Medical Imaging Based on Magnetic Fields and Ultrasounds

Medical Imaging Based on Magnetic Fields and Ultrasounds

Edited by
Hervé Fanet

iSTE

WILEY

First published 2014 in Great Britain and the United States by ISTE Ltd and John Wiley & Sons, Inc.

ISTE Ltd
27-37 St George's Road
London SW19 4EU
UK

www.iste.co.uk

John Wiley & Sons, Inc.
111 River Street
Hoboken, NJ 07030
USA

www.wiley.com

Library of Congress Control Number: 2013956559

British Library Cataloguing-in-Publication Data
A CIP record for this book is available from the British Library
ISBN 978-1-84821-502-3

MIX
Paper from responsible sources
FSC® C013604

Printed and bound in Great Britain by CPI Group (UK) Ltd., Croydon, Surrey CR0 4YY

Table of Contents

Foreword

After almost a century of quiescence, medical imaging has experienced considerable progress over the past thirty years. This progress has resulted primarily from the convergence of major innovations in the fields of detection, information processing and instrumentation. It should be stressed from the very outset that this convergence would never have happened without the extraordinary progression of computing power, the use of which is an absolute necessity nowadays, in view of the enormous increase in the volume of data needing to be handled. Although originally, X-rays, nuclear medicine, MRI and ultrasounds represented mutually independent "monospectral" methods, it is evident today that another convergence is coming about: that of multispectral imaging, stemming from the combination of two imaging techniques in the same piece of equipment. The best example of this would be positron emission tomography – computed tomography (PET-CT). Let us also point out that the capabilities offered by this convergence go beyond the diagnosis stage – the device can also be used in a therapeutic capacity: a very typical example is focused high-energy ultrasounds under MRI.

Information sciences and the development of "physiological" models have opened up functional imaging for the use of methods initially used for their morphological properties: the extraction of circulatory parameters from dynamic scanner or MRI sequences has become an essential tool in the study of tumor response.

Initially developed with a view to studying the whole body, high-resolution imaging techniques are beginning to emerge: an example is optical imaging, but the weakness of this technique is the small sample size. However, its use on humans, particularly in endoscopy and likely, in tomorrow's world, in imaging-guided biopsy methods, appears to have a very bright future.

While imaging is the subject of very intense intrinsic research, it is also considered a crucial tool in physiological and metabolic research, or even cognitive research by integration of physiological signals with imaging data. Thus, methods for magnetic tattooing of cardiac muscle have opened up numerous avenues for the physiology of heart contraction; the study of aortic distendibility is now considered an early sign of aging. In addition, these imaging methods have become absolutely essential in pre-clinical trials on animals: thus, the development of new medicines benefits greatly from such methods. More generally, platforms for imaging of small animals have developed in the context of multidisciplinarity, and demonstrate the interconnection of imaging with the physical sciences, information sciences, chemical sciences and biological sciences.

In this context, the development and fine-tuning of markers and tracers represent a common goal for all imaging methods; the molecular imaging which has already been developed in the field of nuclear medicine should see future progress with other imaging techniques, both in animals and in humans. Substances with diagnostic and therapeutic properties are beginning to come to light, and are developing with no difficulty at all. Thus, progress is also being made in imaging, by way of the advances made in the field of chemistry.

The aim of this book is to provide an overview of the progress made in the various domains of imaging from various different angles: there can be no doubt that readers will find it enriching.

Guy FRIJA

January 2014

Chapter 1

Ultrasound Medical Imaging

1.1. Introduction

Ultrasound imaging accompanies each of us, from several months before we are born, throughout our lives. To monitor our development, 17 million ultrasound examinations are performed each year in France in the private sector, and around twice that if we add those done in the public sector. Thus, ultrasound imaging is the most widely used type of imaging for diagnostics after radiography. The world market for ultrasonography is still growing, and is worth an estimated 4.9 billion dollars (data from 2009). Ultrasonography plays a central role, both in hospitals and in doctors' surgeries. The reasons for its ever-growing success are mainly based on its portability, its reasonable cost in comparison with other methods, its performances in terms of yielding results in real time and the fact that it uses non-ionizing waves. Historically, ultrasonography is associated with specialties in obstetrics for pregnancy-monitoring and cardiology. Today, ultrasound medical imaging covers a far wider range of specialties like no other type of imaging. Imaging of the digestive system, breasts, liver with elastography, thyroid or prostate, are examples of the most commonly performed operations [HTT 09].

In the future, significant progress is expected which will enable a doctor to reach a more certain diagnosis in a shorter period of time. For this to happen, technological innovations will be accompanied by methodological developments in signal- and

---

Chapter written by Didier VRAY, Elisabeth BRUSSEAU, Valérie DETTI, François VARRAY, Adrian BASARAB, Olivier BEUF, Olivier BASSET, Christian CACHARD, Hervé LIEBGOTT, Philippe DELACHARTRE. All the authors of this chapter are members or close collaborators of the "Ultrasound Imaging" team at CREATIS: University of Lyon; CNRS UMR5220; Inserm U1044; INSA-Lyon; University of Lyon 1, France.

image processing. 3D or 4D imaging techniques, particularly in the area of cardiovascular care, are likely to emerge in the near future, with the development of new probes and new modes of acquisition, e.g. using sparse sampling techniques. At the same time, progress in modeling, simulation and image processing will be at the heart of new quantitative analysis software built into ultrasound machines. The diagnosis of myocardial infarction, the replacement of the heart valves or the detection of atherosclerosis are examples of medical exams for which these innovations will be essential (Figure 1.1).

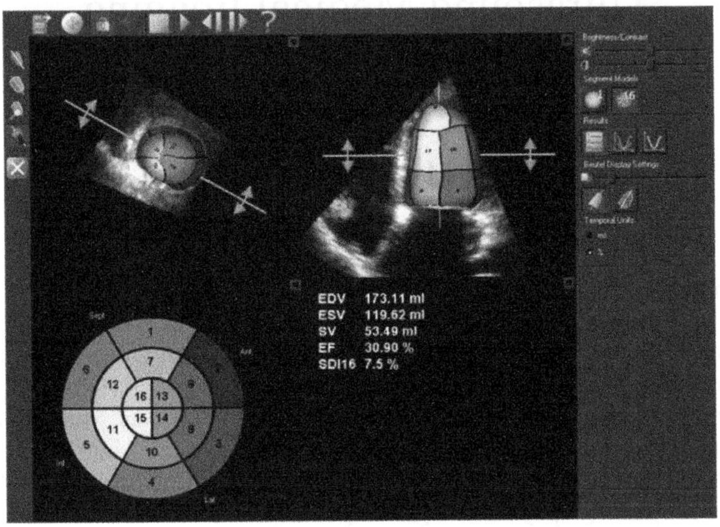

Figure 1.1. *Program for measuring the volume of the left ventricle of the heart, running on a Vivid 6 clinical echogram machine (GE HealthCare). For a color version of this figure, see www.iste.co.uk/fanet/medimagnet.zip*

Other means of imaging based on estimation of the physical properties of healthy and diseased tissues will also supplement conventional imaging tools. Different modes of quasi-static, dynamic or transient elastography should be able to quantify the elasticity of tissues for diagnosing liver disorders or quantify the development of cancer, for instance.

In this chapter, we are going to present both the physical basics of ultrasound (US) imaging and the main advances expected of the echography of tomorrow. The chapter begins with a presentation of the physical principles upon which ultrasound imaging is based. Then, we shall detail the different modes of imaging in ultrasound systems: B-mode, M-mode, Doppler modes, contrast and harmonic imaging. The hardware aspects will also be touched upon, with a discussion of the different types of linear or sectorial probes used in clinical practice, depending on the compromise between resolution and penetration. We shall then return to statistical analysis of the

US image using the properties of the ultrasound speckle. This knowledge will help to simulate realistic images and sequences which are useful to validate the image formation models and processing methods. The final part of this chapter will be given over to the advances in ultrasound image acquisition and processing which will serve as a diagnostic aid to doctors. We shall present the most recent probe technologies which are based on new materials to transmit and receive ultrasounds in 2D and 3D with sensor matrices. These probes can be based on innovative methods of image formation using the methods of synthetic aperture, “tagging” or sparse sampling. We shall also illustrate the contribution made by new techniques such as elastography, nonlinear imaging or parametric imaging, and the performances of real-time tracking methods or motion estimation more generally. The end of the chapter will be devoted to multimodality imaging. Using the example of bi-modal US/Optical imaging, we shall show that the combination of the anatomical information provided by the US image with the functional or metabolic information provided by the other modes of imaging facilitates a more effective aid to diagnosis and monitoring of the evolution of diseases.

1.2. Physical principles of echography

1.2.1. *Ultrasound waves*

US waves are pressure waves whose frequency is greater than the maximum audible frequency of 20 kHz. These US waves are mechanical vibrations. They need a source to give rise to them, and a support medium (a solid, liquid or gas) in order to propagate. If the vibration generated by the source is oscillating, the particles of the medium initially at rest will oscillate around their equilibrium position when the US wave passes through. There are two modes of oscillation: a longitudinal mode where the particles oscillate along the direction of the wave’s propagation, forming a longitudinal or compression wave, and a transversal mode, where the particles oscillate in the direction transversal to the direction of the wave’s propagation, forming a transversal or shear wave [SZA 04].

In most diagnostic echogram exams, it is soft tissues which are explored, and the range of frequencies of oscillation of the wave is between 2 and 20 MHz for the most common external applications and 30 to 50 MHz for internal explorations, such as intra-vascular imaging. At the frequencies used for diagnostic imaging, the shear waves are greatly attenuated and can therefore be neglected. In addition, as soft tissues are composed primarily of water, the propagation of the US wave in these tissues is very similar to its propagation in liquids. In a liquid, the particles oscillate along the direction of propagation of the wave, forming a longitudinal wave. In water, for the applications in question here, the movement of the particles is roughly a few tens of nanometers and their velocity is a few cm/s, whereas the phase velocity c at which the wave propagates in the medium is around 1500 m/s.

When a sinusoidal disturbance propagates in a liquid medium, regions of compression and dilatation form in the medium when the wave passes through. This phenomenon of compression and dilatation is periodic and is observable in two different ways: at a given time or at a given position. It is due to the displacement of the particles. Particle displacement is greater in the dilatation zone than in the compression zone. The periodicity of the displacement is called the wavelength λ if we observe that displacement as a function of the position at a given time, and period T if we observe it as a function of time at a given position. These two values are linked by the phase velocity $\lambda = cT$ or indeed $c = \lambda f$ when $f = 1/T$, representing the frequency of the oscillations emitted by the source.

In order to study the propagation of a US wave, we need to distinguish two types of media: isotropic and anisotropic media. An isotropic medium is one which exhibits no single prevailing direction of propagation. This means that it is the source generating the initial disturbance which imposes the direction of propagation, rather than the medium. Conversely, with an anisotropic medium it is the medium which imposes the direction of propagation. Such is the case, for instance, in most tissues with an oriented fibrous structure.

1.2.2. *Wavefronts*

As we have just seen, there are two types of waves which are distinguished by their mode of propagation. These waves are also differentiated by the shape of what is called their *wavefront,* i.e. the set of points of a medium which simultaneously experience the same change in pressure as the wave passes. Thus, waves can be classified into three wave shapes:

– spherical waves, which are generated over 4π steradians by a source whose dimension is very small in relation to the wavelength emitted;

– cylindrical waves, which are generated by a cylindrical source of infinite length, vibrating in a radial direction;

– plane waves, which are generated by a flat (or "plane") surface such as a piston of infinite dimensions.

In reality, any given wave is a combination of the three forms described above. For reasons of simplicity, we shall only examine the interaction of a plane wave with the medium. By way of example, consider the configuration represented in Figure 1.2. A point source generates a spherical wave. Target 1, which is "near" to the source, is then subjected to the influence of that spherical wave. That is, if the dimension of that target is larger than the wavelength, not all the points on the anterior face are subject to the same pressure at the same time. On the other hand, as regards Target 2, which is a long way from the source, the wavefront which reaches this target can be considered to be plane.

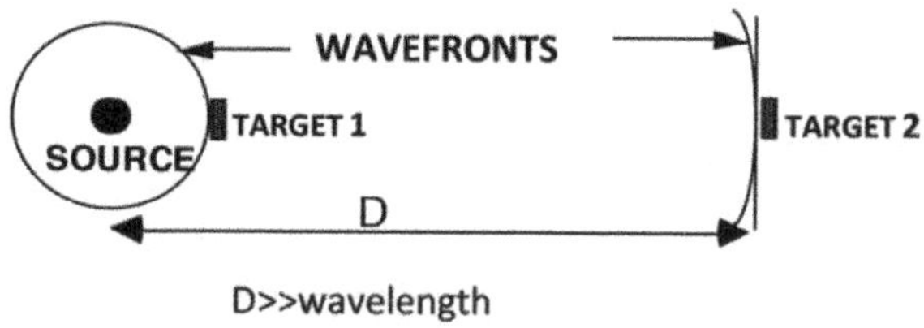

Figure 1.2. *The source emits a spherical wave. Target 1 receives a spherical wave whereas Target 2, which is further away, receives a plane wave*

1.2.3. *Stress/Strain relation*

We consider a perfectly elastic medium, i.e. a medium which, when subjected to a stress, deforms with no internal friction (with no losses) and regains its original form exactly when the stress is removed. As the wave propagates in this elastic and isotropic medium, it locally imposes a stress or pressure causing a displacement of the particles as the wave passes, and a deformation of the medium. In echography, the pressure generated by the source is approximately 1 MPa, causing a local displacement of the particles of a few tens of nanometers. Hence, the displacements are sufficiently small for us to consider the stress/strain relation to be linear. That is, for a wave propagating in direction z, it is written as follows:

$$K_{zz} = (\nu + 2\mu)\varepsilon_{zz}, \quad K_{yz} = \mu\varepsilon_{yz}, \quad K_{xz} = \mu\varepsilon_{xz}$$

where K_{zz} is the longitudinal stress in direction z, and K_{yz} and K_{xz} are the shear stresses in directions y and x. ε_{zz} is the longitudinal strain in direction z, and ε_{yz} and ε_{xz} the shear strains in directions y and x.

The strains are calculated on the basis of the particle displacements (U,V,W) in the three directions (x,y,z) : $\varepsilon_{zz} = \dfrac{\partial W}{\partial z}$ $\varepsilon_{yz} = \dfrac{\partial V}{\partial z}$ and $\varepsilon_{xz} = \dfrac{\partial U}{\partial z}$.

The constants ν and μ are the Lamé parameters. They depend on the propagation medium and are linked to usual values such as Young's modulus (E in N/m² or Pa), the bulk modulus (K in N/m² or Pa) or the compressibility coefficient (β in m²/N or Pa^{-1}),

$$E = \frac{\mu(3\nu+2\mu)}{\nu+\mu}, \quad K = \nu + \frac{2}{3}\mu, \quad \beta = \frac{3}{3\nu+2\mu} = \frac{3}{E}\frac{\mu}{\nu+\mu}$$

1.2.4. *Propagation equation*

The wave equation in direction z is obtained by applying Newton's second law; the sum of the forces is equal to the product of the mass and the acceleration caused by that force. Thus, by summing the volumetric forces in direction z ,we obtain:

$$\frac{\partial K_{zz}}{\partial z} + \frac{\partial K_{yz}}{\partial y} + \frac{\partial K_{xz}}{\partial x} = \rho \frac{\partial^2 W}{\partial t^2}$$

where t is the time and ρ the density of the medium.

If we take into account the same initial remark as in the case of the application of US waves to soft tissues – that the shear waves are greatly attenuated and are therefore negligible – the wave equation is reduced to the equation of the compression wave:

$$\frac{\partial K_{zz}}{\partial z} = \rho \frac{\partial^2 W}{\partial t^2} \quad \Leftrightarrow \quad \frac{\partial^2 W}{\partial^2 z} = \frac{\rho}{\nu+2\mu}\frac{\partial^2 W}{\partial t^2} \quad \Leftrightarrow \quad \frac{\partial^2 W}{\partial^2 z} = \frac{1}{c^2}\frac{\partial^2 W}{\partial t^2}$$

where $c = \sqrt{\dfrac{\nu+2\mu}{\rho}} = \sqrt{\dfrac{K+\frac{4}{3}\mu}{\rho}} \approx \sqrt{\dfrac{K}{\rho}}$

In soft tissues, $K \approx$ 2.2 GPa and $\mu \approx$ 0.1 MPa. Note that the phase velocity depends only on the compressibility ($1/K$) and density of the medium.

In the case of a sinusoidal particle displacement of amplitude W_0 with angular frequency $\omega = 2\pi f$, the solution to the wave equation is:

$$W(z,t) = W_0 e^{j(\omega t \pm kz)}$$

with $k = \omega / c$ being the wavenumber. The sign (–) indicates a progressive plane wave, moving in the direction $+z$, and vice versa.

1.2.5. *Acoustic impedance*

Here, we shall retain the hypothesis of a sinusoidal excitation source giving rise to a plane wave, propagating in direction z. The particle velocity is obtained from the time-derivative of the particle displacement: $u_z = \frac{\partial W}{\partial t} = j\omega\, W$. The velocity is therefore out of phase by 90° in relation to the displacement. The stress K_{zz} expressed above is linked to the pressure p by the relation: $p = -K_{zz} = -(\nu + 2\mu)\frac{\partial W}{\partial z}$

If we differentiate the displacement of the particles generated by a compression wave, we have: $p = \pm jk(\nu + 2\mu)W = \pm j\omega\rho c W = \pm \rho c u_z$. Note that the pressure is linked to the particle displacement by the term $z = \rho c$, which represents the acoustic impedance expressed in kg.m^{-2}.s^{-1} or "Rayleigh", in tribute to Lord Rayleigh for his work on acoustic waves. Acoustic impedance is an important parameter in ultrasound imaging because it determines the amplitude of the echoes registering on the echogram.

1.2.6. *Acoustic intensity*

The acoustic intensity is the total energy, per unit time, of a US wave traversing the unit of surface perpendicular to the direction of propagation of the wave. The intensity is expressed in W/m² or in mW/cm² for medical applications. In the case of a sinusoidal plane wave, we express the average intensity over a period as a function of the maximum pressure p_0:

$$I = \int_0^T pu dt = \frac{1}{\rho c}\int_0^T p^2 dt = \frac{p_0^2}{\rho c}\int_0^T e^{2j(\omega t - kz)} dt = \frac{p_0^2}{2\rho c}$$

In echography, the waves are usually brief pulses comprising a few oscillations, and the pressure field varies greatly depending on the spatial position, i.e. the distance $\vec{d}$ between the emitting source and the point of observation. In general, the pressure field is emitted by a focused source, and the maximum intensity is experienced at a distance $\vec{d}_{\max}$ corresponding to the focal point. For these reasons, other definitions of intensity are used.

By defining an instantaneous intensity, dependent on time and on the position $\vec{d}$ in the pressure field, $I_i(t,\vec{d}) = \vec{p}^2(t,\vec{d})/\rho c$, the intensity can be averaged over the pulse repetition frequency time T_{PRF} – this is I_{spta} (*spatial peak temporal averaged intensity*): $I_{spta} = \frac{1}{T_{PRF}}\int_0^{T_{PRF}} I_i(t,\vec{d}_{\max})dt$

The intensity can be averaged over the duration τ of the emission I_{sppa} (*spatial peak pulse averaged intensity*): $I_{sppa} = \frac{1}{\tau}\int_0^{\tau} I_i(t,\vec{d}_{\max})dt$

The maximum intensity I_m can be calculated between two times around the maximum pressure. Thus, for a sinusoidal pulse, I_m is calculated over a half-period around the maximum. $I_m = \frac{I_0}{0.5 f_0}\int_0^{0,5/f_0} \sin^2(2\pi f_0 t)dt = \frac{I_0}{2} = \frac{p_0^2}{2\rho c}$

EXAMPLE.– for five periods of oscillations at 5 MHz, the pulse duration is τ = 1 µs. The pulse repetition frequency (T_{PRF}) is limited by the time taken for the wave to propagate in both directions. For a there-and-back distance of 15 cm, the propagation time is 100 µs, considering a velocity c = 1500 m/s. The pulse repetition frequency time can therefore be no less than 100 µs in this example. For a pulse repetition frequency of 5 kHz (T_{PRF} = 200 µs) and a probe emitting a rectangular sinusoidal plane wave with pulse duration $\tau = 3$ µs at 5 MHz and a maximum pressure of 500 kPa, the intensities are $I_m = \frac{I_0}{2} = \frac{p_0^2}{2\rho c}$ = 1.65 W/cm².

$$I_{spta} = \frac{1}{T_{PRF}}\int_0^{\tau} \frac{p_0^2}{2\rho c}dt = \frac{\tau}{T_{PRF}}\frac{p_0^2}{2\rho c} = 25 \text{ mW/cm}^2$$

and $$I_{sppa} = \frac{1}{\tau}\int_0^{\tau} \frac{p_0^2}{2\rho c}dt = \frac{p_0^2}{2\rho c} = I_m = 1.65 \text{ W/cm}^2,$$

with the pulse envelope being rectangular in the example: $I_m = I_{sppa}$. In practice, the envelope is often Gaussian in form, which yields $I_m > I_{sppa}$.

Approval of an ultrasound system by the "United States Food and Drug Administration" (F.D.A.) is based on a series of intensity values (Table 1.1 and Table 1.2).

The acoustic *power* P is linked to the intensity by the relation $i(t) = dP / dA = p(t)u(t)$, where A represents the unit of surface area.

Use	I_{spta} (mW cm^{-2})		I_{sppa} (W cm^{-2})		I_m (W cm^{-2})	
	in situ	water	*in situ*	water	*in situ*	water
Cardiac	430	730	65	240	160	600
Peripheral vessels	720	1500	65	240	160	600
Ophtalmology	17	68	28	110	50	200
Fetal exam	46	170	65	240	160	600

Table 1.1. *Maximum acceptable values for the acoustic field emitted by ultrasound systems*

Type of imaging	Max pressure (MPa)	I_{SPTA} (mW/cm^2)	I_{SPPA} (mW/cm^2)	Power (mW)
B-mode	1.7	19	174	18
M-mode	1.7	73	174	4
Pulsed Doppler	2.5	1140	288	31
Color Doppler	2.6	234	325	80

Table 1.2. *Characteristic values of the acoustic field emitted for different types of imaging*

1.2.7. *Mechanical Index*

Ultrasound systems refer to the Mechanical Index (MI) to quantify the level of emission. This parameter is defined as the ratio of the maximum amplitude of the pressure pulse emitted (in MPa) to the square root of the frequency (in MHz): $\mathrm{MI} = p_0 / \sqrt{f}$. It usually varies between 0 and 2, with a value of 2 corresponding to an intensity I_{spta} of 720 mW/cm^2.

1.2.8. *Generation, emission*

A US imaging system requires a sensor – usually *piezoelectric* in nature – to convert electrical signals into US waves and vice versa. Thus, the sensor generates a US *beam*, and converts the US pressure field it receives into electrical signals. To acquire 2D or 3D ultrasound data, the beam needs to be moved in one or more directions. In echography, the displacement of the beam is controlled electronically, meaning that the piezoelectric elements are grouped into subsets, with each active subset able to emit a US beam and receive waves.

For instance, with a probe comprising 128 elements (channels or beams), the matrix is commonly grouped into sets of 32, with a sweep of 1 element to acquire each new scan line. This means we need to acquire $32 \times 97 = 3104$ *raw signals*[1] to form an image comprising 97 *radio-frequency (RF) signals*.

The choice of a probe is determined by its *resonant frequency and its bandwidth*. A US pulse has a broad spectrum which will shift toward low frequencies during the course of propagation because the acoustic medium behaves like a low-pass filter. In view of the medium being observed, we need to use a probe with an appropriate resonant frequency and bandwidth. For instance, when exploring the breast or the heart, the resonant frequency of the transducer is respectively 7.5 MHz and 3.5 MHz.

1.2.9. *Resolution*

Three other parameters are very important for appreciating the quality of an echographic probe: the *axial, lateral and azimuth resolutions* of the US beam.

– The *axial resolution* is defined as the minimum distance (Δz) perceptible by the probe between two sufficiently close reflective structures in planes perpendicular to the direction of propagation of the ultrasound wave (Figure 1.3).

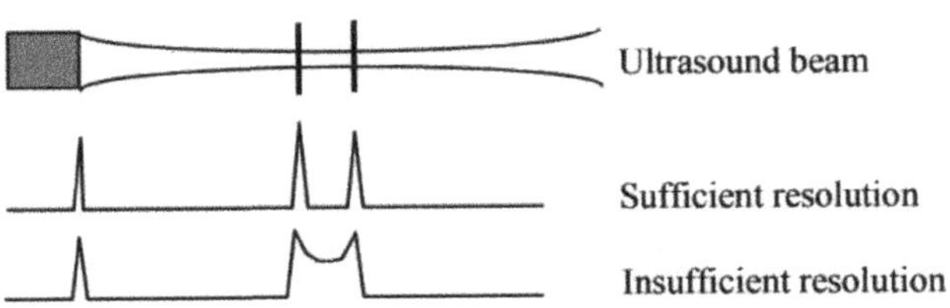

Figure 1.3. *Axial resolution of an echographic probe*

Let n be the number of periods of the sinusoidal pulse emitted, and λ its wavelength; in this case, we have: $\Delta z = n\lambda/2$. This axial resolution is sub-millimetric in the field of echography. It depends on the shape and duration of the

1 The raw signals are those captured by the echographic probe and the RF signals are obtained by linear combination (beamforming) of the raw signals.

US wave, and therefore on the impedance matching of the probe; on the wavelength of the acoustic signal but also on the bandwidth of the probe for probes with broad frequency bands. For instance, with a probe emitting 4 sinusoidal periods at 5 MHz, the axial resolution is $\Delta z = 0.6$ mm. One way of increasing the axial resolution is to increase the frequency of the emitted wave but, as we shall see later on, the attenuation of the wave is greater when the frequency increases.

– *The lateral and azimuth resolutions* are among the most significant factors affecting the quality of an ultrasound image. They are defined by the capacity of the probe and of the ultrasound system to distinguish between two nearby structures situated in the same plane perpendicular to the axis of the beam. Their values depend on the width of the US beam. Therefore, one way of increasing the lateral resolution is to work with a focused beam. The width of the beam at the focal point W_b is linearly dependent on the wavelength of the signal emitted: $W_b = f_\# \lambda$, where $f_\#$ is the aperture number (also commonly called the "*f* number"), which is the relation between the focal length and the aperture of the beam. In the case of linear array, where the transducers are rectangular, the lateral resolution expresses the resolution in a direction parallel to the width of the transducers. The azimuth resolution (or elevation resolution) expresses the resolution in a direction parallel to their length – it expresses the thickness of the imaging plane (Figure 1.4). The lateral and azimuth resolutions are of approximately a millimeter.

The lateral resolution depends on the geometry of the transducer, the frequency used and the focusing of the beam. For instance, with a linear probe at 5 MHz focused at 4 cm with an aperture comprising 32 active elements and a pitch (distance between two elements) of λ, we have $W_b = 1.3$ mm.

These three resolutions determine an *elementary resolution volume*.

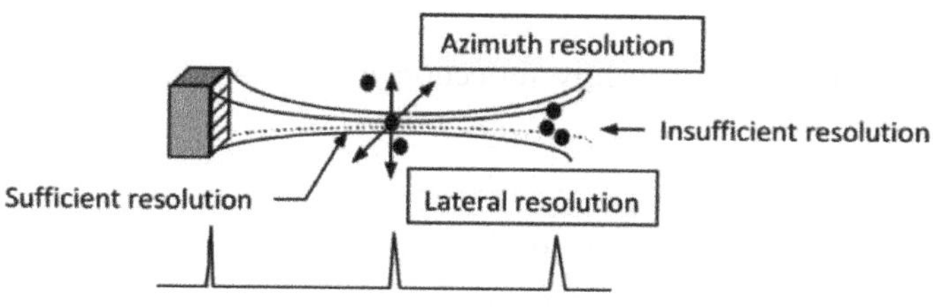

Figure 1.4. *Lateral and azimuth resolutions*

1.2.10. *Propagation of a plane wave in a finite isotropic medium*

When an incident plane compression wave encounters a plane interface separating two isotropic elastic media (1 and 2) with different acoustic impedances

($z_1 \neq z_2$), some of the energy is reflected, propagating at the same velocity (c_1) as the incident wave, and the rest of the energy is transmitted into medium 2, within which the wave will propagate with a different velocity (c_2).

In writing the following equations, we consider a longitudinal incident wave. If we mark the incident, reflected and transmitted waves as *I*, *R* and *T* respectively, the boundary conditions at the interface *S* are written thus:

$$\Sigma \text{ displacements medium 1} = \Sigma \text{ displacements medium 2} \Leftrightarrow \{\ u_I + u_R = u_T\ \}_S$$

$$\Sigma \text{ pressures medium 1} = \Sigma \text{ pressures medium 2} \Leftrightarrow \{\ p_I + p_R = p_T\ \}_S$$

We can show that the reflected and transmitted waves have the same frequency as the incident wave, and that they are situated in the plane of incidence. We can also deduce the Snell–Descartes law, which can be used to determine the directions of propagation of the reflected and transmitted waves, i.e. the reflection- and transmission angles as a function of the angle of incidence and of the celerities in the two media. These angles are defined between the direction of propagation and the normal to the interface. By definition, they vary between 0 and π/2: $\frac{\sin\theta_I}{c_1} = \frac{\sin\theta_R}{c_1} = \frac{\sin\theta_T}{c_2}$

We therefore define the coefficients of amplitude reflection *R* and transmission *T*, obtained by continuity of the pressure and particle velocity across the interface.

$$R = \frac{p_R}{p_I} = \frac{z_2 \cos\theta_I - z_1 \cos\theta_T}{z_2 \cos\theta_I + z_1 \cos\theta_T} \qquad T = \frac{p_T}{p_I} = \frac{2\, z_2 \cos\theta_I}{z_2 \cos\theta_I + z_1 \cos\theta_T}$$

or indeed the coefficients of energy reflection *R* and transmission *T*, obtained by $I = \frac{p^2}{2z}$

$$\alpha_R = \frac{I_R}{I_I} = \left(\frac{z_2 \cos\theta_I - z_1 \cos\theta_T}{z_2 \cos\theta_I + z_1 \cos\theta_T} \right)^2$$

$$\alpha_T = \frac{I_T}{I_I} = \frac{4 z_1\, z_2 \cos\theta_I}{(z_2 \cos\theta_I + z_1 \cos\theta_T)^2}$$

Note that if the impedance of medium 2 is negligible in comparison to that of medium 1, which is the case when medium 2 is air, $\alpha_R \approx 1$ and $\alpha_T \approx 0$. That is to say that all the energy is reflected; there are no transmitted waves. Table 1.3 gives a few values for the reflection coefficients for a wave with normal incidence angle.

INTERFACE	Ratio of amplitude Ar / Ai	Reflected energy (%)
Fat/Muscle	0.10	1.08
Muscle/Blood	0.03	0.07
Bone/Fat	0.69	48.91
Soft biological tissues/Water	0.05	0.23
Soft biological tissues/Air	0.9995	99.9
Skin/piezoelectric element	0.89	80

Table 1.3. *Reflection coefficients in the case a normal incidence angle*

1.2.11. *Propagation of a plane wave in a non-homogeneous medium*

In a homogenous, non-absorbent medium, a plane wave maintains constant amplitude and direction. In reality, in biological tissues, propagation no longer obeys these principles. For example, the intensity of the wave decreases as it penetrates deeper into the medium. Similarly, the plane wave is not preserved (the direction and/or shape of the wave is altered). This *attenuation* of the US wave has various causes, which we are going to study separately although they are interrelated. It may result from the absorbent nature of the medium. The phenomenon of *absorption* transforms the incident ultrasound energy into heat. This energy transformed into heat constitutes a *bona fide* loss. Attenuation may also be caused by the inhomogeneities of the medium.

Diffusion, in the broadest sense, occurs when a wave propagates through a non-uniform medium. Some of the energy is redirected and appears separately to the initial wave. Either it is merely delayed or its actual direction is altered. The simplest case is that of a plane interface at a normal incidence angle, discussed in the previous section. This is an example of reflection and transmission of the wave, which can be easily resolved by looking at the specific impedances of the two media. Remember that the plane interface theory applies when the dimensions of the object are larger than the wavelength. In the field of medicine, such interfaces are rare, and the discontinuities are very variable in shape, size, position and orientation. Diffusion exists when the size of the inhomogeneity is small in comparison to the wavelength of the incident wave. The inhomogeneity then behaves like a point source, and the energy is re-emitted throughout the whole of the space in the form of a spherical wave.

Attenuation covers all losses, i.e. energy which is not transmitted through the medium and would not be picked up by a receiver facing the emitting probe. Thus, reflection, refraction, diffusion, diffraction and absorption all contribute to attenuation. The pressure of a monochrome plane wave propagating in direction z decreases exponentially as a function of the distance covered: $p(z) = p(z=0)e^{-\alpha z}$, where $p(z=0)$ is the pressure at $(z=0)$ and $\alpha = \frac{1}{z} Ln\left(p(z=0)/p(z)\right)$ is the coefficient of attenuation of the pressure, expressed in nepers per centimeter or, more usually in medical applications, in decibels per centimeter $\alpha(dB/cm) = 20(\log_{10} e)\alpha(np/cm) = 8,686(np/cm)$.

Biological tissues and materials	Temperature (°C)	Density (kg.m^{-3})	Celerity (m.s^{-1})	Acoustic impedance (kg m^{-2} s^{-1})	Attenuation (dB/cm^{-1}) at 1 MHz
Air	0 20	1.28 1.20	331 344	430	12
Water	20	1000	1480	1.52 × 106	0.0022
Blood	20	1060	1530	1.62 × 106	0.18
Fat		920	1410 to 1470	1.34 to 1.39 × 106	0.6
Muscle		1070	1550 to 1640	1.65 to 1.7 × 106	1.2
Liver		1060	1550	1.65 × 106	0.9
Kidney		1040	1560	1.62 × 106	
Bone	20 to 37	1200 to 1810	2700 to 4100	3.2 to 7.8 × 106	20
Lung	35		650	0.26 × 106	40
Skin			1519	1.58 × 106	
Blood Vessels		1080	1530	1.65 × 106	
Soft tissues			1540 on average	1.63 × 106	
Breast			1440 to 1550		1.68 ± 0.48
Quartz			5750	1.52 × 107	

Table 1.4. *Acoustic characteristics of biological tissues and certain materials*

This attenuation coefficient is itself proportional to the frequency. Consequently, attenuation is much greater at high frequencies. In biological tissues, attenuation is roughly 1 dB cm^{-1} MHz^{-1}. In order to illustrate the effect of this attenuation, let us take a probe working at 3 MHz and compare the amplitude of the echoes of two identical targets at a distance of 10 cm (a round trip of 20 cm). In these conditions, the attenuation is $1 \times 20 \times 3 = 60$ dB, i.e. there is a ratio of 1000 between the amplitude of the two echoes. If we consider a probe working at 6 MHz, the attenuation is 120 dB (a ratio of 1,000,000). It is for this reason that high frequencies are not used to image organs that are close to the probe. In echography, we are led to strike a balance between a high frequency – which is to be desired if we want good image definition – and a lower frequency to obtain good penetration. Table 1.4 shows the acoustic characteristics of biological tissues in comparison with those of other materials.

1.2.12. *Speckle*

An elementary cell whose resolution is defined by the characteristics of the transducer (axial, lateral and azimuth resolutions), corresponds to the smallest volume within which point targets cannot be individually resolved, but whose presence contributes to the creation of an echo. The dimension of these reflectors is smaller than the wavelength emitted (Figure 1.5). The US wave reaches the point scatterers at different times, depending on how far they are from the transducer; they in turn emit spherical waves and the pressure at each point of the acoustic medium traversed by the echoes is the result of the summation of the wave emitted by each reflector.

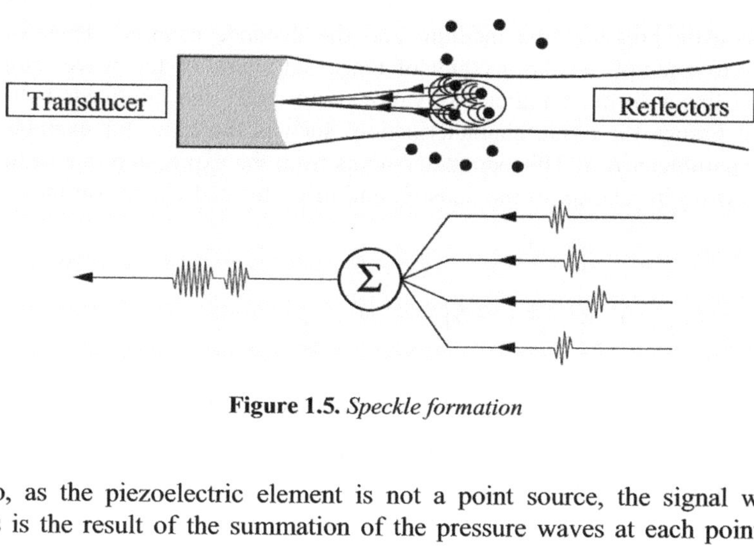

Figure 1.5. *Speckle formation*

Also, as the piezoelectric element is not a point source, the signal which it delivers is the result of the summation of the pressure waves at each point on its surface. If the difference between the arrival times of the echoes is less than the duration of the emitted pulse, it is not possible to distinguish the contribution of each reflector, and the high-frequency signal gives us no indication about the number of position of the targets. Interference, resulting from the summation of the reflected waves, then occurs. The point corresponding to the projection of the elementary resolution cell in the image plane will appear brighter if the dominant interferences are constructive; if not, the point will be darker.

This set of points lends the appearance of granite to the echogram image – this characteristic is known as *texture noise* or ***b***. For certain forms of diagnosis, speckle is considered to be parasitic noise, and image processing methods have been developed to reduce it (filtering, spatial composition, frequency composition, etc.). On the other hand, with other forms of diagnosis (e.g. with liver disease), the texture of the image is a piece of information which can be used to characterize different tissues. Speckle depends on parameters relating to the structure of the tissues (density, size, composition, distribution of the reflectors), so healthy and diseased tissues may well exhibit different textures.

1.2.13. *Nonlinear waves*

The phenomena discussed hitherto have been limited to US waves of low intensity, for which the variation in density of the medium is slight. Ultrasound systems use focused pressure fields to achieve local pressures of several MPa. In such cases, the acoustic pressure is no longer negligible in comparison to the pressure in the medium at rest. Thus, high levels of acoustic pressure generate waves whose propagation is nonlinear. The shape of these waves depends on the amplitude of the acoustic pressure, the medium and the distance covered. Therefore, it is necessary to attempt to take account of the exact shape of the wave, rather than simply its linear form. Another motivation is to study the nonlinear behavior of biological tissues for tissue characterization: such is the case, for example, when using the parameter *B/A*. This parameter stems from the expression in equation form of the pressure in relation to the density, and of its limited expansion into a Taylor series.

$$p = p(s_0, \rho_0) + A\left(\frac{\rho-\rho_0}{\rho_0}\right) + \frac{1}{2}B\left(\frac{\rho-\rho_0}{\rho_0}\right)^2 + \ldots$$

where ρ_0 and s_0 are the density and entropy at the equilibrium point with $A = \rho_0[\partial p(s_0, \rho_0)/\partial \rho]$ and $B = \rho_0^2[\partial^2 p(s_0, \rho_0)/\partial \rho^2]$. Table 1.5 gives the values of the B/A ratio for different media.

Medium	B / A
Water (30°C, 1 atm)	5.2
Blood	6.3
Liver	7.6
Spleen	7.8
Fat	11.1

Table 1.5. *Value of the ratio B/A for different media [BAM 80]*

1.2.14. *Contrast agents*

While the use of contrast agents in other modes of imaging has been common practice for many years, the development and commercialization of contrast agents designed specifically for ultrasound imaging only occurred very recently. As with other types of imaging, the injection of contrast agents during an ultrasound exam is intended to facilitate the detection and diagnosis of specific diseases. Contrast-enhanced ultrasound imaging is based on the backscattering of ultrasound wave by microbubbles. These microparticles are injected into the bloodstream intravenously, in pellet or drip form. These particles cannot pass through the endothelial wall, so they show up perfused areas in contrast to non-perfused areas. Many different solid, liquid or gaseous particles have been tested as ultrasound contrast agents (UCAs), but gaseous particles are most effective because the liquid/air interface presents a significant discontinuity in terms of acoustic impedance. UCAs comprise microbubbles – either unprotected or encapsulated to increase their lifetime – of a few microns in diameter.

The microbubbles do not only behave as simple scatterers, they vibrate in a nonlinear fashion. An acoustic wave is composed of an alternating pattern of high and low pressures. When this acoustic wave interacts with a microbubble, alternately the microbubble is compressed during the pressurization phase and expands during the depressurization phase. As the bubble can expand more easily than it can contract, the change in its diameter is not symmetrical. Instead of producing a sinusoidal wave, it generates a non-symmetrical wave. This asymmetry causes harmonics, meaning that the diffused waves are broad band. In addition to the emitted frequency (the fundamental frequency) they contain harmonics of that frequency (mainly the second harmonic, which is double the emitted frequency). In addition, these microbubbles are resonating systems, meaning that the amplitude of the backscattered wave is greater when the incident wave is at a frequency near to its own resonant frequency. The product of the resonant frequency of the bubble by the radius of that bubble is around 3. Therefore, by happy coincidence, the resonant frequency of these bubbles, which have a radius of a few microns, is within the frequency range used in diagnostic echography.

It is this nonlinear behavior of the microbubbles and that property of resonance that are exploited by new imaging techniques developed for contrast-enhanced imaging.

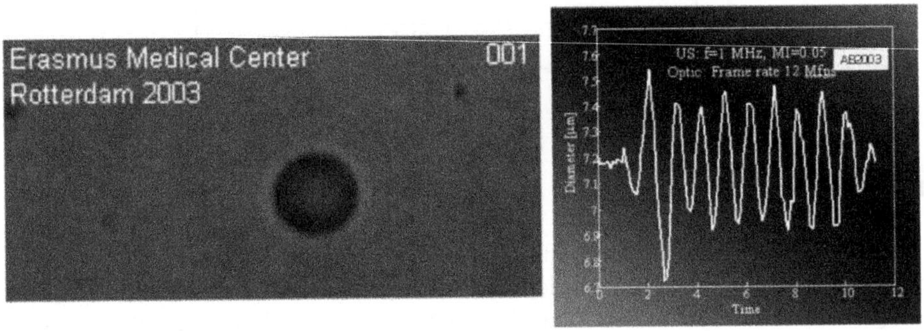

Figure 1.6. *Left, image of a microbubble of contrast agent filmed by a Brandaris camera; right: change in diameter of a microbubble over time under the influence of a US wave. With a low MI, the vibration is quasi-linear. With the courtesy of N. de Jong, Erasmus Medical Center, Rotterdam*

1.3. Medical ultrasound systems

1.3.1. *Principle*

Pulse echo imaging techniques are based on the determination of the amplitude and delay of an ultrasound signal reflected by a medium, and thereby generate a representation of its structure. Generally, a pulsed echo system can be represented by a block diagram of different functions around a main element which is the converter. The transmitter converter is periodically excited by an electrical pulse. An acoustic wave is then caused, which interacts with the medium within which it is propagating. Reflection and diffusion waves are formed. The portion of that wave which is reflected and backscattered in the direction of the receiver converter is converted into an electrical signal: the *RF signal*. The RF signal is then processed to extract the data necessary to create an image. An RF signal provides one-dimensional (1D) spatial information which indicates the depth of the interfaces. To construct the complete image of a cross-section of an acoustic medium, it is necessary to move the US beam within the plane of the cross-section. This imaging system enables the velocity of the ultrasounds to remain constant within the acoustic medium being sounded (c = 1540 m/s in soft tissues) and each transducer element to be unidirectional.

1.3.2. *The different stages in image formation*

1.3.2.1. *Emission*

The controller of the *beamformer* is a crucial element. It synchronizes the emitted and received signals, knowing the width and depth of the area being explored. This region is transposed into a number of rows to be scanned and a number of focal points per row. The controller determines which piezoelectric elements to activate for the row in question, and the delay to apply to each element depending on the depth and orientation of the desired focal point. The controller begins with the first row, feeding voltage (±100 V) and current (±2A) to the piezoelectric elements involved in the active part of the antenna. The high-voltage electrical signal emitted passes through a transmission/reception hub to protect the receiver part, wherein the voltage levels are of the order of a micro- or millivolt. Before being transmitted, this electrical pulse is properly adjusted at each active piezoelectric element, to produce a focused beam, directed at the desired focal point, for the row under investigation.

1.3.2.2. *Receiving*

After propagation and interaction with the medium being explored, the return wave is received by the elements and converted into a low-voltage signal. This signal is then amplified by one or more VCAs (voltage controlled amplifiers) and then digitized by an ADC (analog-to-digital converter) whose sampling frequency is set at around 4 times the central frequency of the probe. The VCA is configured in such a way that the received signal is amplified as a function of the time (i.e. of the depth) so as to compensate for the attenuation of the wave within the tissues. The number of VCAs and ADCs depends on the number of active channels used for beamforming.

1.3.2.3. *Beamforming upon receipt*

The signals are passed through the beamformer when they are received. Each signal is differentiated and weighted, and then the phases of all the signals are added together ("coherent signal summing") to form the RF signal corresponding to a row of the image. All the above operations, from emission to reception, are repeated to form each row of the image. The number of rows may vary greatly, but generally an image will contain around a hundred rows. The beamforming operations are performed by a specialized device – an FPGA (field programmable gate array), a DSP (digital signal processor) or a combination of the two. The choice of hardware architecture depends on the number of channels used for digital beamforming.

The more the number of channels increases, the more the computation time per beam needs to decrease so as not to slow down the image cadence. This image cadence, which can range from 30 to 300 frames per second, is a major asset in echography.

1.3.2.4. *Processing of received signals in different types of imaging*

After beamforming upon receipt, the RF signal thus constituted is filtered to reduce the noise outside of the bandwidth. The treatments performed after beamforming depend on the mode or modes of imaging selected. For instance, the classic display of ultrasound images in grayscale corresponds to *B-mode* ultrasound, with B representing Brightness. B-mode uses demodulation, envelope detection and logarithmic compression, to aid the viewing of slightly echogenic structures against highly echogenic structures. These operations may be followed by 2D noise-reduction or contrast-enhancement treatments. The last operation is scan conversion, which enables the image to be displayed in a format appropriate for the means of data acquisition. For instance, with an abdominal probe acquiring data by sectorial geometry (as is done with fetal ultrasounds), the scan converter ensures the final image is displayed in this same format.

Doppler mode can be used to study movement – mainly the movement of the bloodstream – in a region which is covered by B-mode. In this mode, various types of display are possible: a spatial map of velocities with a *color Doppler*, or the computation of a *sonogram* to visualize the change in speed of the flow over time. Sonogram computation is done by way of a local Fourier transform of the demodulated RF signal as a function of the depth for a given direction in the image, and repeating this process for multiple emissions and receptions.

All the post-treatments performed after beamforming are done in real time. The choice of components such as FPGA, DSP or GPU (graphics processor unit) is guided by the constraints of portability and low power consumption, or of computation power for video display with superposition of images in the case of a color Doppler.

There is one more mode of emission/reception – that of a continuous wave (*CW mode*), used to measure velocities by the Doppler Effect. This mode offers a more accurate measurement of velocities but no longer facilitates the spatial localization of those velocities. In this specific mode, a separate emitter and a receiver are necessary, and the chain of acquisition is analog: an analog beamformer followed by a demodulator. Indeed, the dynamic of amplitude of the continuous signals received and modern ADCs are not, at present able to digitize these signals with a sufficient signal-to-noise ratio (SNR) to estimate the velocity. After beamforming and demodulation, the signal can then be sampled at lower frequencies (around a kHz) and digitized over a greater dynamic of amplitude. Figure 1.7 illustrates all the functions, from the acquisition of the signals to the display of the ultrasound image.

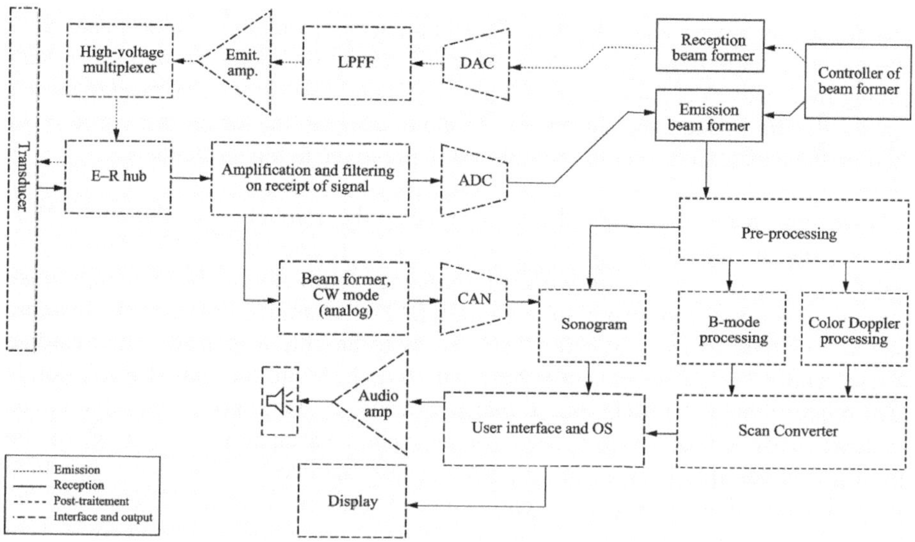

Figure 1.7. *Block diagram of an ultrasound imaging system.*

1.3.3. *Ultrasound imaging probe*

The probe is the crucial part in an ultrasound imaging system, because it is the component which generates the US wave and receives the echoes. Initially, they comprised one or more piezoelectric elements performing mechanical scanning of the medium; today, however, the majority of ultrasound probes comprise a set of piezoelectric elements performing electronic scanning of the medium.

Electronic scanning consists either of successively activating a set of elements in order to enable the beam thus created to move around, or to reorientate the beam in different directions using delay laws on the elements, on emission and on reception.

The choice of central frequency for the elements used, of their shape, geometric dimensions and ordering, is strongly linked to the intended application.

1.3.3.1. *Frequency*

When we speak of the frequency of a probe, we are in fact speaking of the frequency of the elements which make it up. In general, an indication will also be given of the bandwidth of the elements.

The frequency and bandwidth have a direct impact on the physical resolution of the system. Indeed, in order to have good axial resolution, we need to be able to transmit signals of very short duration. This implies rapid variations in the signal, and, consequently, high frequencies. A good axial resolution therefore implies a high imaging frequency. Yet a high central frequency is not enough; we also need to have a large bandwidth. Indeed, the narrower the band of a signal, the longer that signal is. In addition, the lateral and azimuth resolutions are improved at higher frequency.

However, there are drawbacks to increasing the frequency of the probe. For instance, the attenuation of the US waves is also greater when the frequency is higher. Hence, the choice of working frequency will always be a compromise between resolution and depth of exploration. For example, for external cardiac imaging, we tend to work at a few MHz so as to be able to see all of the cardiac muscle, whose largest dimension is between 10 and 15 cm. the spatial resolution is much poorer than at 10 MHz, but at that frequency, it would not be possible to see the heart in its entirety. Conversely, for peripheral vascular imaging (e.g. of the carotid), it is possible to work at 10 MHz because the depth of the vessel in which we are interested in only a few centimeters. Therefore, it is possible to see the vascular wall with good spatial resolution. If good resolution is needed for a deep internal organ, we need to try to get the probe as close as possible. This is what is done, for instance, in intravascular, transvaginal or transrectal imaging.

Another issue of which we need to be wary when increasing the frequency is the control of the position of the network lobes. These are secondary lobes which exist because of the composition of the probe, comprising regularly-spaced elements rather than a single continuous element. These network lobes are equivalent in antenna processing to the emergence of aliases when we run a Fourier transform of the sampled signals. In order to have network lobes beyond the field of view, we need to space the elements at least a distance equal to half the wavelength apart. Increasing the frequency leads to a decrease in the wavelength, so we need to decrease the distance between the elements if we wish to limit the influence of the network lobes.

1.3.3.2. *Geometry*

The most intuitive way of arranging the elements is to line them up to form what we call a linear probe. In this case, if the translational motion of the active part is occurring in a direction perpendicular to the axis of the US beam, we obtain a rectangular image. This is the type of probe that is used, for instance, in peripheral vascular imaging. Unfortunately, the usefulness of this approach is limited if we want to obtain an image with lateral dimensions greater than a few centimeters.

In this case, we arrange the elements across a convex surface. The beam is perpendicular to the anterior face of the probe, so translational motion of the active part yields a sectorial scan of the medium. With this scanning technique, we obtain an image whose dimensions are greater than those obtained with linear probes. This type of probe is used for abdominal imaging, and particularly for monitoring fetal development.

Finally, there is an even more restrictive geometric configuration. If we wish to image the heart, it is important to have a fairly broad sector of an image, but the window of view is reduced because the heart is behind the ribs. This being the case, the probe used is a linear alignment of elements across a reduced surface area, and all of the elements are used simultaneously. Delays are introduced in the interests of electronic disalignment of the ultrasound beam.

1.3.3.3. *Examples*

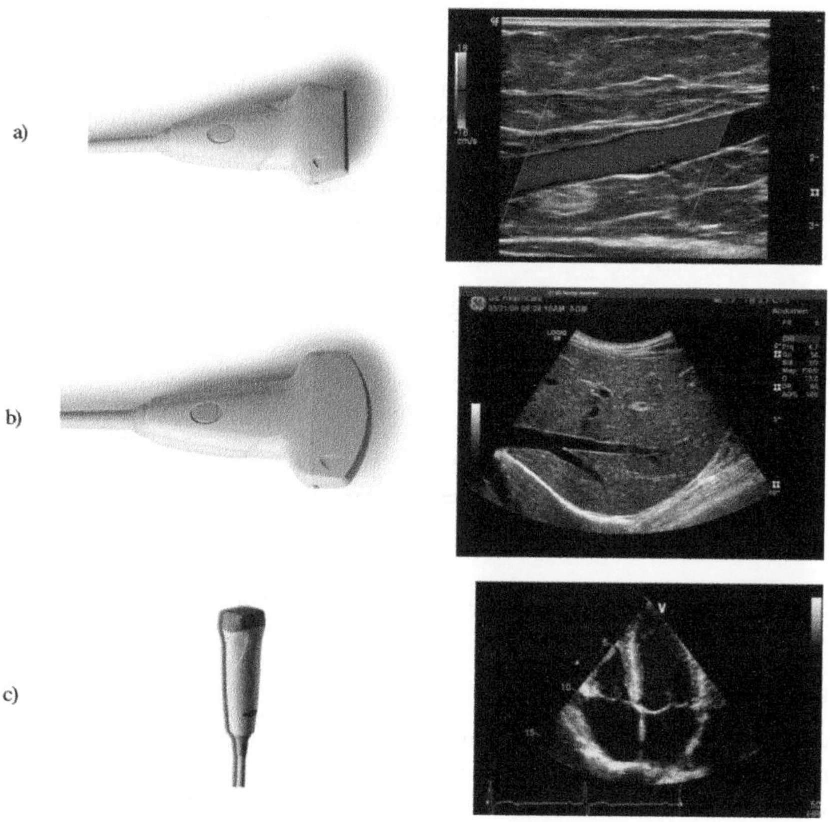

Figure 1.8. *a) Linear probe and corresponding image with rectangular geometry. This is an image of the carotid with the image of the bloodstream superposed upon it; b) Convex probe and corresponding image with sectorial geometry, showing a blind spot at the top of the image. This is an image of the liver; c) Phased array cardiac probe. Corresponding image of a heart – apical view of the four chambers. ESAOTE probes and image from GE Healthcare. For a color version of this figure, see www.iste.co.uk/fanet/medimagnet.zip*

1.3.4. *Modes of imaging, B-mode and M-mode, and harmonic imaging modes*

Pulse echo imaging techniques are based on the representation of the amplitude on the basis of the delay of an ultrasound signal reflected by a medium to represent its structure.

In order to deal with the requirements of the different domains of application, there are variants in the use of this signal to create a set of modes of ultrasound imaging which are made available to the doctor. The modes of imaging commonly used for echography are: B-mode, M-mode, harmonic mode and pulse inversion mode.

1.3.4.1. *A-mode (Amplitude)*

The oldest and simplest mode of the pulse echo method is "A-mode" (or "A-scan"). This technique uses the emission of ultrasound wave and reception of the echo along a single line of propagation. The vertical deflection corresponds to the amplitude of the RF signal (Figure 1.9). The horizontal deflection is a linear timebase which can be converted into a distance scale if the speed of propagation of the acoustic waves is known.

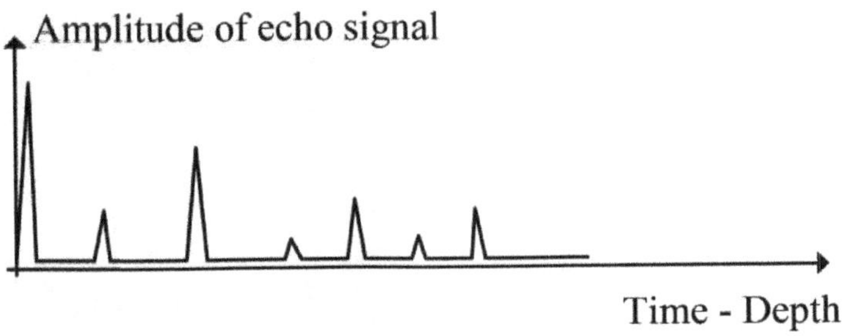

Figure 1.9. *Representation of a signal in A-mode*

1.3.4.2. *M-mode (Motion)*

M-mode or TM (Time Motion) mode is an offshoot from A-mode. It is used to visualize the motion of structures whose position or shape varies over time. M-mode shows the evolution over time of an A-type signal. The transducer is stationary. Vertical scanning of the screen shows the position of the echoes in terms of depth. The amplitude of these echoes is represented by a modulation of the spot intensity. A slow horizontal timebase can be used to juxtapose successive A-type signals. In particular, this mode is used to study the motions of the heart. In M-mode, an immobile echogenic structure is represented by a straight line. If the structure is moving, we obtain a curve of displacement over time which, in the case of the heart, has the period of the echocardiogram (see Figure 1.10).

1.3.4.3. *B-mode (Brightness)*

B-mode is at present the most widely used of medical acoustic imaging systems. It is also derived from A-mode. The two-dimensional (2D) image is constructed by juxtaposition of a large number of rows, each of which expresses an A-mode echogram. The different rows are obtained either by moving the transducer so that the propagation paths of the ultrasounds always remain in the same plane or by using an array of transducers, which enables us to explore several rows without moving the array. On the viewing screen, each row corresponds in position and direction to the trajectory of the ultrasound waves for each position of the transducer. The amplitude of the signals (after treatment) is represented on screen by modulation of the intensity of the trace. The resulting image is therefore a 2D representation of the distribution of discontinuities in acoustic impedance over a cross-section of the object being sounded. The imaging plane is formed by the direction of propagation of the ultrasound wave and the direction of displacement of the transducer or the axis of the array otherwise (Figure 1.10).

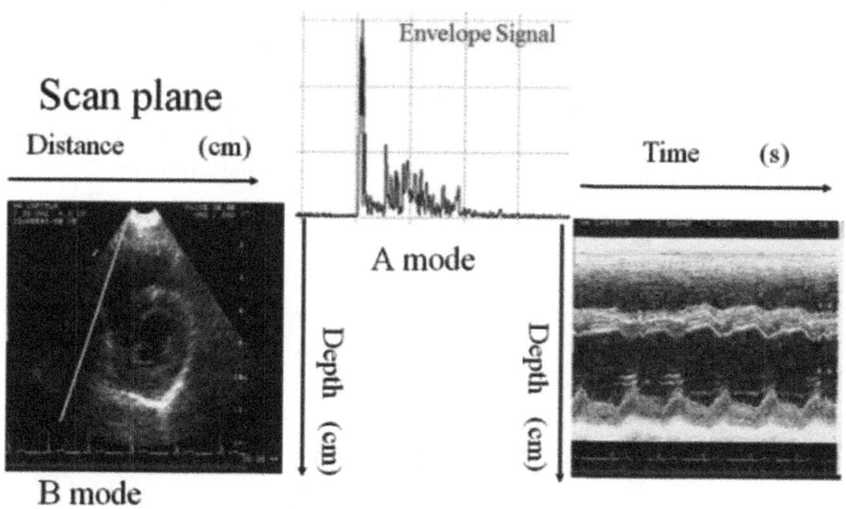

Figure 1.10. *Top center: A-mode, envelope of the RF signal along a particular direction of observation; left: B-mode, imaging plane obtained in the direction of the plane of observation of the probe; right: TM mode, juxtaposition of the data in one direction of observation over time*

1.3.4.4. *C-mode (Constant depth)*

C-mode provides 2D images at "constant depth", i.e. in a plane perpendicular to the direction of propagation of the waves. The sweeping motion of the transducer is such that the point located at a constant distance along the propagation path (axis of the beam) creates a plane. In practical terms, a time portal selects the signal corresponding to the reflection on this plane (delay constant in comparison to the emission). The luminosity of the spot is proportional to the amplitude of the echo in question.

1.3.4.5. *Harmonic imaging mode*

Harmonic ultrasound imaging is based on selection by filtering of the component at double the transmission frequency. The technique was first developed in the 1990s to use nonlinear vibration of microbubbles acting as contrast agents, working with the hypothesis that propagation in tissues was linear and the harmonics were generated only by the bubbles. In reality, the tissues also exhibit nonlinear behavior which deforms the wave transmitted over the course of propagation, and causes the apparition of harmonics. As harmonic imaging uses double the transmission frequency, the resolution is improved by the same ratio.

Harmonic imaging is restricted by the bandwidth of the echographic probes. Indeed, it is crucial to prevent any overlap between the bandwidth of the transmission signal and the bandwidth of its harmonics (Figure 1.11). This is done by limiting the transmission band and choosing it to be in the lower part of the bandwidth of the probe, which results in a decreased spatial resolution. A compromise needs to be found in terms of choosing the emission frequency.

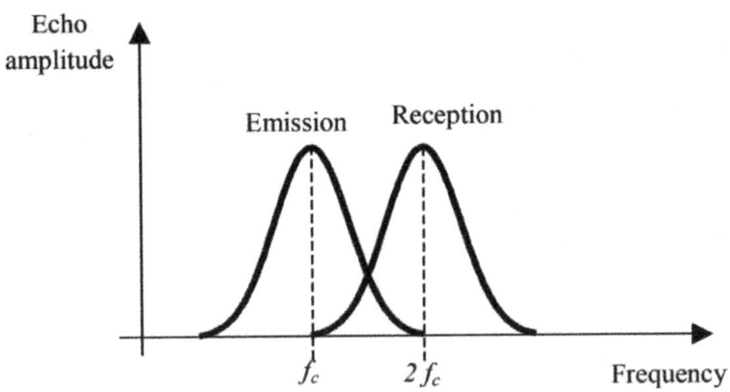

Figure 1.11. *Overlap of the bandwidth between transmission and reception*

1.3.4.6. *Pulse inversion mode*

The limitation of imaging with the second harmonic is the frequency overlap of the bandwidth of the transmission signal and that of the second harmonic. Pulse inversion imaging helps combat this effect. In pulse inversion imaging, a sequence of two pulses in phase opposition is transmitted successively into the medium. The principle of pulse inversion exploits the asymmetry of the response of the scatterers of these two pulses transmitted in phase opposition. Indeed, in the case of a linear scatterer, the sum is zero, whereas in the case of a nonlinear scatterer, the result of the summing is a non-zero value (Figure 1.12). Pulse inversion uses the whole bandwidth of the echoes received and provides a higher-resolution image. The drawback to this mode is the halving of the imaging frequency.

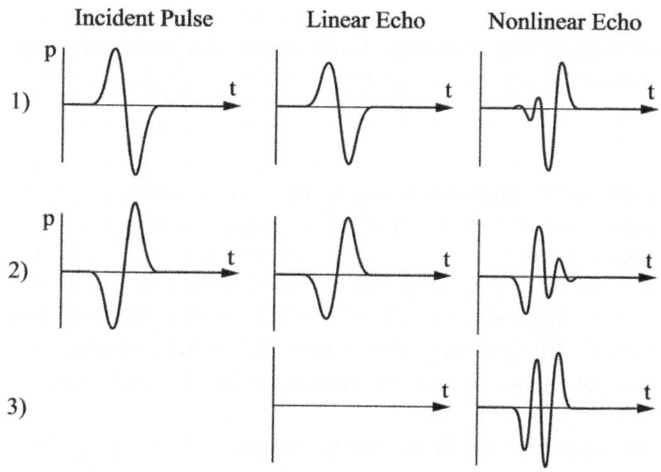

Figure 1.12. *Principle of pulse inversion [SIM 99]. 1) First emitted pulse and its echoes corresponding to the linear and nonlinear scatterers; 2) second emitted pulse; 3) summing s(t) of the two echoes with phase readjustment. The value of s(t) will be non-null only in the case of nonlinear scattering. For a color version of this figure, see www.iste.co.uk/fanet/ medimagnet.zip*

1.3.5. *Doppler imaging*

1.3.5.1. *Physical principle*

We have all had the experience of perceiving the horn of a train passing by us on the railroad track. The sound seems higher-pitched as the train is approaching, and drops in pitch as it moves away. Figure 1.13 gives a schematic illustration of this phenomenon. We suppose that the emitter E produces a sound at frequency f0, represented by the concentric circles. The sound is received by the receiver R. In a), E and R are stationary, so the observer perceives the sound at frequency f0. In b), E is coming nearer to R, who therefore perceives a higher-pitched sound (the period of repetition of the circles appears smaller for R). The frequency of the sound has increased by fD, which is called the Doppler shift. R perceives f0 + fD. In c), the reasoning is similar, but this time R perceives f0 – fD.

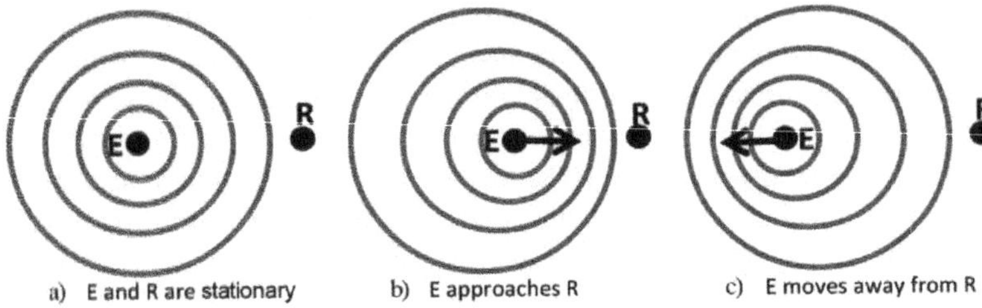

Figure 1.13. *Doppler Effect. The sound emitted by a moving object (E) is perceived by an observer (R) with a frequency shift that depends on the relative velocity between E and R*

The same physical phenomenon governs the Doppler mode in ultrasound imaging to measure the movement of blood or tissues. In this case, the wave emitted by the probe plays the part of an emitter at the nominal frequency f0 of the probe. Red blood cells, or more generally the moving components of biological tissues, backscatter a wave whose frequency is shifted by the Doppler shift in relation to the nominal frequency. By analyzing this wave, the system is able to calculate the Doppler shift frequency and deduce the velocity of the tissues being examined.

Consider the case of a single scatterer, moving at velocity v in a direction that forms an angle θ with the axis of the ultrasound beam (Figure 1.14).

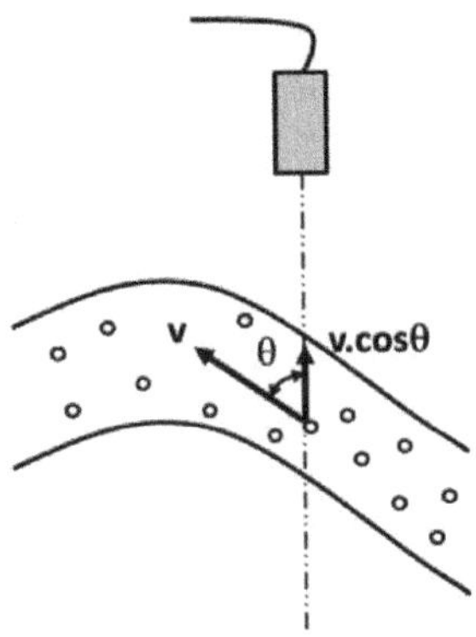

Figure 1.14. *Principle of measurement of axial velocity using the Doppler Effect*

The Doppler frequency, which corresponds to the shift (either positive or negative) in relation to the frequency of the US emission, is given by the formula:

$$f_D = -2 f_0 \frac{v}{c} \cos\theta$$

Note that when the angle θ is equal to 90°, then fD is 0. This means that it is not possible to estimate the velocity of a reflector whose motion is perpendicular to the firing axis of the ultrasound probe. In practice, we choose an angle of less than 60°. In any case, it is necessary to know the angle θ in order to estimate the speed of blood-flow in a vessel; this is one of the constraints with Doppler imaging. The velocity of propagation of the ultrasound waves is taken to be constant in biological tissue, c = 1540 m/s, which is not always absolutely true. Despite these limitations, the Doppler modes are essential to doctors in vascular imaging to measure flows ranging from a few cm/s to several m/s with some diseases.

1.3.5.2. *Continuous-wave Doppler*

As its name suggests, continuous-wave Doppler imaging is based on the continuous emission of a US wave by a transducer. A second transducer continuously receives the signal backscattered by the moving tissues in the field of emission/reception. Thus, continuous-wave Doppler imaging gives us overall information about the distribution of the velocities in all areas traversed by the beam. Technically, the received signal is demodulated by the emission signal and then put through a low-pass filter. Spectral analysis of the received signal is performed by a Fourier transform, and yields a frequency distribution corresponding to the distribution of the velocities and the direction of the flow over time. It can be used to record very high-speed flows and thus precisely analyze the maximum speed of the flow with great precision. The main limitation relates to the location of the flow being analyzed, which is not hugely accurate. Figure 1.15 gives an example of a Doppler analysis of the carotid.

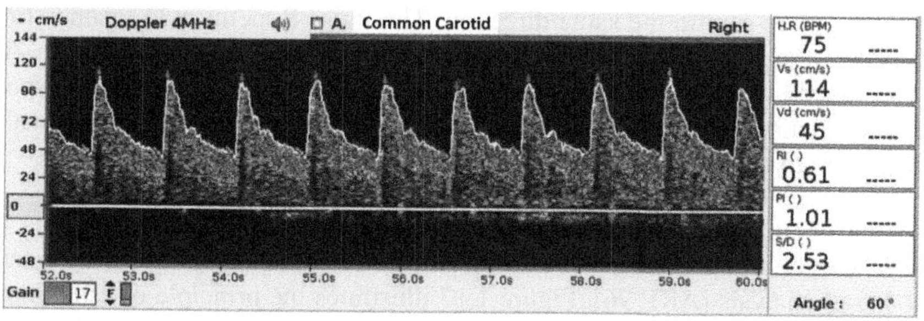

Figure 1.15. *Example of a continuous-wave Doppler image of the carotid. Data from Atys Médical, www.atysmedical.fr. For a color version of this figure, see www.iste.co.uk/fanet/medimagnet.zip*

Continuous-wave Doppler mode requires two separate transducers. They can both be included in a small device, which can easily be put in position, e.g. to measure venous blood flow or to monitor cerebral blood flow.

1.3.5.3. *Pulsed Doppler*

Pulsed Doppler mode helps to circumvent one of the limitations of continuous-wave mode by enabling the user to choose a sample volume within which to perform the velocity distribution analysis. The size, depth and location of the sample volume are controlled by the user on the basis of the B-mode image (see Figure 1.16).

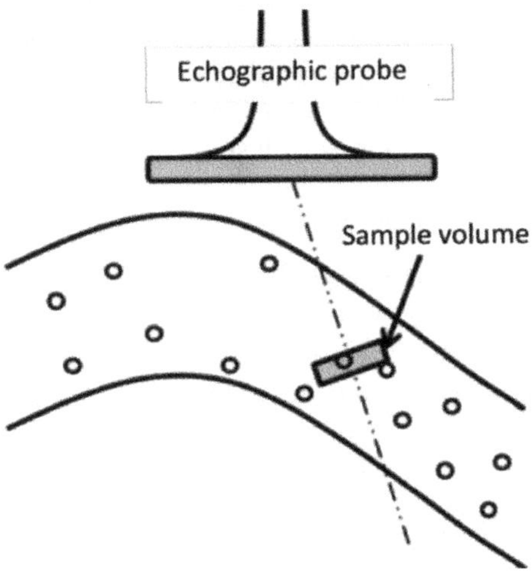

Figure 1.16. *Schematic representation of the location of the sample volume for pulsed Doppler imaging*

The principle behind pulsed Doppler mode is the analysis of data from several successive pulses sent in the same direction. The Pulse Repetition Frequency (PRF) is limited so the echoes of one pulse are not confused with those of the next. Also, the PRF needs to be sufficiently high to estimate the highest maximum speeds. Indeed, the sampling theorem limits the Doppler frequencies estimated without spectral overlap to the frequency PRF/2.

At the depth determined by the sample volume, analysis of the signals thus acquired can be performed in several equivalent ways, by temporal, frequential or phase-based analysis [ARE 96]. Figure 1.17 illustrates the principle of estimation of the Doppler frequency at a given depth on the basis of the Doppler pulses at the PRF.

A low-pass anti-aliasing filter prevents overlap due to excessive velocities, and a high-pass filter is used to eliminate the influence of low motions due to breathing or other movements of the tissues which we do not want to take into account.

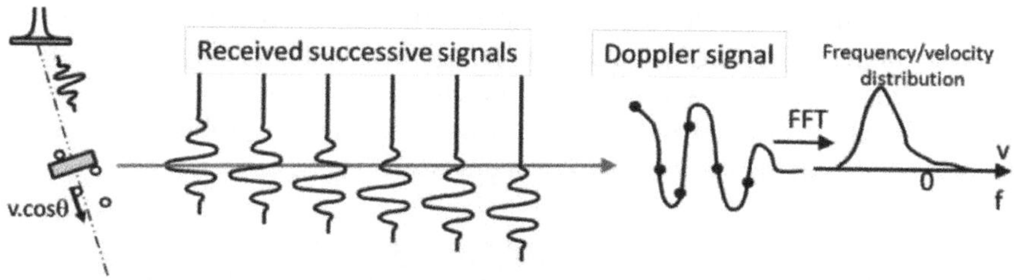

Figure 1.17. *In pulsed Doppler mode, various successive pulses are sent in the same direction at the PRF*

Finally, the velocity distribution in the sample volume is computed by the Fourier Transform (Figure 1.17). Figure 1.18 shows an example of blood flow analysis in Pulsed-Wave (PW) Doppler mode, with the sonogram that is displayed during the cardiac cycle. Note that the frequency of the B-mode images is inevitably reduced by the superposition of the Doppler pulses, but modern US imaging systems are able to smoothly interweave the different modes to preserve the fluidity of the visualization.

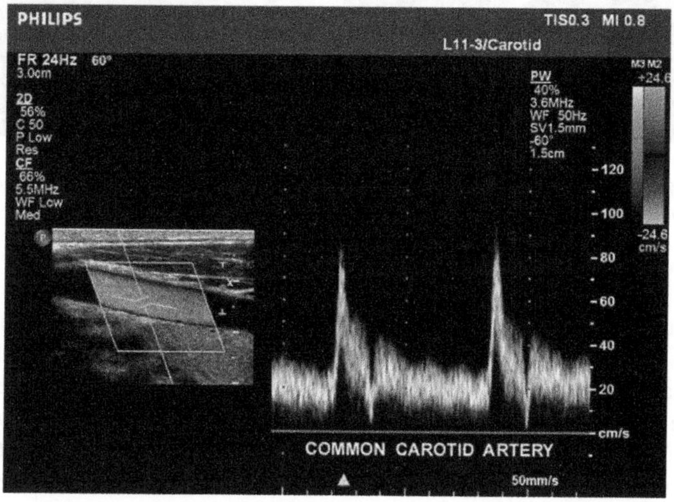

Figure 1.18. *Example of analysis of the blood flow in the common carotid artery. On the left, the sample volume is chosen in B-mode. The PW Doppler mode, whose results are displayed on the right, shows the evolution of the velocities in the carotid artery over the course of a cardiac cycle. For a color version of this figure, see www.iste.co.uk/fanet/medimagnet.zip*

1.3.5.4. *Color Doppler and Power Doppler*

The modes of Doppler imaging which are Color Doppler, Power Doppler or Color M-Mode Doppler enable us to see bloodstreams in real time in the form of a color image. Thus, we obtain velocity-related information for the whole of a region of interest, rather than simply for a sample volume. Color coding is used for the scale of speed of the flows and also their direction. The color is blue when the flow is getting further away and red when it is approaching the probe. Color Doppler imaging mode is superposed on B-mode in grayscale.

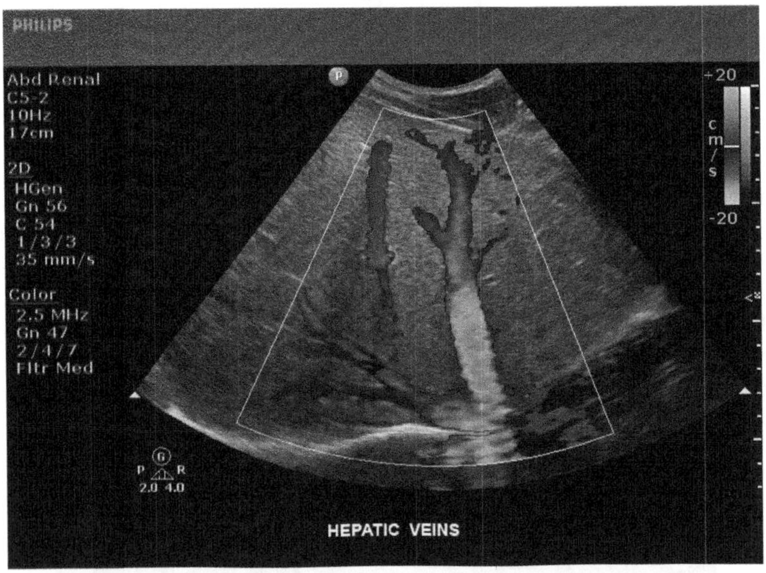

Figure 1.19. *Example of Color Doppler imaging, showing the ascending and descending flows of the vascularization of the liver. For a color version of this figure, see www.iste.co.uk/fanet/medimagnet.zip*

The principle used to estimate the speed at each pixel of an image is the same as that set out in the discussion of PW Doppler, but numerous methodological and technological developments have been necessary since the 1980s to get us to the point where today, we can obtain real-time visualizations of flow maps such as those shown in Figure 1.19. This developmental progress has been so great that now, multiple modes can be shown simultaneously on clinical systems. The evolution in terms of electronics and computing has largely contributed to these advances. New signal- and image-processing algorithms have also been developed, e.g. for computing a Fourier Transform, estimating frequency or filtering. Color Doppler mode is now an indispensable tool in the diagnosis of vascular and cardiovascular diseases. However, the estimation of speed is less accurate than with PW Doppler mode, because at each pixel, it is based on a limited number of samples. Doctors also need to be aware of the Doppler angle, which is identical throughout the region of interest, whereas the vessels may have various different orientations.

A variant of Color Doppler mode is Power Doppler mode, which uses a color scale to represent the amplitude of the Doppler signal, irrespective of the direction. Power Doppler performs best when imaging flows in small vessels, even when the flow is slight. Figure 1.20 shows 3D vascularization of the placenta in Power Doppler mode.

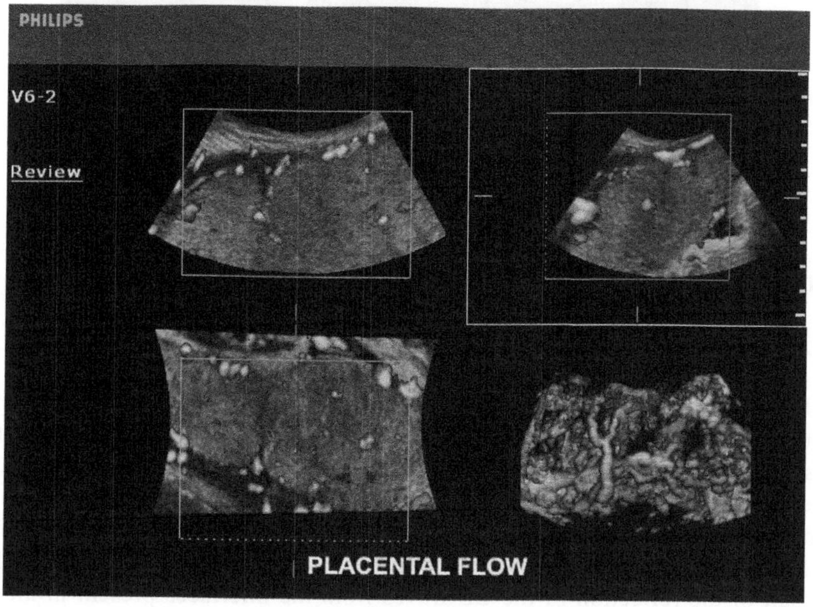

Figure 1.20. *Example of Power Doppler imaging, giving a 2D and 3D view of the vascularization of the placenta. For a color version of this figure, see www.iste.co.uk/fanet/medimagnet.zip*

The field of Doppler imaging is still changing at present, and new modes are being developed to serve practitioners' needs. For instance, one might cite Color M-mode Doppler, which enables us to view a Doppler map on a single row over time, or Tissue Doppler, whereby we can view the movement of tissues rather than of flows. Thus, cardiac diseases can be diagnosed by looking at and measuring faults with the contraction of the myocardium (Figure 1.21).

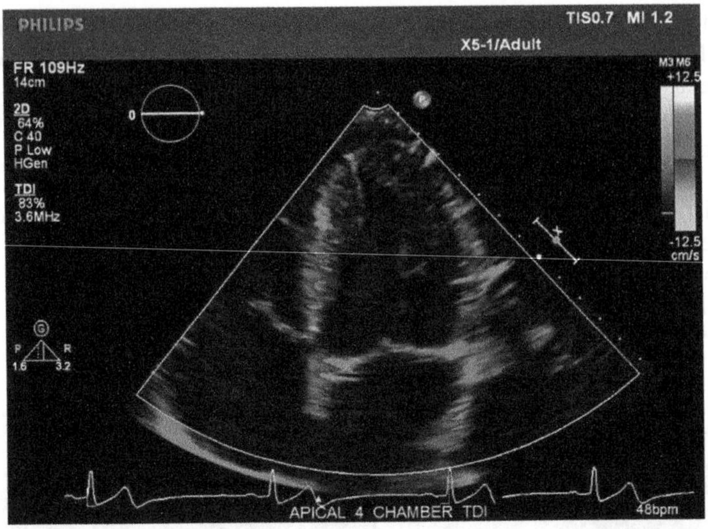

Figure 1.21. *Example of Tissue Doppler imaging of the heart – apical 4-chamber view. For a color version of this figure, see www.iste.co.uk/fanet/medimagnet.zip*

1.4. The US image

1.4.1. *Properties of speckle, echostructure and statistical laws*

The RF signal, generated by the interaction between the wave propagating in the medium and the constituents of that medium, contains information about the acoustic properties of the tissues, such as attenuation or backscatter capacity. The conventional ultrasound image is the result of the convolution of the RF signal with the transfer function of the imaging system. The image formed, called the envelope image, makes only partial use of the information available in the RF signal. It shows the result of the reflection of the US waves from the interfaces present in the zone being investigated (specular echoes) and backscattering on the microscopic structure of the medium (speckle). Speckle reduces the contrast between the different tissues in ultrasound images and makes it difficult to identify the contours of the different regions. Therefore, segmentation techniques using contour detection have problems in handling ultrasound images. However, speckle is related to the makeup of the tissues, and by studying it, it is possible to discover certain characteristics of the medium being explored.

1.4.1.1. *Tissue composition and speckle*

Speckle is a consequence of the diffusing behavior of the multitude of scatterers present in the ultrasound beam. This depends on a number of factors:

1) the difference in acoustic impedance between the scatterer and the surrounding medium: the greater the difference, the greater the diffused energy;

2) the spatial distribution of the inhomogeneities, which may be totally random (e.g. in blood) or more ordered (in muscles, for instance);

3) the density of scatterers, which influences the intensity of the backscattered wave;

4) the size of the inhomogeneities and the distance between them. When these dimensions are smaller than the wavelength λ of the ultrasounds, we observe Rayleigh scattering; when they are the same as the wavelength, diffraction occurs and finally, when they are much greater than the wavelength, the object behaves like a mirror, and geometric phenomena such as reflection and refraction predominate.

Speckle appears in images when the characteristic dimension of the reflectors is smaller than or similar to the wavelength λ, and when the elementary resolution cell contains multiple reflectors. Each point in the reconstructed image corresponds to the result of the backscattering of the incident wave within the resolution cell.

Over the propagation path of the ultrasound pulse through the tissue, the resolution cell shifts, bringing different sets of reflectors into play. A random distribution of reflectors gives a random realization of the RF signal, determined each time by the particular arrangement of the reflectors within the resolution cell.

1.4.1.2. *Tissue model and statistics of the image envelope*

Depending on the density of scatterers in the resolution cell and their distribution principle throughout the tissue being imaged, the statistical behavior of the image envelope varies. When the scatterers are smaller than the US wavelength, randomly distributed and in a sufficient concentration, we observe an echo signal whose statistical properties can be described by a Rayleigh probability density function p(A):

$$p_A(A) = \frac{A}{\sigma^2} \exp\left\{-\frac{A^2}{2\sigma^2}\right\}; \qquad A \geq 0$$

σ^2 is a parameter which depends on the quadratic mean of the amplitude scattered by the particles. Therefore, it matches the energy of the signal. This distribution (Figure 1.22) has the property of having a constant SNR of 1.91.

It has been shown that the average level of amplitude is proportional to the square root of the density of scatterers [THI 03] and therefore that the average intensity is proportional to that density.

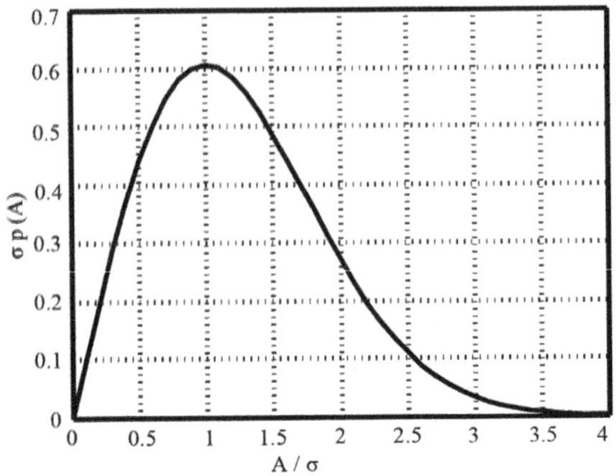

Figure 1.22. *Rayleigh probability density function*

It should be noted that this model assumes the presence of a large number of scatterers, randomly distributed throughout a resolution cell. If this hypothesis is verified, we then speak of "fully developed" speckle.

When the scatterers are also small but are distributed regularly throughout the medium (muscle tissue, presence of vessels, etc.), a coherent component appears in the echo signal. The distributions of the resulting signals are then described by a Rice probability density function.

The Rayleigh probability density and its extension to the Rice probability density constitute a good approach to modeling of the statistics of the image envelope. However, the constraint of a high number of scatterers in the medium limits their domain of applicability. When the hypothesis of a large number of scatterers is not vindicated, the probability density of the backscattered signal envelope is given by the K-distribution. This distribution is described by a parameter α which characterizes the density of scatterers in the medium.

The Nakagami distribution, put forward more recently in the literature, exhibits the advantage of modeling all configurations of scatterers that are found in tissues. Two parameters, m and Ω, are used to represent distributions of reflectors which may or may not show organization, and with no constraint on the density. The Nakagami probability density function is given by:

$$p_X(x) = \frac{2m^m x^{2m-1}}{\Gamma(m) \cdot \Omega^m} \cdot \exp(-\frac{mx^2}{\Omega})$$

where $\Gamma(.)$ is the gamma function. $m > 0$ is a shape parameter and $\Omega > 0$ is a scale factor. m and Ω can be estimated on the basis of the different moments of the distribution [SHA 01].

Thus, the 2nd-order moment gives Ω, which corresponds to the energy of the signal:

$$E[X^2] = \Omega$$

and the 2nd- and 4th-order moments give m by the equation:

$$m = \frac{E[X^2]^2}{E[X^4] - E[X^2]^2}$$

The Nakagami parameter, m, is directly linked to one of the three distributions traditionally encountered:

$m = 1$ corresponds to a Rayleigh scattering of the scatterers;

$0 < m < 1$ corresponds to a K-distribution arrangement of the scatterers;

$m > 1$ corresponds to a Rician distribution of the scatterers.

Figure 1.23. shows the Nakagami distribution for different values of its parameters.

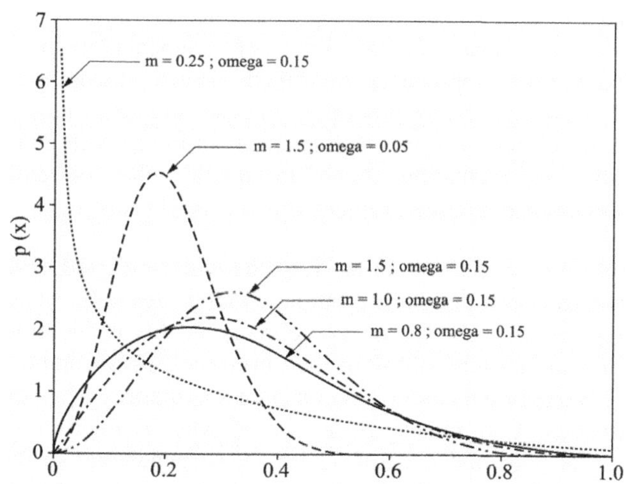

Figure 1.23. *Nakagami distributions*

1.4.1.3. *Logarithmic compression*

The statistics described above are valid for envelope images which have not undergone logarithmic compression of the echo dynamic. However, the dynamic of the signal envelope may be as high as 100 dB, and on a black-and-white image, the human eye can only perceive a dynamic of around 30 dB. For this reason, the ultrasound systems that are routinely used carry out logarithmic compression of the signals to reduce the dynamic of the echoes, thereby facilitating the observation of less echogenic areas. In practice, this entails logarithmic amplification of the signal from the transducer, instead of linear amplification.

The nonlinearity of the amplification obviously changes the statistics of the signal envelope. Thus, the distribution models presented about are no longer valid.

1.4.1.4. *Statistics of the RF image*

The speckle which appears on envelope images and the associated probability densities stem from low-frequency filtering of the RF signal. As research labs increasingly have access to RF signals, it is tempting to model the statistics of the RF signal with an appropriate probability density function. That is what was done by Bernard *et al.* [BER 07], who proposed a generalized Gaussian distribution for segmentation purposes. The advantage to this is that its expression is simple, and it offers a robust estimation of the parameters.

1.4.2. *Segmentation of US images*

The number of articles on the topic in the existing body of literature demonstrates the importance of the issue of segmentation in digital image processing. The various approaches described always attempt to adapt to the peculiarities of the images being processed and the intended outcomes to give an optimum result, which presents the information in an image with a more condensed and easily-usable representation. How to qualify the successfulness of a segmentation technique remains a problem which is rarely solved.

When ultrasound images are processed, the noise-afflicted and low-contrast aspect of the images means that segmentation, which is generally based on grayscale values, is always a tricky step. The mediocre image quality is due to the presence of speckle, to the inconstant attenuation of the ultrasound signals and to the reflection from specular structures which depends on their orientation. Thus, a contour may appear discontinuous. Also, in spite of the use of dynamic focusing techniques, the lateral resolution is poor and structures that are perpendicular to the direction of propagation of the US waves appear blurred.

As well as the poor quality of the images, we are also faced with the fact that, in most cases, the structures we are trying to detect have similar acoustic characteristics to those of the surrounding tissues. Therefore, the contrast between the different tissues is low, and it is difficult to accurately determine a boundary. The attenuation of the ultrasounds depends on the nature of the tissues being explored. Consequently, a homogeneous tissue does not appear entirely homogeneous on the image. It can be seen with a slight variation in intensity in the direction of propagation, in spite of the correction offered by Time Gain Compensation (TGC), which is constant and independent of the nature of the tissues.

Given that the information provided by the intensity of the pixels is shrouded in speckle, an alternative way to discriminate between different tissues is to use other information such as that gleaned from the measurements taken for tissue characterization. This information is usually based on textural analysis performed on the ultrasound envelope images, or on acoustic measurements of the RF signal.

A further improvement to the segmentation can be made if we have access to three-dimensional (3D) data. The continuity of the data in 3D facilitates more robust detection of the contours of an object. This remains true if the dataset is constructed on the basis of the evolution over time of a dynamic image in the same plane.

Segmentation refers to the notions of difference and similarity perceived by the human visual system. This gives rise to two approaches, commonly called the "region-based" and "contour-based" approach. The aim of the region-based approach is to reveal homogenous regions using a criterion (grayscale, texture, etc.), while the contour-based approach attempts to find contours or boundaries of regions, where there is a sudden variation in the same criterion.

The quality of ultrasound data is insufficient for segmentation on the basis of low-level criteria, which are based solely on the pixels. More complex methods, which facilitate the combination of the constraints and *a priori* knowledge in the process of segmentation, are necessary.

1.4.2.1. *Contour-based approaches*

Boundary approaches are amongst the most classic of methods for image segmentation. These methods generally require an *a priori* model of the discontinuities being sought, and operate on a very localized level. The techniques most commonly used in ultrasound imaging are based on variational approaches, such as deformable models, which produce closed contours. They take account of overall information about the contour, which is generally gained from an *a priori* contour model. A deformable model is a geometric structure which evolves iteratively so as to adjust to the boundaries of the objects sought. 2D deformable models such as active contours are represented by a closed curve. Segmentation is formulated as an optimization problem, with the goal being to determine the curve which minimizes an energy functional. Level sets have also been suggested as an alternative to active contours.

1.4.2.2. *Region-based approaches*

Unlike boundary approaches, which look for dissimilarities, region-based approaches look for similarity over contiguous collections of pixels. These similarities can be used to generate a map of bounded regions. However, the localization of the boundaries is generally still not hugely accurate. Of the various region approaches, we can essentially identify three types of methods.

Classification methods: the classes are constructed on the basis of the attributes of their pixels. Each pixel is assigned a unique class. The most conventional methods of segmentation of ultrasound data use Bayesian classifications [BOU 03] or neural methods [HUI 98]. Segmentation by classification techniques falls down on the problem of the choice of the number of classes (with unsupervised approaches) and choice of the attributes.

Region growing methods: these methods implicitly integrate spatial information into the segmentation process. The regions are created by way of an initialization phase which corresponds to the choice of a new seed (starting point for a new region) and an iterative phase which adds neighboring pixels to the seed using a homogeneity criterion until convergence.

"Split–merge" methods: after an initialization stage, the process of segmentation is iterative, alternating between two phases: a phase of splitting of all non-homogeneous regions and a phase of merging of all adjacent regions such that the resulting region always fulfills the criterion of homogeneity.

1.4.2.3. *Example of clinical applications*

In echocardiographic imaging, segmentation is based on methods using active contours to distinguish the left ventricle or the different structures, right to the epicardium [MIS 03, MIG 01, YAN 03]. Ultrasounds are also used to detect focal lesions in ultrasound images of the breast (Figure 1.24). In this case, segmentation is done by a region-based approach using a classification tool (such as a neural network) based on local readings [CHE 02]. Segmentation of the contours of the prostate is also a first step in guiding a biopsy needle. The introduction of *a priori* information about the shape and nature of speckle in the segmentation process, with a view to automatic segmentation, has been suggested [ABO 04]. Segmentation of vessels in endovascular images affords us access to quantitative data about the potential presence, size, shape and nature of atherosclerotic lesions. In this domain, many methods, generally based on deformable models, have been put forward

[BRU 04]. In obstetrics, image segmentation provides measurements which can be used to evaluate a fetus' growth and detect any malformations.

In many applications, semi-automatic methods can be used. *A priori* information is input by an operator, who manually positions the seeds or initial contours of the regions sought.

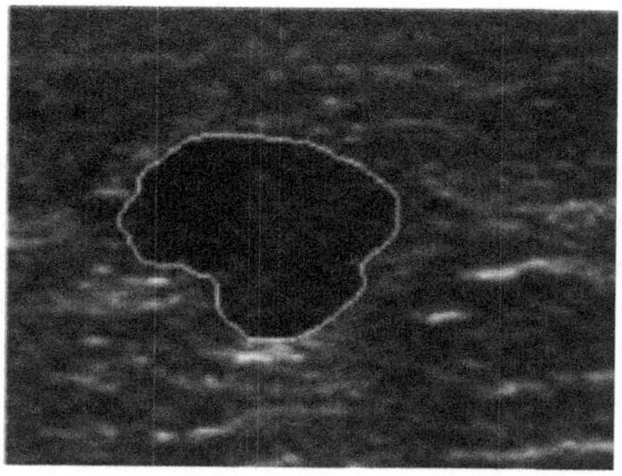

Figure 1.24. *Bayesian segmentation of an ultrasound image of a breast lesion*

1.4.3. *Simulation of US images*

1.4.3.1. *Context*

In ultrasound imaging, as with other types of imaging, simulation of images in research projects is a crucial step. Firstly, it helps to perfectly understand and grasp the different stages in the creation of the image. It can be used to simulate all envisageable situations in a clinical context in order to test a set of data-processing methods and algorithms. Thus, different situations are considered and serve as reference points. Recently, new strategies have emerged which enable us to simulate ultrasound images in real time and offer learning aids to doctors.

1.4.3.2. *Methodology*

– Introduction

Two major categories of methods exist to simulate US images. For the most part, these methods have been published within the past thirty years. The first type of approach is based on linear convolution, while the second considers the acoustic aspects of the ultrasound field and the behavior of the probe. All these methods are founded on the representation of the medium as a collection of scatterers which backscatter the US wave towards the probe. The amplitude of the backscattering signal from the medium characterizes the echogenic or non-echogenic nature of a medium.

– Linear model

The linear model was initially advanced by Bamber and Dickinson [BAM 80]. A convolution between all the scatterers in the medium and the pulsed response of the system (point spread function – PSF) is performed in order to generate the US image. However, as the PSF is considered to be constant throughout the medium, the resolution of the ultrasound image does not vary with depth, as it does in clinical imaging.

– Acoustic model

The acoustic model aims to resolve the approximation made with the PSF in the linear model. Thus, the ultrasound field created by the active elements of the probe is considered at all depths in order to take account of the aspects of focusing of the US waves. These efforts, begun in the 1970s, were continued by Jensen with the creation of the Field II software suite [JEN 92]. There are other tools in existence, such as DREAM [PIW 99], Ultrasim [HOL 01] and CIVA [CAL 06].

1.4.3.3. *Existing software*

– Field II

Field II is the reference software in the ultrasound community. With it, we can simulate RF images for different configurations of probes. It takes account of the interaction of an acoustic wave propagating through the medium with the scatterers which make up that medium. The main weakness of Field II is the very long computation time needed to construct an image. However, it has been shown to be possible to reduce that computation time by distributing the computation of each US row between parallel computers. Also, only linear propagation is considered with Field II.

– CREASIMUS

CREASIMUS, based on the linear convolution model [MAR 09], can simulate US images of vessels and model an arbitrary flow. The method used and its implementation have enabled users to very rapidly simulate ultrasound sequences of several seconds.

– CREANUIS

CREANUIS is a recent nonlinear RF-image simulation tool [VAR 13]. CREANUIS combines a tool for simulation of the nonlinear US field [VAR 11] with an image reconstruction algorithm similar to that used by Field II. The field of propagation can therefore be considered for media with inhomogeneous nonlinearity, and it has been implemented on a GPU to reduce the computation time for the US image as far as possible [VAR 11]. Thereby, the computation time has been reduced by a factor of between 2 and 4 in comparison to Field II, also providing a harmonic image.

Figure 1.25 shows images simulated with the different simulators.

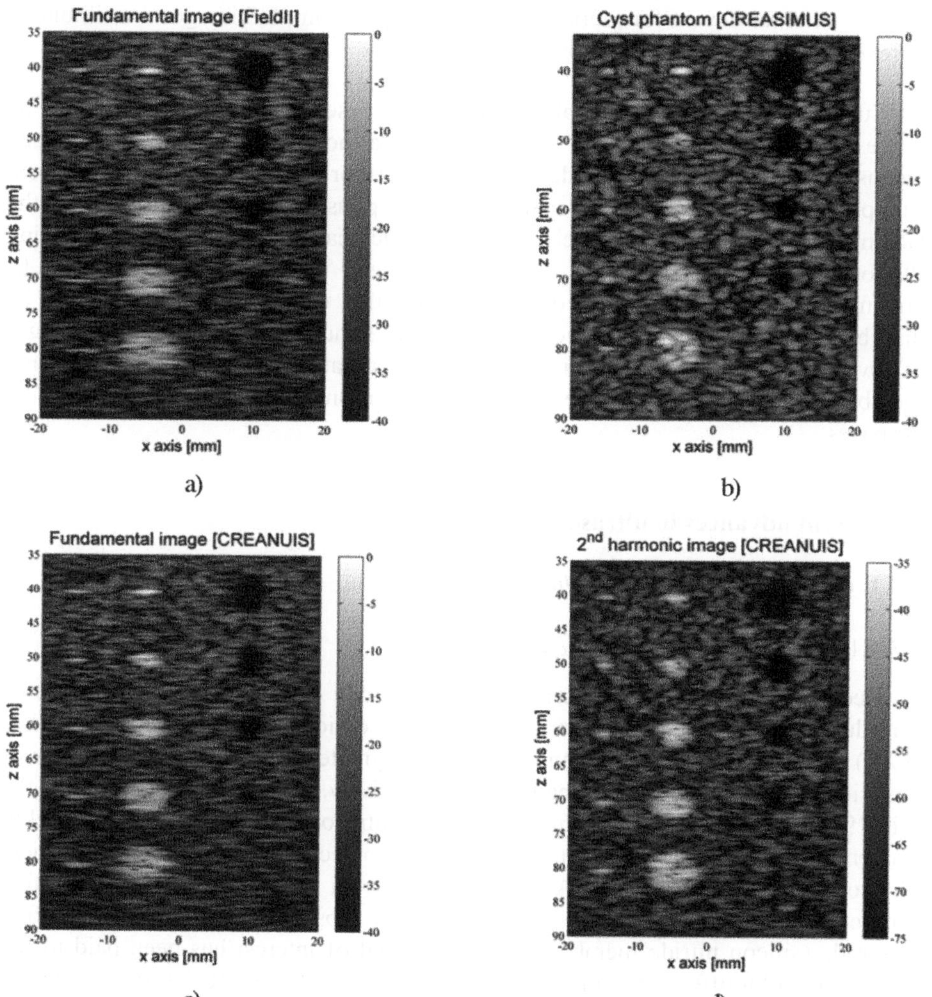

Figure 1.25. *Illustration of the images obtained with different simulators: Field II a), CREASIMUS b) and CREANUIS (c-d). The constant resolution in the CREASIMUS image b) is visible when compared to the acoustic models a) c) and d). The computation time is less than one second with image b), 2 hours with image a) and 33 minutes for the CREANUIS images (c-d)*

1.4.3.4. *Prospects*

The prospects for simulation are many. GPU programming has significantly reduced the computation time, and new tools using the linear convolution model are able to simulate US images extremely quickly [KUT 09]. By coupling this method with an external device for motion-tracking of a phantom, US images are simulated in real time which can be used for learning purposes.

Currently, CREANUIS is able to simulate the second harmonic. It would be interesting to have access to higher-order harmonics. Similarly, US image reconstruction needs to be speeded up, and GPU portage is one possible solution. Computation of the transducer's spatial pulsed response on a GPU would enable the US image to be simulated entirely on the graphics card. VIP (the Virtual Imaging Platform) at the Creatis laboratory (see http://vip.creatis.insa-lyon.fr/documentation/) enables the computation of the different US images to be distributed, and thereby use an acoustic model to simulate sequences of US images of several seconds in length. It is also able to simulate various modes of imaging, thereby extending the usefulness of the simulation to research and validation purposes.

1.5. Recent advances in ultrasound imaging

1.5.1. *Generation/emission of ultrasounds*

1.5.1.1. *Evolution of US probes, cMUT technology*

Recently, a great deal of research has been done on micromachined ultrasonic transducers (MUTs). In comparison with conventional Lead-Zirconate-Titanate (PZT) transducers, this new technology offers a greater bandwidth both in terms of emission and reception, whilst maintaining sensitivity. Since their emergence in the mid-1990s, cMUTs (capacitive micromachined ultrasonic transducers) have quickly proven to be an interesting alternative to piezoelectric transducers for medical ultrasound imaging [CAL 05]. A cMUT comprises a set of capacitive cells (a few μm across) composed of an electrode fixed to a substrate (silicon: Si) and a free electrode (silicon nitride membrane). A great deal of interest has been paid to the large bandwidth offered by this type of transducer (>100%) [HUA 06]. Because of its capacity to receive echoes at frequencies of up to four or five times the transmission frequency, the cMUT probe affords us the possibility of performing nonlinear imaging.

1.5.1.2. *Transverse oscillation (TO)*

At the root of transverse oscillation (TO) methods is the issue of multidimensional motion estimation [LIE 10]. Indeed, whether in terms of blood flow imaging, elastography or echocardiography, which are the main applications for TO methods, in all cases it is the need to estimate motion in many spatial directions which has caused the development of transverse oscillations. For all these applications, the observation is the same: we are able to accurately estimate the apparent displacement along the axis of propagation of the ultrasounds but, either it is impossible to estimate the transverse motion (in Doppler mode), or the estimated motion is of poor quality. The idea of transverse oscillation techniques, therefore, is to attempt to produce, in the direction perpendicular to the axis of propagation of the ultrasounds, a US field with similar characteristics to those of the axial field, i.e. the field that follows the axis of propagation of the ultrasounds.

– Principle of beamforming for TOs

TOs are obtained by beamforming. The shape of the pressure field results from the superposition of the fields in transmission and reception. Unlike with transmission, reception can be dynamic and can therefore be adapted as a function of the depth. The easiest solution to implement, therefore, is to emit a plane wave, which will have little impact on the final shape of the field, and form the TOs on receipt of them. The design of the beamformer on reception is based on the choice of two sets of parameters, the delays and weighting functions for the different elements. When the Fraunhofer conditions are all fulfilled, a simple approximation can be used to link the shape of the field to that of the weighting function. These conditions are fulfilled when the wave in question is focused. This being the case, the lateral profile of the field h(x) can be approximated by the Fourier transform of the weighting function $p(x_e)$ for the elements with lateral position x_e.

$$h(x) = TF\left\{p\left(\frac{x_e}{\lambda z}\right)\right\}$$

Note, however, that this transform depends on the distance from the focal point z, and the weighting function therefore needs to be variable with varying depth. λ is the wavelength of the transmission wave.

In order to obtain a field with a clear shape, we need only focus the wave and choose the inverse Fourier transform of the desired profile as the weighting function. For TOs, we need to produce a field which exhibits oscillations. We can then write the lateral profile of the field:

$h(x) = w(x).\text{mod}(x)$ where w(x) is a weighting window, used to limit the extent of the field, and mod(x) is a sine- or cosine function. If we take a simple example as illustrated in Figure 1.26:

$$h(x) = \exp^{-\pi\left(\frac{x}{\sigma_x}\right)^2} \cos\left(\frac{2\pi x}{\lambda_x}\right)$$

then

$$p(x_e) = TF^{-1}\left\{h\left(\frac{x}{\lambda z}\right)\right\} = K\left\{\exp^{-\pi\left(\frac{x_e - x_0}{\sigma_0}\right)^2} + \exp^{-\pi\left(\frac{x_e + x_0}{\sigma_0}\right)^2}\right\}.$$

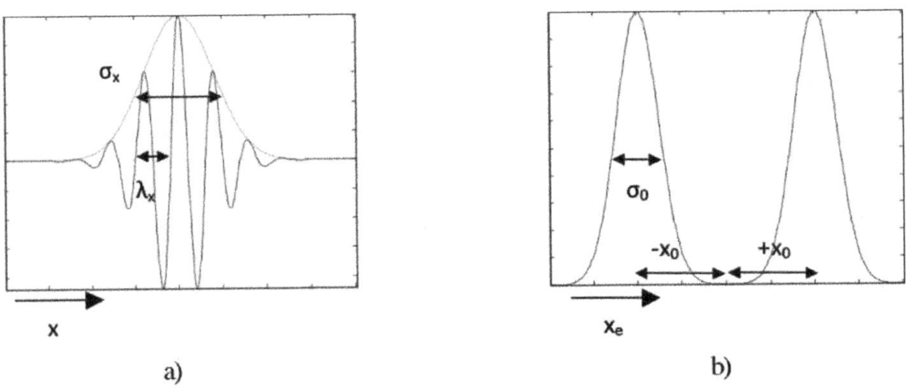

Figure 1.26. *Lateral profile of the PSF a), and associated weighting function b). All of the parameters are represented*

– Applications of TOs

The first application of TOs is blood flow estimation [JEN 98]. This technique is currently built into an ultrasound system made by B&K. Figure 1.27 shows an estimation of blood flow obtained using the TO technique.

TOs have also been applied in the imaging of the elasticity of biological tissues (elastography), where it is also necessary to estimate the 2D displacement [SUM 08, LIE 07]. The most recent studies focus on the application of TOs in sectorial geometry for echocardiography [LIE 08]. In this case, the geometry of acquisition complexifies the beamforming technique.

Finally, it should be noted that with the emergence of matrix probes for 3D imaging, the extension of TOs to a 3D application is envisageable, and is the subject of research in the area of flow.

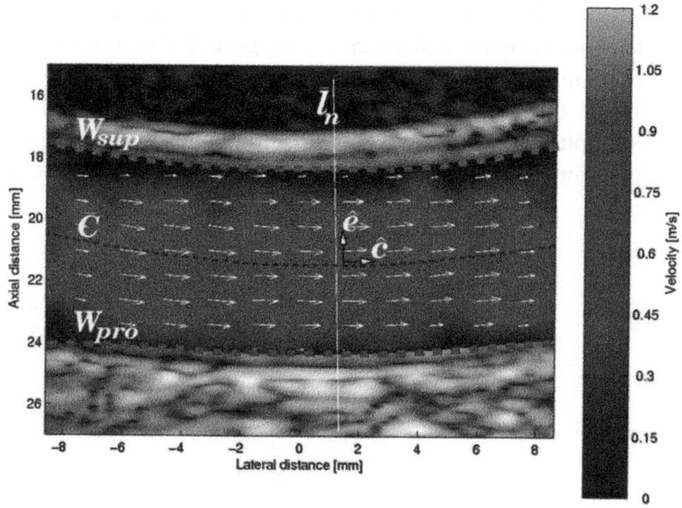

Figure 1.27. *Representation of a flow obtained with the TO technique. Note that this flow, which runs parallel to the surface of the skin, would be difficult to estimate using a conventional Doppler technique [HAN 09]*

1.5.1.3. *Synthetic aperture imaging*

The spatial resolution of ultrasound images is a point on which it is difficult to make any progress using conventional focusing methods (Figure 1.28). With a delay-and-sum approach, dynamic focusing on reception of the signals enables us to focus optimally on each point in space for reception. For transmission, it is possible to have only one focal point. Multiple focusing would necessitate multiple transmissions, which can be done, but to the detriment of the imaging frequency.

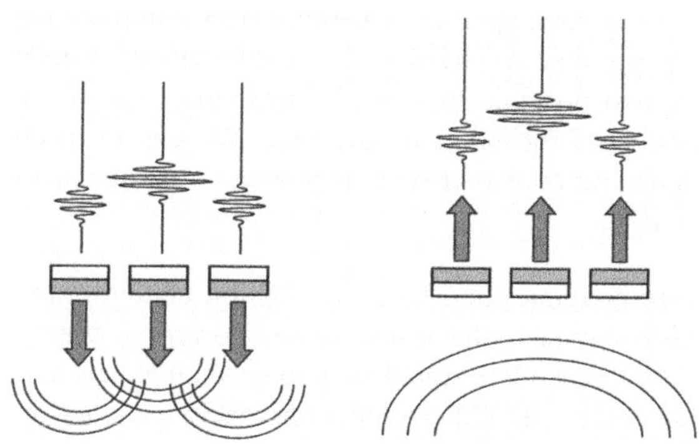

Figure 1.28. *In conventional imaging (illustrated here with three active elements), all the elements transmit and all the elements receive*

A different approach, called synthetic aperture imaging, facilitates dynamic focusing both for transmission and reception (Figure 1.29). For this, the imaging strategy is reviewed in its entirety. Each element in the probe transmits individually, and all of the signals on reception are recorded, with the data being focused during the post-processing phase. A synthetic aperture imaging sequence therefore enables us to exploit data from all of the pulses, from each transmission element to each reception element.

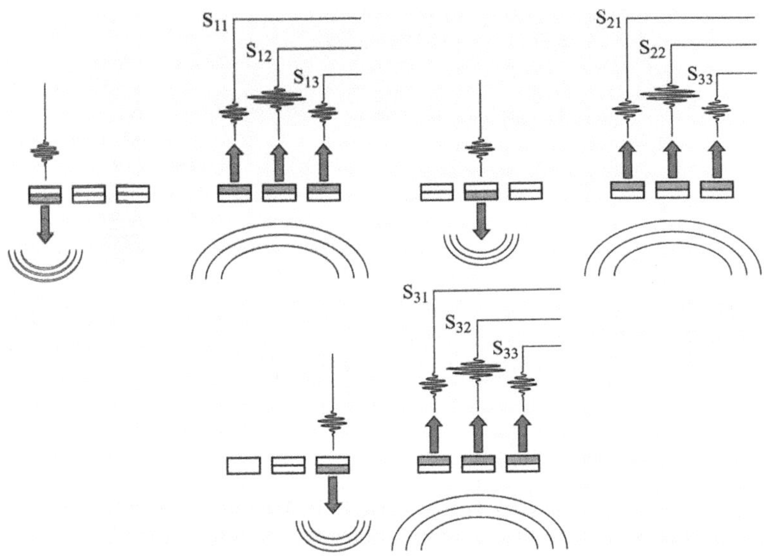

Figure 1.29. *In aperture synthesis, each element transmits separately, and all of the raw signals reaching each reception element are recorded for the post-processing phase*

The technique of synthetic aperture imaging offers complete flexibility. With the post-processing, it is possible to envisage any combination of delays in emission or reception, and it is also possible to vary the weighting functions in emission and reception as a function of the point in question. We can even envisage using the same set of raw signals to construct several images in parallel with complementary beamforming approaches.

Synthetic aperture imaging does, however, exhibit certain drawbacks. Due to the emission by a single element, the transmission power is relatively low, and the resulting SNR in each reception signal is poorer than when all the elements are transmitting simultaneously. In addition, this approach needs all the elements to fire before an image can be formed. Thus, the image rate may be reduced. Certain approaches inspired by channel coding in telecommunications try to circumvent this problem by simultaneously transmitting from all the elements and assigning a different channel to each element. The contribution from each element is then obtained by decoding on reception. Frequency multiplexing approaches or the use of orthogonal codes can help to deliver both an optimum spatial and temporal resolution.

1.5.1.4. *Fast imaging*

In certain medical applications, and particularly those where we need to be able to observe "rapid" phenomena, the rate of acquisition of ultrasound images is no longer sufficient. Transient elastography, where the propagation of a shear wave is imaged and analyzed to extract parameters relating to the local elasticity of the tissue being observed necessitates an imaging frequency of several KHz. Similarly, it is sometimes interesting to be able to simultaneously form two or more RF signals (or a whole image) with a single ultrasound pulse. We can cite flow imaging, where such an approach could create a velocimetry chart in a large region far more quickly.

The most widespread approach for fast imaging is to transmit a plane wave, which is able to illuminate a much larger region of tissue than with a focused pulse. Dynamic beamforming is applied on reception of the echo, which enables us to obtain a new image with every pulse. It is important to note here that the images obtained using this approach have a much poorer spatial resolution than is delivered by a focused transmission. The primary goal here is rapidity, so the degradation of the image is not overly problematic.

This approach is also used for 3D imaging with matrix probes. Indeed, if we wish to construct a 3D volume, the scan time with a conventional approach would be extremely long. The possibility of constructing a whole volume (or at least part of that volume) with a single pulse is therefore potentially of great interest.

1.5.2. *Signal- and image processing*

1.5.2.1. *Elastography, imaging of elasticity of biological tissues*

Ultrasound elastography is an imaging technique using ultrasound data to yield information about the local mechanical properties of a tissue. This technique can provide physicians with useful information for diagnosis. By way of example, cirrhosis, which is a liver disease, is known to significantly alter the elasticity of tissue. Similarly, in a study published in 1998 on mechanical tests done on tissue samples from the breast and the prostate, Krouskop *et al.* [KRO 98] measured, under sufficient tissue precompression conditions, elasticity moduli (stress/strain ratio) that were much higher for cancerous tissues than for healthy tissues. In other words, when subjected to the same level of stress, cancerous tissues deformed less than normal tissues. This also accounts for the fact that, in some cases, the lesion can be detected by palpation.

Various approaches can be distinguished on the basis of the mechanical stimulus used, the elasticity-related parameter evaluated and the estimation method employed [BER 04, HOY 07, NIG 03, OPH 91, SAN 03, XU 10]. In particular, we shall consider herein what is referred to as quasi-static elastography methods, where we image the deformation (strain) of the medium subjected to compression that can be applied with the ultrasound probe, for instance. An example of the result obtained with a test object (CIRS, model 059[2]) specially designed for studies of this type is shown in Figure 1.30, and clearly illustrates the potential of using elastography alongside echography.

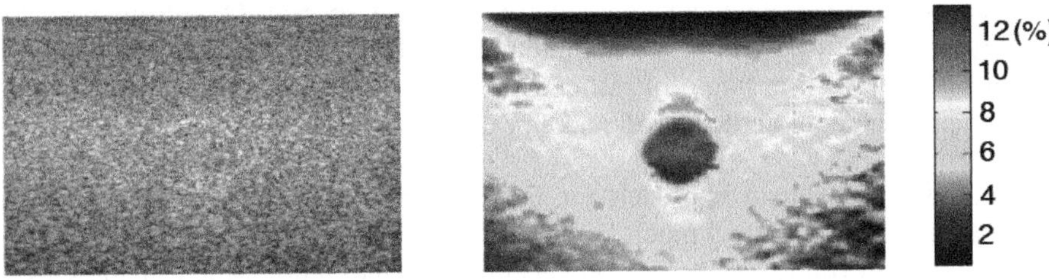

Figure 1.30. *Result obtained on a breast-mimicking phantom which contains several spherical inclusions to simulate the presence of lesions. Conventional ultrasound image of a plane containing one such inclusion, which in this example is difficult to detect (left). Image of the corresponding axial strains or elastogram, showing medium compression (in %, right) and clearly revealing the presence of the inclusion. For a color version of this figure, see www.iste.co.uk/fanet/medimagnet.zip*

The earliest techniques, which offered 1D digital processing of ultrasound data, later evolved to offer 2D or even 3D processing.

Generally speaking, locally estimating tissue strain is performed by dividing the image into multiple regions of interest (ROIs). For each of these regions, the modifications induced by the compression of the medium are modeled and estimated, taking account of effects of varying degrees of complexity. Most techniques assume that tissue compression engenders a 2D translational movement of the ROIs [CHE 09, DOY 01, ZHU 02]. This description suggests a rigid motion of the different ROIs making up the medium. Using this model, improved techniques have been put forward, notably involving multi-level or multi-scale approaches, and/or regularization to obtain continuous displacement fields [PEL 04, SHI 07]. Strains are then deduced from the gradient of estimated displacements.

Other methods consider more complex transformations of the ROIs during the deformation of the tissue. In addition to a translational movement, terms describing

2 http://www.cirsinc.com/products/all/83/breast-elastography-phantom/.

the axial, lateral and shear strains can also be integrated into the model and thereby directly estimated [BRU 08, LIU 09, MAU 04].

An example of results obtained on patients exhibiting breast lesions is shown in Figure 1.31 [BRU 11, BRU 13].

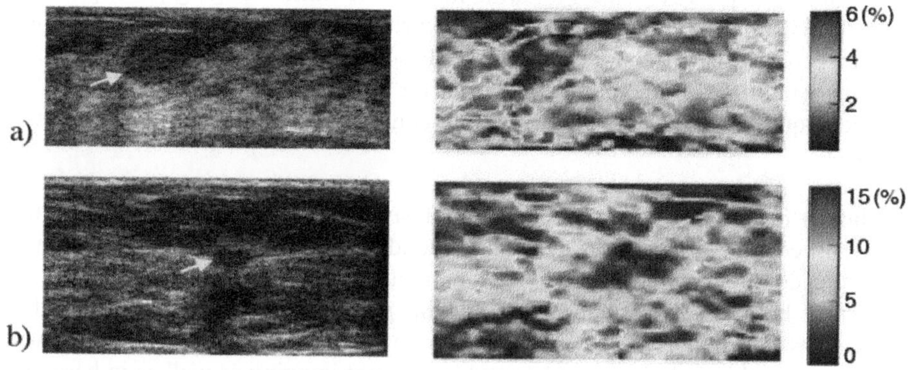

Figure 1.31. *Results of quasi-static ultrasound elastography, obtained on breast lesions, a) in a patient with a fibroadenoma (benign lesion), b) in a patient suffering a grade II infiltrating ductal carcinoma. In both cases, the conventional ultrasound image is presented on the left, with the lesion marked by an arrow, and the axial strain image (in %) on the right. For a color version of this figure, see www.iste.co.uk/fanet/medimagnet.zip*

This example is illustrative of the results found in the existing body of literature, showing notably that cancerous breast lesions (case b) often appear on elastograms as low-strain regions of significantly larger size than the size observed with echography [ITO 06, HAL 03].

Finally, modules devoted to the imaging of tissue deformation are now available in commercial ultrasound scanners, and can be used to evaluate the diagnostic contribution of this technique on a larger scale.

1.5.2.2. *New modes of harmonic imaging*

– Mono-pulse techniques

B-mode imaging and second harmonic imaging use the transmission of a single pulse. In addition to these conventional techniques, many techniques derived from harmonic imaging have been developed to increase the CTR (contrast to tissue ratio): sub-harmonic imaging by reception at half the transmitted frequency [FOR 00] or super-harmonic imaging by low-frequency emission (0.8 MHz) and receiving the higher-order harmonics (third, fourth, fifth, etc.) [BOU 02, BOU 03].

Other techniques attempt to increase the CTR of the second harmonic by modifying the transmitted signal to reduce the second harmonic generated by the tissue. The difference between these techniques lies in the way in which the source signal is created. With HCS (harmonic cancellation system), the shape of the wave at the focal point is predicted, and the transmitted signal is composed of the inversion of the predicted second harmonic and the initial imaging pulse [KRI 96, KRI 98, CHR 99, KRI 08]. The harmonic contribution of the tissues is reduced only within a certain area around the focal point. Pasovic *et al.* deal with this problem by defining a multiple-component signal from the second harmonic signal [PAS 10]; this technique is called mcSHRS (multiple-component second-harmonic reduction signal).

Shen *et al.* propose to simultaneously transmit a fundamental component and a phase-shifted third harmonic wave. The second harmonic component is decreased when the component of the sum of the frequencies and the component of the difference between the frequencies are in phase opposition. When the two components are in phase, the second harmonic component is improved. Consequently, this technique can be used in contrast imaging as well as in harmonic tissue imaging [SHE 07a, SHE 07b, SHE 08a, SHE 08b, SHE 10, SHE 11].

– Multi-pulse techniques

In order to increase the sensitivity for detection of harmonic echoes, a longer pulse (with a duration of several periods with a narrow-band pulse), needs to be transmitted, which unfortunately decreases the image resolution [CHI 03, FRI 00].

To deal with the compromise between harmonic detection and resolution in conventional harmonic imaging, or discrimination between the zones perfused with contrast agent from un-perfused tissues, various multi-pulse techniques have been developed. The transmission sequences comprise multiple transmission pulses with different parameters: frequency, amplitude, phase, duration, number and repetition frequency. The received signals are processed with a view to reinforcing these characteristics, by weighted summing or filtering. These techniques are offshoots from pulse inversion (PI), the first multi-pulse sequence to be put forward, which is a sequence of two opposed-phase pulses [SIM 99, HWA 99].

Pasovic *et al.* suggested transmitting a sequence of two pulses with 90° phase-shift, and summing together the two received signals. As the second harmonic component has a quadratic relation to the fundamental component, the phase-shift between the two second harmonic components received is 180°, so the second harmonic is cancelled successfully in the tissue, whilst good preservation is noted in the region with contrast agent, and therefore the CTR is increased. This method is called SHI (second harmonic inversion) [PAS 11].

There are also a variety of methods to increase the nonlinear response of microbubbles. Brock-Fisher *et al.* use amplitude modulation [BRO 96]. The linear echoes are cancelled, whilst the odd-numbered harmonic components are preserved. The combination of phase- and amplitude modulation in a more general nonlinear detection process was proposed by Haider and Chiao [HAI 99]. Phillips altered this technique, with multiple sequences called contrast pulse sequences (CPS) [PHI 01, PHI 04]. Eckersley *et al.* [ECK 05] propose to combine amplitude modulation and pulse inversion; the second pulse is inverted and its amplitude is half that of the first pulse. The response to the second pulse is doubled before being added to the response to the first pulse. Wilkening *et al.* suggest transmitting five pulses with equidistant phases. A weighted sum treatment of the received echoes then eliminates the selected harmonics [WIL 00, WIL 01]. Lin *et al.* proffered a generalization of these multi-pulse sequence transmission techniques.

1.5.2.3. *Motion estimation*

Motion estimation in ultrasound imaging has numerous applications such as elastography, heart motion monitoring or flow estimation.

– Block matching

The most widely used method in tissue motion estimation in echography is "block matching", also known as "speckle tracking" in the literature in ultrasound imaging (the term "block matching" is used in general image processing). The principle behind this method is to divide the image into blocks of pixels (usually rectangular) and suppose that all the pixels in a block have experienced the same displacement. In order to estimate this displacement (rigid translational motions), the block in an image at time t is sought in the image at time $t+1$ in a search area, using a similarity function (correlation, normalized correlation) or a cost function (sum of absolute differences (SAD) or sum of squared differences (SSD)). The position of the block most closely resembling the original is used to compute an estimation of the motion of the pixels involved. The choice of the size of the search area is generally dictated by *a priori* knowledge about the maximum amplitude of motion between times t and $t+1$. However, a compromise needs to be found between too small a search area (with the inherent danger of missing the appropriate resembling block) or too large a search area (danger of confusion between several similarly resembling blocks). In order to deal with this problem, *a priori* knowledge about the motion being estimated has to be exploited; for an example see [BAS 08] in ultrasound elastography.

In many applications, given the spatial sampling of ultrasound images, subpixel precision in the estimation of the displacement may be required. In this case, it is necessary to refine the search area, using interpolation techniques. As well as a significant increase in computation time, these interpolation techniques may also give rise to estimation errors. Furthermore, the choice of interpolation factor is often empirical, and in turn largely influences the quality of the estimation.

One final point which must be stressed regarding block-matching methods is the choice of size of the blocks. Here too, a compromise needs to be found: too small a size may mean that the blocks do not contain enough information and are therefore difficult to identify in the search area, whereas too large a size may invalidate the hypothesis of identical displacement of all the pixels in a block. One strategy to partially resolve this problem is to use models of more complex parametric motions than simple rigid translational motions, on the scale of a block. Examples using an affine model [SUH 05] or a bilinear one [BAS 08, BAS 09] can be found in the existing literature.

Figure 1.32 illustrates the principle of block-matching methods.

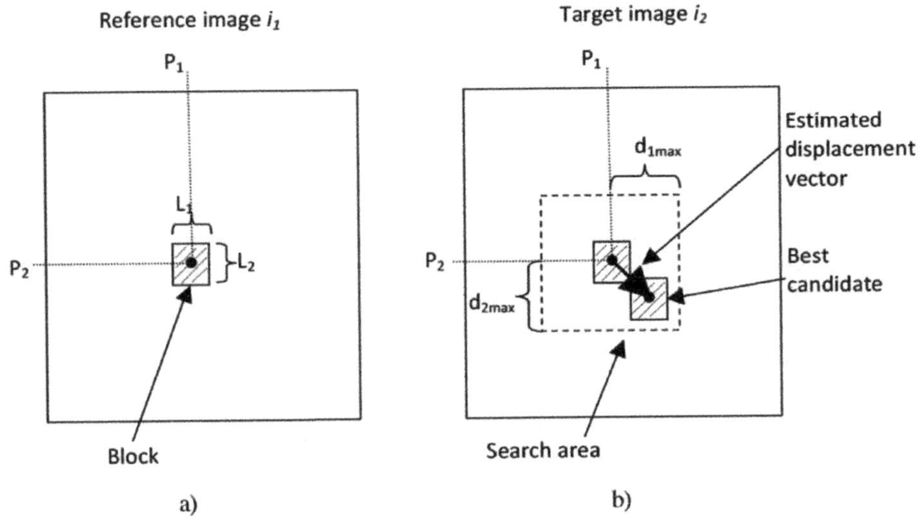

Figure 1.32. *Principle of block-matching methods. a) A block of size $L_1 \times L_2$ is considered in the reference image. b) A search area is considered in the target image. The block is sought within that search area and the best candidate (that which most closely resembles the reference block, on the basis of a given criterion) is chosen. Note that the size of the search area imposes the maximum displacement detectable for the block. The search area can also be interpolated to ensure subpixel accuracy of the estimation*

By way of illustration, we shall now show an example of motion estimation in ultrasound elastography extracted from [BAS 09]. A series of ultrasound images of a thyroid presenting a malignant tumor was recorded while light pressure was exerted with the US probe. The motion caused by this pressure was estimated using the deformable block-matching method introduced in [BAS 08]. The result obtained is shown in Figure 1.33. This motion was also used to generate a parametric map, wherein the contrast between two regions belonging to the healthy and disease parts of the thyroid was improved by a factor 3 in comparison to the echograph image.

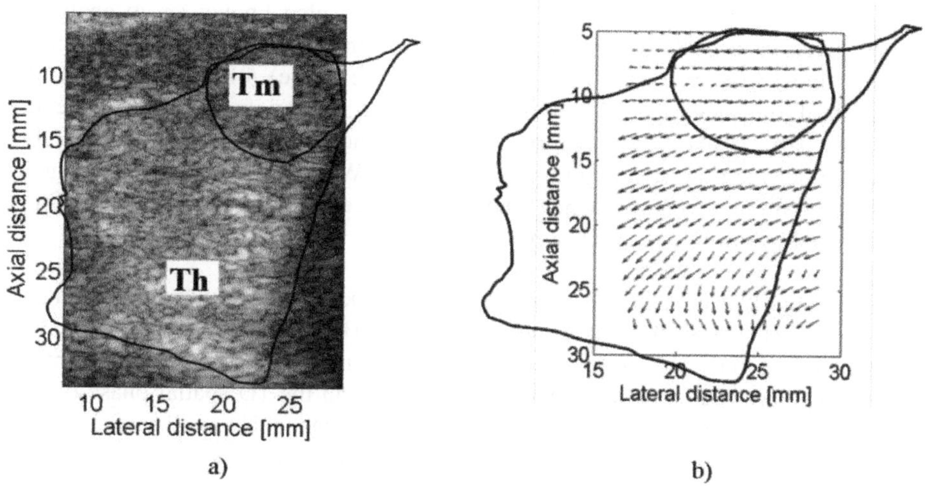

Figure 1.33. *a) an ultrasound image used for motion estimation, b) 2D field of the estimated motion. The outlines of the thyroid (Th) and tumor (Tm) were traced by a doctor on the B-mode ultrasound image*

– Differential methods

A second category of motion estimation methods, which are widely used in optical imaging, is differential methods. They express a strict hypothesis of pixels' conservation of light intensity over time, in an equation known as the optical flow equation. Given that a single equation is insufficient to find the two unknowns (the two components of the motion), two approaches are possible: local, considering that neighboring pixels will have the same displacement, or global, regularizing the estimation on the scale of the whole image.

In ultrasound imaging, differential methods have had less success, notably because of the hypothesis of conservation of the level of gray of the pixels over time, which is not or is only slightly respected because of the intrinsic nature of these images. However, they have the advantage of producing a subpixel estimation of the motion without using interpolation of the images. For these two reasons, we have used differential approaches in ultrasound imaging not on the images themselves, but rather on images giving structural data about the images, which shall be explained later on. Thus, we obtain a subpixel estimation without the need for interpolation, and which is unaffected by the variations in light intensity of the pixels over time.

– Phase image

The complex representation of a 1D real signal, also known as an analytical signal, gives us access to local data such as the instantaneous amplitude, spatial phase or local frequency (the time-derivative of the phase). In one dimension, the real part of the analytical signal is composed of the original signal, while the imaginary part is its Hilbert transform. It should be noted that there is a considerable difference between the spatial and spectral phases. Whereas the latter gives angular information for a given frequency, the former gives structural information for a given spatial position in the real signal.

The spatial phase of the RF signals has largely been used in ultrasound imaging to estimate 1D axial displacements [PES 99, EBB 06, FRO 07], or the two components of the motion for lateral oscillation images [LIE 07]. Thus, many publications have demonstrated that methods based on the 1D spatial phase offer a gain in precision compared to conventional methods using 1D similarity functions such as the correlation- or cost function, as done with the SAD or SSD method.

In two dimensions (2D signals or images), many attempts to generalize the 1D analytical signal can be found in the literature. Here, we classify these approaches into two categories: multi-dimensional analytical signals (total, partial, single-orthant) and monogenic signal.

– Multi-dimensional analytical signals

Multi-dimensional analytical signals are obtained, similar to the 1D case, by combining the original signal with its total Hilbert transform and/or its partial Hilbert transforms [HAN 96, BUL 01]. For ease of reading, here we consider the 2D case, and give the expressions of the Hilbert transforms in the Fourier domain. Thus, for an image notated as $s(x_1,x_2)$ an its frequential representation $S(u_1,u_2)$, the 2D Hilbert transforms are defined thus:

Total: $S_H(u_1,u_2) = -\mathrm{sgn}(u_1)\mathrm{sgn}(u_2)S(u_1,u_2)$

Partial: $S_H^1(u_1,u_2) = -j\,\mathrm{sgn}(u_1)S(u_1,u_2)$ and $S_H^2(u_1,u_2) = -j\,\mathrm{sgn}(u_2)S(u_1,u_2)$

where $sgn(u)$ is the function “sign” with its usual definition.

The two partial transforms presented here correspond to the axial and vertical directions. It should be noted that partial transforms can be constructed for any direction in space.

By combining the original signal with the three Hilbert transforms above, we obtain two so-called "single-orthant" analytical signals (notated s_1 and s_2). This name comes from the fact that they are obtained by cancelling out three of the four orthants in the Fourier domain.

$$S_1(u_1,u_2) = S(u_1,u_2) - S_H(u_1,u_2) + j\left(S_H^1(u_1,u_2) + S_H^2(u_1,u_2)\right)$$

$$S_2(u_1,u_2) = S(u_1,u_2) + S_H(u_1,u_2) + j\left(S_H^2(u_1,u_2) - S_H^1(u_1,u_2)\right)$$

– Monogenic signal

In spite of their interesting properties, multi-dimensional analytical signals are not a true generalization of the 1D analytical signal, essentially in the sense of isotropy and capacity to separate the image into local amplitude information and local phase information depending only on the structures, and invariant with the local variations in energy.

In order to deal with this shortcoming, Felsberg and Sommer in 2001 introduced the monogenic signal [FEL 01]. Presented in the form of an incomplete quaternion (two imaginary parts), the monogenic signal combines the initial image and its Riesz transform, which represents a generalization of the Hilbert transform for vectors. The better to understand this generalization, an analogy is made in [UNS 09] the Riesz transform is to the Hilbert transform as the gradient is to the differential operator. However, a hypothesis has to be made about the nature of the images for which the monogenic signal is appropriate: the local structures must have a single prevailing orientation. The two imaginary parts are obtained by filtering the initial image with $H_1(u_1,u_2) = \frac{u_1}{\sqrt{u_1^2 + u_2^2}}$ and $H_2(u_1,u_2) = \frac{u_2}{\sqrt{u_1^2 + u_2^2}}$. Prior filtering of the image is usually employed to ensure that it has a limited frequency band and to obtain a multi-scale analysis. Log-Gabor- or Poisson-type filters are those which are most commonly encountered in the existing literature.

Based on the monogenic signal (s, s_{H1}, s_{H2}), where $S_{H1,2}(u_1,u_2) = S(u_1,u_2) \times H_{1,2}(u_1,u_2)$, local measurements of orientation ($\theta(x_1,x_2)$) and of phase ($\varphi(x_1,x_2)$) can be obtained by using the formulae given in [FEL 01]:

$$\theta(x_1, x_2) = \operatorname{atan}\left(\frac{s_{H2}}{s_{H1}}\right)$$

$$\varphi(x_1, x_2) = \text{atan}\left(\frac{s}{s_{H1}\cos(\theta) + s_{H2}\sin(\theta)}\right)$$

These local measurements are used to estimate the motion between two [MAL 10] or more [BAS 11] ultrasound images, as shown by the summary diagram that is Figure 1.34.

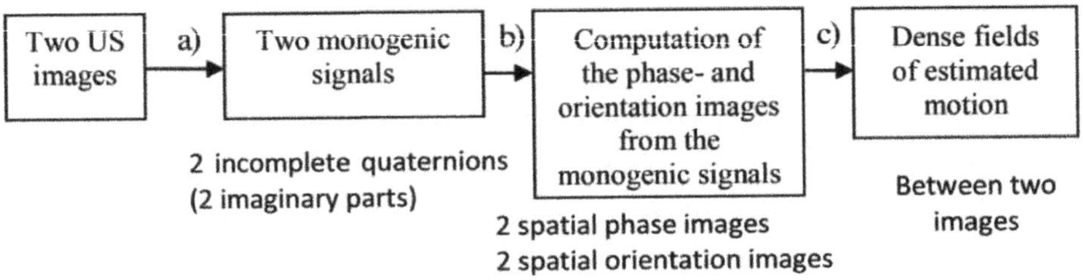

Figure 1.34. *Summary diagram of the motion estimation method based on the monogenic signal and applied to ultrasound imaging presented in [MAL 10] and [BAS 11]: a) computation of the monogenic signals associated with the two US images; b) extraction of the two phase and spatial orientation images; c) motion estimation using the estimator presented in [MAL 10]*

– Results

In order to illustrate the advantage of using local measurements such as the orientation and phase obtained by using the monogenic signal in motion estimation, we shall show a result extracted from [MAL 10]. This result was gained in a classic experimental framework in ultrasound elastography, when a homogeneous medium containing a harder spherical insert is deformed with the ultrasound probe. A B-mode ultrasound image of the sequence thus acquired is shown in Figure 1.35. The deformation map obtained after having estimated the motion using a conventional block-matching method and that given in [MAL 10] are also shown in Figure 1.35. Although they are qualitatively quite similar, it should be noted that the classic result required a 5-fold interpolation of the ultrasound images, in both spatial directions to obtain subpixel precision. As the method presented in [MAL 10] naturally ensures such precision, it was applied directly to the original images.

1.5.2.4. *Compressive sensing*

"Compressive sensing" (CS) is a nascent signal processing technique which can, under certain conditions, enable us to reconstruct a complete signal from a number of samples that is much lower than the limit set by the Shannon theorem. The goal of CS may be i) to improve the quality of signals reconstructed using the same quantity of information; or ii) to accelerate acquisition by limiting the quantity of data acquired whilst preserving the quality of the reconstructed data.

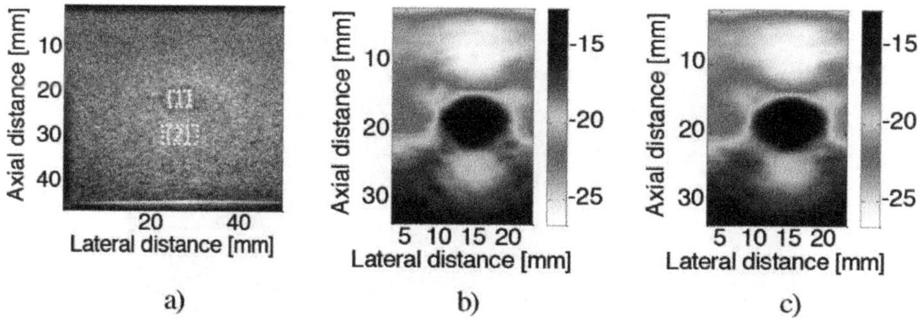

Figure 1.35. *Result of elastography on a test object containing a spherical insert (1) which is harder than the surrounding medium (2), but is very hard to see in the B-mode image: a) echographic image; b) elastogram obtained using a conventional block-matching method and a factor 5 interpolation of the images; c) elastogram obtained using a motion estimation based on the monogenic signal without interpolation of the images. For a color version of this figure, see www.iste.co.uk/fanet/medimagnet.zip*

Ultrasound imaging is a domain of application which could greatly benefit from CS, not only in terms of improving the quality of the images from a visual point of view, but also in terms of speeding up the acquisition process. For instance, it would be very advantageous to be able to construct a 3D volume on the basis of the same number of elements as in 2D (128 elements), whereas a complete 3D probe would actually require $64 \times 64 = 4096$.

In order to be applicable, this theory requires that two conditions be satisfied:

– the signal being reconstructed must have a sparse representation in a certain base. This means that the number of non-zero coefficients in that base must be low;

– the representation base and the acquisition base must be incoherent. This is usually ensured by using random sampling or a set of random projections of the data.

Ultrasound imaging using CS is a very recent notion. If successful, it could represent a real conceptual and methodological breakthrough in ultrasound imaging.

By way of illustration, a study showed the feasibility of reconstructing a good-quality 2D ultrasound image using only 20% of the signals originally sampled. Various reconstruction bases have been tested, and the "wave atoms" base used in geophysics to represent oscillating phenomena has been able to deliver the best results.

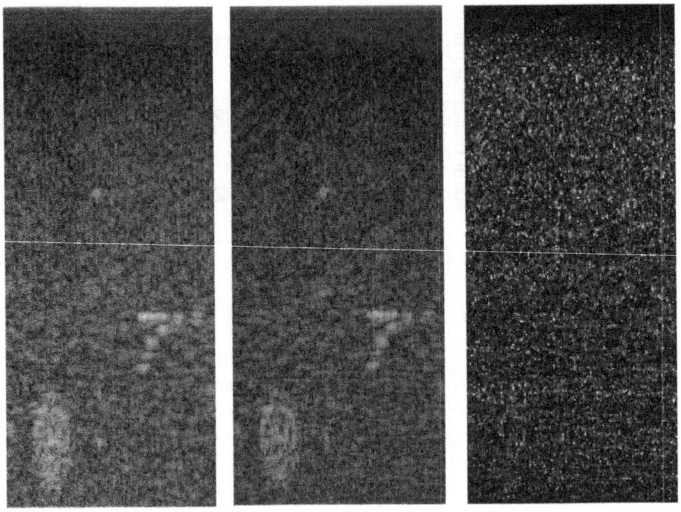

Figure 1.36. *Reference image constructed using all the data (left). Image reconstructed with 20% of the initial data using compressive sensing (center). Error image (right)*

1.5.3. *Multimodal imaging*

1.5.3.1. *Magnetic resonance elastography versus US elastography*

Biological tissues behave like viscoelastic media, meaning that they simultaneously possess the viscous properties of a fluid and the elastic properties of a solid. The propagation of mechanical waves is directly linked to these properties. The study of these waves using MRI therefore seems advantageous to characterize biological tissues. Indeed, it is well known that palpation helps identify certain tumoral processes or other disease-induced mechanical changes. However, palpation remains subjective, and limited to regions that can be explored with the hand. Magnetic Resonance Elastography (MRE) is based on the feature offered by MRI to measure the displacements generated by the propagation of mechanical waves in tissues. This method of characterization of biological tissues was developed in the 1990s (Muthupillai *et al.* [MUT 95]). It consists of visualizing the propagation of shear waves (or transverse waves) at low frequency (generally 50 Hz to 500 Hz) in the tissues, the shear wave speed being linked to the shear modulus. Compression waves or longitudinal waves, which are much faster, are difficult to measure with this technique. If focused ultrasound can be used (Souchon *et al.* [SOU 08]), the shear waves are mainly generated using a mechanical or acoustic driver placed near the region of interest. Visualization of the shear waves is obtained by adding, to the acquisition sequence, motion-encoding gradients (MEGs), whose duration is conditioned by the period of the mechanical excitation. The most widely-used acquisition sequences are gradient-echo sequences or Echo Planar Imaging (EPI) sequences. The images obtained are phase images which, when analyzed, inform us about the viscoelastic properties of the tissues, such as the shear modulus and viscosity. MRE requires synchronization of the excitation with the MSGs in the sequence, and whose oscillation frequency is imposed by the frequency of the mechanical excitation applied. Note that the most widely used "dynamic" excitation, and which consists in applying harmonic mechanical excitation, does not require knowledge of the stress applied to the tissues to estimate the value of the shear modulus, which is a major advantage. Nevertheless, the estimated values are dependent upon the excitation frequency. The shear modulus measured may vary in a wide range of values between healthy and diseased tissues. Thus, this "remote palpation" technique offers new prospects in early diagnosis of diseases such as breast cancer, certain visceral conditions and liver fibrosis [MAN 01, YIN 07].

Two illustrations of MRE are shown in Figures 1.37 and 1.38. The first is drawn from an examination of a patient suffering from a cholangiocarcinoma. The second shows the results of an experiment on a test object, specifically designed for elastographic studies.

The test object examined (CIRS, model 049) is presented in the form of a parallelepiped in which are included eight spheres with different mechanical properties. The ERM result was obtained for a plane containing four of these spheres, with respective Young's moduli of E = 6 kPa, 17 kPa, 54 kPa and 62 kPa, with the Young's modulus of the surrounding medium being around 29 kPa.

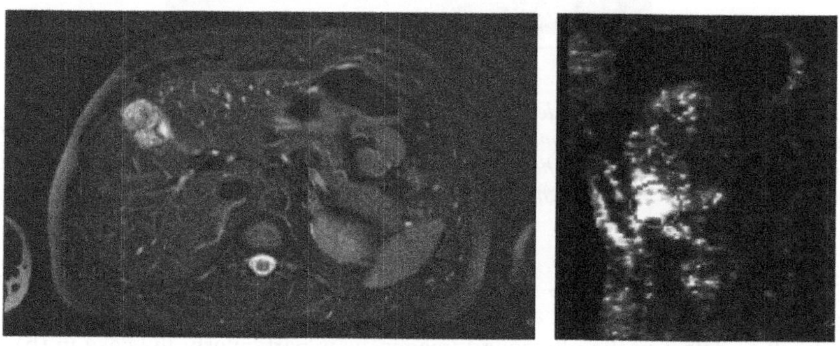

a) b)

Figure 1.37. *a) T2-weighted axial image of the liver of a patient with a cholangiocarcinoma. b) Result of magnetic resonance elastography in the form of a map of the shear modulus obtained along an oblique coronal plane. The tumor attains a value of around 31 kPa, as compared to the normal liver tissue with a value of around 1 kPa. For a color version of this figure, see www.iste.co.uk/fanet/medimagnet.zip*

a)

b)

c)

Figure 1.38. *Images of amplitude a) and phase b) obtained with an ERM gradient echo sequence on the CIRS 049 phantom with an excitation frequency of 177 Hz. Map c) in kPa of the shear modulus estimated on the basis of eight phases. For a color version of this figure, see www.iste.co.uk/fanet/medimagnet.zip*

For comparative purposes, the results obtained on the same physical phantom using ultrasound elastography are presented in Figure 1.39. While the spheres may be difficult to detect in the ultrasound images, as is particularly the case in (b), they are clearly shown in the strain images. The local value for the Young's modulus is obviously not accessible, but the deformation is consistent with the known mechanical properties of the medium, and is able to reveal the spherical inclusions.

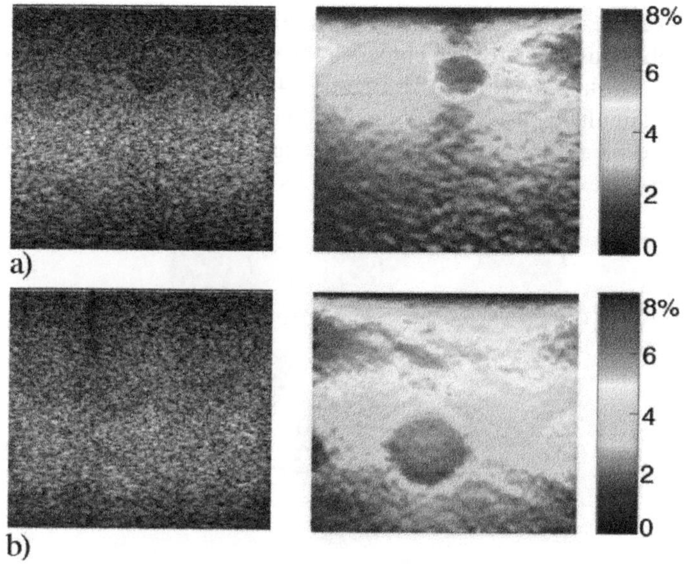

Figure 1.39. *Ultrasound elastography results obtained a) on a region containing a sphere with a Young's modulus of 6 kPa, and b) on a region containing a sphere with a Young's modulus of 54 kPa. In both cases, we have given the conventional ultrasound image (left) and the strain image (in %, right). For a color version of this figure, see www.iste.co.uk/fanet/ medimagnet.zip*

1.5.3.2. *US/Optic, photoacoustics*

The coupling of different modes of imaging is based on a simple idea: each mode provides additional information which will help doctors in reaching their diagnoses. In concrete terms, it is interesting to combine a mode which provides anatomical information with a functional or metabolic mode of imaging. Already in hospitals we can find devices combining CT/PET or MRI/PET, particularly for cancer diagnosis. The combination of ultrasound imaging with a different modality is also very promising. Research currently being done on imaging of the prostate or the liver, for instance, is aimed to placing an ultrasound imaging plane into a previously-acquired MRI volume scan. One imagines that in thc future, doctors will be able to acquire US data in real time and view them, superposed on the 3D MRI data for the patient.

The coupling between US and optical technology has recently emerged, and initial clinical trials are currently being carried out to test the contribution of this type of bi-modal imaging and thereby improve prostate cancer diagnosis. For this purpose, optical imaging is based on specific marking of tumor tissues with fluorescent tracers. One of the advantages of the US/optical combination is that the two imaging devices can be combined in a single probe, handheld by the doctor. A prototype probe was recently developed in the context of a research consortium [ANR TecSan ProstaFluo] [PRO 12]. Figure 1.40 shows the bi-modal probe, which is capable of simultaneously acquiring US images and (3D) optical tomographic data for time-resolved fluorescence imaging. Figure 1.41 shows a fluorescent insert reconstructed on the basis of optical data, and its matching with the US image.

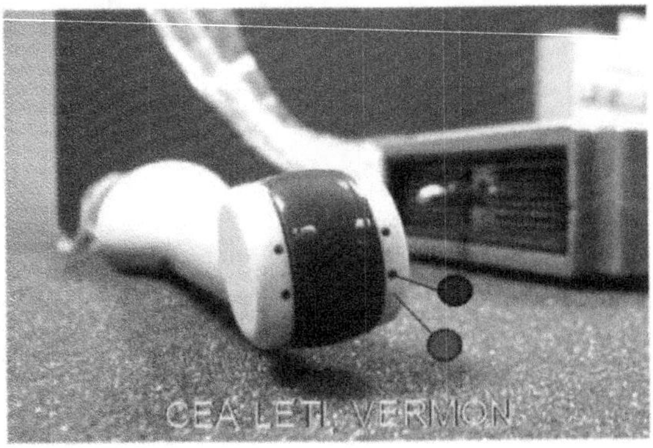

Figure 1.40. *Bimodal US/optical probe containing 6 excitation fibers, 4 reception fibers and, in the center, a US probe of 128 elements at 6.7 MHz [BOU 09]. For a color version of this figure, see www.iste.co.uk/fanet/medimagnet.zip*

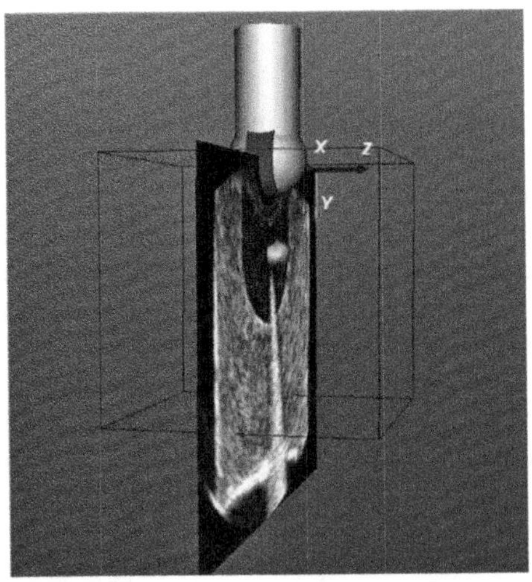

Figure 1.41. *Registration and visualization of US and optical data, showing on a phantom image the correspondence between the echographic location and the tomographic reconstruction of a fluorescent marker (yellow). For a color version of this figure, see www.iste.co.uk/fanet/medimagnet.zip*

1.6. A bright future for ultrasound imaging

To begin with, in this chapter, we recapped the fundamental physical principles underlying the emission and reception of US waves. We then went on to describe the main functionalities of ultrasound systems. Filled with electronics and computer technology, these systems offer doctors many different modes of imaging to explore the human body. At a rate of several dozen or even several hundred images per second, in 2D and/or 3D, ultrasound systems are at the heart of a doctor's diagnosis – particularly in cardiology and obstetrics. The Doppler modes provide additional information to quantify the speeds of blood flow and tissue movement.

The second part of the chapter illustrated the constant evolution of this mode of imaging, which caters for a very broad clinical spectrum in a way no other mode does. Innovations are being made at all stages of the process of image formation. New probes are emerging which improve the spatial resolution, facilitate 3D imaging, or enable us to image over a wider range of frequency, such as with harmonic imaging. New image formation techniques are becoming available, to increase the frame rate or generate specific types of waves. Treatments performed on the raw signals' output by the sensors enable us to extrapolate information about the elasticity of the tissues (elastography) or to sample above the Shannon limit, which seemed an impossibility only ten years ago. Software suites for image segmentation and tracking of sequences of images provide doctors with quantitative data about surface area, volume and displacement over time. Also, more recently, the combination of ultrasound imaging with MRI or optical techniques has opened the door to functional and metabolic imaging. In conclusion, today, ultrasound imaging plays a central role in clinical diagnostics, and will undoubtedly, in the future, see its fields of specialty expand and witness the development of new systems integrating the innovations discussed in this chapter.

1.7. Bibliography

[HTT 09] http://www.ameli.fr/, section Statistiques et publications, N°31, CCAM en secteur libéral -2009.

[ABO 04] ABOLMAESUMI P., SIROUSPOUR M.R., An interacting multiple model probabilistic data association filter for cavity boundary extraction from ultrasound images, *IEEE Trans. Med. Imag.*, vol. 23, no. 6, pp. 772–784, 2004.

[ARE 96] ARENDT JENSEN J., Estimation of Blood Velocities Using Ultrasound, Cambridge University Press, 1996.

[BAM 80] BAMBER J.C., DICKINSON R.J., "Ultrasonic B-scanning: a computer simulation", *Phys Med Biol*, vol. 25, no. 3, pp. 463–79, 1980.

[BAS 08] BASARAB A., LIEBGOTT H., MORESTIN F., *et al.*, "A method for vector displacement estimation with ultrasound images and its application for thyroid nodular disease", *Medical Image Analysis,* Elsevier, vol. 12, no. 3, pp. 259–274, 2008.

[BAS 09] BASARAB A., LYSHCHIK A., GRAVA C., *et al.*, "Ultrasound image sequence registration and its application for thyroid nodular disease", *Journal of Signal Processing Systems,* Springer, vol. 55, no. 1–3, pp. 127–137, April 2009.

[BAS 11] BASARAB A., VIDAL F., ABBAL R., *et al.*, "Motion estimation in ultrasound imaging applied to the diagnostic of pelvic floor disorders", *IEEE Engineering in Medicine and Biology Society Conference (EMBC 2011)*, Boston, pp. 8058–8061, 2011.

[BER 04] BERCOFF J., TANTER M., FINK M., "Supersonic Shear Imaging: a new technique for soft tissue elasticity mapping", *IEEE Transactions on Ultrasonics, Ferroelectrics and Frequency Control,* vol. 51, no. 4, pp. 396–409, 2004.

[BER 07] BERNARD O., TOUIL B., D'HOOGE J., *et al.*, Statistical modeling of the radio-frequency signal for partially and fully developed speckle based on a generalized Gaussian model with application to echocardiography, *IEEE Transactions on Ultrasonics, Ferroelectrics and Frequency Control*, vol. 54, no. 10, pp. 2189–2194, 2007.

[BOU 02] BOUAKAZ A., FRIGSTAD S., TEN CATE F.J., *et al.*, "Super harmonic imaging: a new imaging technique for improved contrast detection", *Ultrasound in Medicine & Biology*, vol. 28, no. 1, pp. 59–68, 2002.

[BOU 03] BOUAKAZ A., DE JONG N., "Native tissue imaging at superharmonic frequencies" *IEEE Transactions on Ultrasonics, Ferroelectrics and Frequency Control*, vol. 50, no. 5, pp. 496–506, 2003.

[BOU 03] BOUKERROUI D., BASKURT A., NOBLE A., *et al.*, "Segmentation of ultrasound images – Multiresolution 2D and 3D algorithm based on global and local statistics" *Pattern Recognition Letters*, vol. 24, no. 4-5, pp. 779–790, 2003.

[BOU 09] BOUTET J., DEBOURDEAU M., HERVÉ L., *et al.*, "A bi-modal ultrasound and fluorescence approach for prostate cancer diagnosis", *Journal of Biomedical Optics*, vol. 14, pp. 064001–064007, 2009.

[BRO 96] BROCK-FISHER G.A., POLAND M.D., RAFTER P.G., Means for increasing sensitivity in non-linear ultrasound imaging systems, US Patent 5, 577, 505, November 1996.

[BRU 04] BRUSSEAU E., DE KORTE C.L., MASTIK F., *et al,*. "Fully automatic luminal contour segmentation in intracoronary ultrasound imaging – A statistical approach", *IEEE Trans. Med. Imag.*, vol. 23, no. 5, pp. 554–566, 2004.

[BRU 08] BRUSSEAU E., KYBIC J., DEPREZ J.F., *et al.*, "2-D locally regularized tissue strain estimation from radio-frequency ultrasound images: theoretical developments and results on experimental data", *IEEE Trans. Medical Imaging*, vol. 27, no. 2, pp. 145–160, 2008.

[BRU 11] BRUSSEAU E., DETTI V., COULON A., *et al.*, "A two dimensional locally regularized strain estimation technique: preliminary clinical results for the assessment of benign and malignant breast lesions", *Medical Imaging 2011*: *Ultrasonic Imaging, Tomography, and Therapy, Proceedings of SPIE*, 7968, 2011.

[BRU 13] BRUSSEAU E., DETTI V., COULON A., *et al.*, In vivo response to compression of 35 breast lesions observed with a two-dimensional locally regularized strain estimation method, *Ultrasound in Medicine & Biology*, 2013.

[BUL 01] BULLOW T., SOMMER G., "Hypercomplex signals–A novel extension of the analytic signal to the multidimensional case", *IEEE Trans. Signal Process.*, vol. 49, no. 11, pp. 2844–2852, 2001.

[CAL 05] CALIANO G., CAROTENUTO R., CIANCI E., *et al.*, "Pappalardo, Design, fabrication and characterization of a capacitive micromachined ultrasonic probe for medical imaging", *IEEE Transactions on Ultrasonics, Ferroelectrics and Frequency Control*, vol. 52, pp. 2259–2269, 2005.

[CAL 06] CALMON P., MAHAUT S., CHATILLON S., *et al.*, "CIVA: an expertise platform for simulation and processing NDT data", *Ultrasonics*, vol. 44, no. 1, pp. e975-9, 2006.

[CHE 02] CHEN D.R., CHANG R.F., KUO W.J., *et al.*, Diagnosis of breast tumors with sonographic texture analysis using wavelet transform and neural networks, *Ultrasound in Medicine & Biology*, vol. 28, no. 10, pp. 1301–1310, 2002.

[CHE 09] CHEN L., TREECE G.M., LINDOP J.E., *et al.*, "A quality-guided displacement tracking algorithm for ultrasonic elasticity imaging", *Med Image Anal*, vol. 13, no. 2, pp. 286–296, 2009.

[CHI 03] CHIAO R.Y., HAO X.H., "Coded excitation for diagnostic ultrasound: a system developer's perspective", *IEEE Ultrasonics Symposium*, vol. 52, no. 2, pp. 160–170, 2003.

[CHR 99] CHRISTOPHER T., "Source prebiasing for improved second harmonic bubble-response imaging", *IEEE Transactions on Ultrasonics, Ferroelectrics and Frequency Control*, vol. 46, no. 3, pp. 556–563, 1999.

[DOY 01] DOYLEY M.M., BAMBER J.C., FUECHSEL F., *et al.*, "A Freehand elastographic imaging approach for clinical breast imaging: system development and performance evaluation", *Ultrasound in Medicine & Biology*, vol. 27, no. 10, pp. 1347–1357, 2001.

[EBB 06] EBBINI E.S., "Phase-coupled two-dimensional speckle tracking algorithm", *IEEE Transactions on Ultrasonics, Ferroelectrics, and Frequency Control*, vol. 53, no. 5, pp. 972–990, 2006.

[ECK 05] ECKERSLEY R.J., CHIN C.T., BURNS P.N., "Optimising phase and amplitude modulation schemes for imaging microbubble contrast agents at low acoustic power", *Ultrasound in Medicine & Biology*, vol. 31, no. 2, pp. 213–219, 2005.

[FEL 01] FELSBERG M., SOMMER G., "The monogenic signal", *IEEE Trans. Signal Process.*, vol. 49, no. 12, pp. 3136–3144, 2001.

[FOR 00] FORSBERG F., SHI W. T., GOLDBERG B., "Subharmonic imaging of contrast agents", *Ultrasonics*, vol. 38, no. 1, pp. 93–98, 2000.

[FRI 00] FRINKING P.J.A., BOUAKAZ A., KIRKHORN J., *et al.*, "Ultrasound contrast imaging: current and new potential methods", *Ultrasound in Medicine & Biology*, vol. 26, no. 6, pp. 965–975, July 2000.

[FRO 07] FROMAGEAU J., LIEBGOTT H., BRUSSEAU E., *et al.*, "Estimation of time-scaling factor for ultrasound medical images using the hilbert transform", *Eurasip J Adv Signal Proces*, vol. 2007, pp. Article ID 80735, 13 pages, 2007.

[HAI 99] HAIDER B., CHIAO R.Y., "Higher order nonlinear ultrasonic imaging", *IEEE Ultrasonics Symposium*, vol. 2, pp. 1527–1531, 1999.

[HAL 03] HALL T.J., ZHU Y., SPALDING C.S., "In vivo real-time freehand palpation imaging", *Ultrasound in Medicine & Biology,* vol. 29, no. 3, pp. 427–435, 2003.

[HAN 09] HANSEN K. L., UDESEN J., THOMSEN C., *et al.*, "In vivo validation of a blood vector velocity estimator with MR angiography", *IEEE Transactions on Ultrasonics, Ferroelectrics and Frequency Control*, vol. 56, pp. 91–100, 2009.

[HAN 96] HAHN S.L., *Hilbert Transforms in Signal Processing,* Boston, Artech House, 1996.

[HOL 01] HOLM S., Ultrasim – a toolbox for ultrasound field simulation, in Nordic Matlab Conference, 2001.

[HOY 07] HOYT K., PARKER K.J., RUBENS D.J., "Real-time shear velocity imaging using sonoelastographic techniques", *Ultrasound in Medicine & Biology*, vol. 33, no. 7, pp. 1086–1097, 2007.

[HUA 06] HUANG Y., HAEGGSTROM E., BAYRAM B., *et al.*, "Comparison of conventional and collapsed region operation of capacitive micromachined ultrasonic transducers", *IEEE Transactions on Ultrasonics, Ferroelectrics and Frequency Control*, vol. 53, no. 10, pp. 1918–1933, 2006.

[HUI 98] HUISMAN H.J., THYJSSEN J.M., "Adaptative tissue features extraction with application to ultrasonic image analysis", *Ultrasonic Imaging*, vol. 20, no. 2, pp. 132–148, 1998

[HWA 99] HWANG J., Two pulse technique for ultrasonic harmonic imaging, US Patent vol. 5, no. 951, p. 478, 1999.

[ITO 06] ITOH A, UENO E, TOHNO E, *et al.*, "Breast disease: clinical application of US elastography for diagnosis", *Radiology*, vol. 239, no. 2, pp. 341–350, 2006.

[JEN 92] JENSEN J.A., SVENDSEN N.B., "Calculation of pressure fields from arbitrarily shaped, apodized, and excited ultrasound transducers", *IEEE Transactions on Ultrasonics, Ferroelectrics and Frequency Control*, vol. 39, no. 2, pp. 262–267, 1992.

[JEN 98] JENSEN J.A., MUNK P. "A new method for estimation of velocity vectors", *IEEE Transactions on Ultrasonics, Ferroelectrics and Frequency Control*, vol. 45, pp. 837–851, 1998.

[KRI 08] KRISHNAN K.B., THOMENIUS K.E., "Improved contrast ultrasound with tissue harmonic minimizing pulse", *IEEE Transactions on Ultrasonics, Ferroelectrics and Frequency Control*, vol. 55, no. 1, pp. 249–253, 2008.

[KRI 96] KRISHNAN S., O'DONNELL M., "Transmit aperture processing for nonlinear contrast agent imaging", *Ultrasonic Imaging*, vol. 18, no. 2, pp. 77–105, 1996.

[KRI 98] KRISHNAN S., HAMILTON J.D., O'DONNELL M., "Suppression of propagating second harmonic in ultrasound contrast imaging", *IEEE Transactions on Ultrasonics, Ferroelectrics and Frequency Control,* vol. 45, no. 3, pp. 704–711, 1998.

[KRO 98] KROUSKOP TA, WHEELER TM, KALLEL F., *et al.*, "Elastic moduli of breast and prostate tissues under compression", *Ultrasonic Imaging*, vol. 20, pp. 260–274, 1998.

[KUT 09] KUTTER O., SHAMS R., NAVAB N., "Visualization and GPU-accelerated simulation of medical ultrasound from CT images", *Comput. Methods Prog. Biomed.*, vol. 94, no. 3, pp. 250–266, 2009.

[LIE 07] LIEBGOTT H., WILHJELM J., JENSEN J. A., *et al.*, "PSF Dedicated to Estimation of Displacement Vectors for Tissue Elasticity Imaging with Ultrasound", *IEEE Transactions on Ultrasonics, Ferroelectrics and Frequency Control*, vol. 54, pp. 746–756, 2007

[LIE 08] LIEBGOTT H., MARINCAS S., BASARAB A., *et al.*, "Tangential oscillations for motion estimation in echocardiography", in Proc. *IEEE Ultrasonics Symposium*, Beijing, China, pp. 1761–1764, 2008.

[LIE 10] LIEBGOTT H., BASARAB A., GUETH P., *et al.*, "Transverse oscillations for tissue motion estimation", *Ultrasonics*, vol. 50, no. 6, pp. 548–555, 2010

[LIU 09] LIU K., ZHANG P., SHAO J., *et al.*, "A 2D strain estimator with numerical optimization method for soft-tissue elastography", *Ultrasonics*, vol. 49, pp. 723–732, 2009.

[MAL 10] MALTAVERNE T., DELACHARTRE P., BASARAB A., "Motion estimation using the monogenic signal applied to ultrasound elastography", *IEEE Engineering in Medicine and Biology Society Conference (EMBC 2010)*, Buenos Aires, Argentina, pp. 33–36, 2010.

[MAN 01] MANDUCA A., OLIPHANT T.E., DRESNER M.A., *et al.*, "Magnetic resonance elastography: non-invasive mapping of tissue elasticity.", *Med Image Anal*, vol. 5, no. 4, pp. 237–54, 2001.

[MAR 09] MARION A., VRAY D., "Toward a real-time simulation of ultrasound image sequences based on a 3-D set of moving scatterers", *IEEE Transactions on Ultrasonics, Ferroelectrics and Frequency Control,* vol. 56, no. 10, pp. 2167–79, 2009.

[MAU 04] MAURICE R.L., OHAYON J., FRÉTIGNY Y., *et al.*, "Noninvasive vascular elastography: Theoretical framework", *IEEE Trans Med Imaging*, vol. 23, no. 2, pp. 164–180, 2004.

[MIG 01] MIGNOTTE M., MEUNIER J., "A multiscale optimization approach for the dynamic contour-based boundary detection issue", *Comput. Med. Imag. Graph.*, vol. 25, no. 3, pp. 265–275, 2001.

[MIS 03] MISHRA A., DUTTA P.K., GHOSH M.K., "A GA based approach for boundary detection of left ventricle with echocardiographic image sequences", *Image Vis. Comput.*, vol. 21, pp. 967–976, 2003.

[MUT 95] MUTHUPILLAI R., LOMAS D.J., ROSSMAN P.J., *et al.*, "Magnetic resonance elastography by direct visualization of propagating acoustic strain waves", *Science 269*, no. 5232, pp. 1854–7, 1995.

[NIG 03] NIGHTINGALE K., MCALEAVEY S., TRAHEY G., "Shear-wave generation using acoustic radiation force: in vivo and ex vivo results", *Ultrasound in Medicine & Biology*, vol. 29, no.12, pp. 1715–1723, 2003.

[OPH 91] OPHIR J., CESPEDES I., PONNEKANTI H., *et al.*, "Elastography: a quantitative method for imaging the elasticity of biological tissues", *Ultrasonic Imaging*, vol. 13, pp. 111–134, 1991.

[PAS 10] PASOVIC M., DANILOUCHKINE M., MATTE G., *et al.*, "Broadband reduction of the second harmonic distortion during nonlinear ultrasound wave propagation", *Ultrasound in Medicine & Biology*, vol. 36, no. 10, pp. 1568–1580, 2010.

[PAS 11] PASOVIC M., DANILOUCHKINE M., FAEZ T., *et al.*, "Second harmonic inversion for ultrasound contrast harmonic imaging", *Physics in Medicine and Biology*, vol. 56, no. 11, pp. 3163–3180, June 2011.

[PEL 04] PELLOT-BARAKAT C, FROUIN F, INSANA MF, *et al.*, "Ultrasound elastography based on multiscale estimations of regularized displacement fields", *IEEE Trans Med Imaging*; vol. 23, no. 2, pp. 153–163, 2004.

[PES 99] PESAVENTO A., PERREY C., KRUEGER M., *et al.*, "A time-efficient and accurate strain estimation concept for ultrasonic elastography using iterative phase zero estimation", *IEEE Transactions on Ultrasonics, Ferroelectrics, and Frequency Control*, vol. 46, no. 5, pp. 1057–1067, 1999.

[PHI 01] PHILLIPS P.J., "Contrast pulse sequences (CPS): imaging nonlinear microbubbles", *IEEE Ultrasonics Symposium*, vol. 2, pp. 1739–1745, 2001.

[PHI 04] PHILLIPS P., GARDNER E., "Contrast-agent detection and quantification", *European Radiology*, vol. 14, pp. 4–10, 2004.

[PIW 99] PIWAKOWSKI B., SBAI K., "A new approach to calculate the field radiated from arbitrarily structured transducer arrays", *IEEE Transactions on Ultrasonics, Ferroelectrics and Frequency Control*, vol. 46, no. 2, pp. 422–40, 1999.

[PRO 12] PROSTAFLUO, Un système d'imagerie bi-modale ultrasons/optique pour l'aide au guidage des biopsies de diagnostic du cancer de la prostate, *Programme ANR TecSan, CEA-LETI Grenoble, CREATIS Lyon, CHU Bordeaux*, Institut d'optique Palaiseau, Amplitude Systèmes Bordeaux, VERMON Tours, 2009–2012.

[SAN 03] SANDRIN L., FOURQUET B., HASQUENOPH J.M., *et al.*, "Transient elastography: a new noninvasive method for assessment of hepatic fibrosis", *Ultrasound in Medicine & Biology*, vol. 29, no. 12, pp. 1705–1713, 2003.

[SHA 01] SHANKAR P.M., "Ultrasonic tissue characterization using a generalized Nakagami model", *IEEE Transactions on Ultrasonics, Ferroelectrics and Frequency Control*, vol. 48, no.6, pp. 1716–1720, 2001.

[SHE 07a] SHEN C.C., WANG Y.C., HSIEH Y.C., "A feasibility study of tissue harmonic generation with 3f0 transmit phasing", *IEEE Ultrasonics Symposium*, pp. 1748–1751, 2007.

[SHE 07b] SHEN C.C., WANG Y.C., HSICH Y.C., "Third harmonic transmit phasing for tissue harmonic generation", *IEEE Transactions on Ultrasonics, Ferroelectrics and Frequency Control,* vol. 54, no. 7, pp. 1370–1381, 2007.

[SHE 08a] SHEN C.C., HSIEH Y.C., "Optimal transmit phasing on tissue background suppression in contrast harmonic imaging", *Ultrasound in Medicine & Biology*, vol. 34, no. 11, pp. 1820–1831, November 2008.

[SHE 08b] SHEN C.C., WANG H.W., CHIU Y.Y., "Contrast harmonic detection with chirp excitation in 3f0 transmit phasing", *Ultrasonic Imaging*, vol. 30, no. 4, pp. 251–261, 2008.

[SHE 10] SHEN C.C., YANG Y.C., CHEN W.S., "Optimal transmit phasing for harmonic-background suppression with bipolar square-wave pulser", *Ultrasonic Imaging*, vol. 47, no. 1, pp. 33– 47, 2010.

[SHE 11] SHEN C.C., SHI T.Y., "Third harmonic transmit phasing for SNR improvement in tissue harmonic imaging with Golay-encoded excitation", *Ultrasonics*, vol. 51, no. 5, pp. 554–560, December 2011.

[SHI 07] SHI H., VARGHESE T., "Two-dimensional multi-level strain estimation for discontinuous tissue", *Phys Med Biol*, vol. 52, no. 2, pp. 389–401, 2007.

[SIM 99] SIMPSON D.H., CHIN C.T., BURNS P.N., "Pulse inversion Doppler: a new method for detecting nonlinear echoes from microbubble contrast agents", *IEEE Transactions on Ultrasonics, Ferroelectrics, and Frequency Control*, vol. 46, no. 2, pp. 372–382, January 1999.

[SOU 08] SOUCHON R., SALOMIR R., BEUF O., *et al.*, "Transient MR elastography (t-MRE) using ultrasound radiation force: theory, safety, and initial experiments in vitro", *Magn Reson Med.*, vol. 60, no. 4, pp. 871–881, 2008.

[STE 71] STEPANISHEN P.R., "Transient radiation from pistons in an infinite planar baffle", *The Journal of the Acoustical Society of America*, vol. 49, no. 5B, pp. 1629–1638, 1971.

[SUH 05] SUHLING, M., ARIGOVINDAN, M., JANSEN, C., *et al.*, "Motion analysis from B-mode echocardiograms", *IEEE Trans. on Image Processing*, vol. 14, no. 4, pp. 525–536, 2005.

[SUM 08] SUMI C., "Displacement vector measurement using instantaneous ultrasound signal phase multidimensional autocorrelation and doppler methods", *IEEE Transactions on Ultrasonics, Ferroelectrics and Frequency Control,* vol. 55, pp. 24–43, 2008.

[SZA 04] SZABO T. L., *Diagnostic Ultrasound Imaging*, Elsevier Academic Press, 2004.

[THI 03] THIJSSEN J. M., "Ultrasonic speckle formation, analysis and processing applied to tissue characterization", *Pattern Recognition Letters*, vol. 24, nos. 4–5, pp. 659–675, 2003.

[UNS 09] UNSER M., SAGE D., VANDEVILLE D., "Multiresolution monogenic signal analysis using the riesz laplace wavelet transform", *Image Processing, IEEE Transactions* on, vol. 18, no. 11, pp. 2402–2418, November 2009.

[VAR 10] VARRAY F., CACHARD C., TORTOLI P., *et al.*, "Nonlinear radio frequency image simulation for harmonic imaging – CREANUIS", *IEEE Ultrasonics Symposium, San Diego*, 2010.

[VAR 11a] VARRAY F., CACHARD C., RAMALLI A., *et al.*, "Simulation of ultrasound nonlinear propagation on GPU using a generalized angular spectrum method", *EURASIP Journal on Image and Video Processing*, vol. 17, no. 1, 2011.

[VAR 11b] VARRAY F., RAMALLI A., CACHARD C., *et al.*, "Fundamental and second-harmonic ultrasound field computation of inhomogeneous nonlinear medium with a generalized angular spectrum method", *IEEE Transactions on Ultrasonics, Ferroelectrics and Frequency Control*, vol. 58, no. 7, pp. 1366–1376, 2011.

[VAR 13] VARRAY F., BASSET O., TORTOLI P., *et al.*, "CREANUIS: A nonlinear radio frequency ultrasound image simulator", *Ultrasound in Medicine and Biology*, pp. 1915–1924, 2013.

[WIL 00] WILKENING W., KRUEGER M., "Phase-coded pulse sequence for non-linear imaging", *IEEE Ultrasonics Symposium*, pp. 1559–1562, 2000.

[WIL 01] WILKENING W., BRENDEL B., JIANG H., *et al.*, "Optimized receive filters and phase-coded pulse sequences for contrast agent and nonlinear imaging", *IEEE Ultrasonics Symposium*, vol. 2, no. 4, pp. 1733–1737, 2001.

[XU 10] XU H., RAO M., VARGHESE T., *et al.*, "Axial shear strain imaging for differentiating benign and malignant breast masses", *Ultrasound in Medicine & Biology*, vol. 36, no. 11, pp. 1813–1824, 2010.

[YAN 03] YAN J.Y., ZHUANG T., "Applying improved fast marching method to endocardial boundary detection in echocardiographic images", *Pattern Recognition Letters*, vol. 24, no. 15, pp. 2777–2784, 2003.

[YIN 07] YIN M., WOOLLARD J., WANG X., *et al.*, "Quantitative assessment of hepatic fibrosis in an animal model with magnetic resonance elastography", *Magn Reson Med*, vol. 58, no. 2, pp. 346–53, 2007.

[ZHU 02] ZHU Y., HALL T.J., "A modified block matching method for real-time freehand strain imaging" *Ultrason Imaging*, vol. 24, pp. 161–176, 2002.

Chapter 2

Magnetic Resonance Imaging

2.1. Introduction

Over the past two decades, the acronym "MRI" has become familiar as being synonymous with medical examination. Much like other types of clinical exploration, it involves a great deal of high-performance technology in the background, which is still a mystery to many. Indeed, behind these clinical installations are hidden devices which we are perfectly able to use, without necessarily understanding their structure, the principles of their operation, all the potential that they offer or indeed the disadvantages which they entail. This introductory remark instantly throws up a number of questions, which we do not intend to fully answer here, given that the subject has become so vast. Its extent, indeed, is so broad, its overspills into all disciplines so numerous and its areas of specialization so varied, that it necessarily involves a great many domains which are sometimes utterly foreign from one another. For instance, if we consider MRI in the category of non-invasive control techniques, we of course find the domain of Medicine and its entourage of Biological and Biomedical Research, but also multiple aspects stemming from Chemical Engineering and Process Engineering. These lead on to industrial control methods, and use installations designed for highly specialized areas: porous media, reactive media, catalysis, polyphasic and microfluidic flows, membrane exchanges, and so on.

Consequently, there is no discipline specifically associated with MRI which, along with MRS (Magnetic Resonance Spectroscopy) – which is widely used in Chemistry, Biochemistry and Biology and therefore Medicine – defines specific

Chapter written by Dominique SAPPEY-MARINIER and André BRIGUET.

orientations for constantly-progressing research stemming from a very diverse group of uses. These uses, which range from the fundamental to the applied, exploit the whole range of values of magnetic fields which we are able to render sufficiently stable, from a value of zero to over 20 teslas. In this case, we usually use the nomenclature "NMR" for Nuclear Magnetic Resonance, as the denomination "MRS" is increasingly reserved for the domain of healthcare for the same reasons of matching of vocabulary, as is the case with the denomination "MRI". Yet it should be remembered that the appellation "NMR" in its own right brings together all the common concepts developed in this discipline which, for over seven decades, from Rabi to Lauterbur and Mansfield, through Bloch, Purcell, Ernst and Wüthrich, has on numerous occasions enjoyed recognition from the Nobel Committee.

In this discussion, we shall limit ourselves essentially to the two states of magnetic energy of the hydrogen nucleus, which is the candidate most often targeted by MRI. The underlying quantum aspect is certainly interesting. However, it is well known, and is widely discussed in numerous monographs, with varying descriptions of it – particularly at the beginning of many theses. Therefore, we shall begin with the notion of nuclear magnetization, whose expression we shall accept. As the notions of MRS which we shall touch upon are greatly marked with the seal of imaging, we do not believe it is necessary to linger over the quantum aspect, given that the spectra of metabolites expressed as MRI images involve only the notion of chemical shift. At present, the notion of the hyperfine structure of nuclear magnetic resonances, which requires a quantum framework upon which to build, refers mainly to multidimensional representations (2D, 3D or more) exploited in Physical Chemistry, Analytical Chemistry and Fundamental Biology, and less frequently in Medicine.

Following this introduction, which is section 2.1 of this chapter, section 2.2 gives a brief overview of the fundamental notions leading to the basic principle behind MRI. Yet throughout this section, we shall linger somewhat on the notion of pulses, using examples, but without going into detail about the fundamentals of the calculations, which are perfectly and clearly explained in a recent book by the late Michel Décorps. As in the rest of the document, we wish to limit the number of formulae presented here, so as not to swamp the reader with excessive mathematical explanations.

Section 2.3 mainly describes the conditions in which the signal is collected, progressively discussing the organization of an MRI installation in a hospital, which differs from an experimental device by a factor of scale, and by various hypertrophies. The dimensions of these installations necessitate the taking of safety measures, which are very briefly discussed in the final part of this section. On that subject, the non-aggressive nature of MRI should be considered an intrinsic property of the interaction between radiation and matter – an interaction which is very slight at the frequencies used in medical examinations. However, the rapid evolution of human research investigation devices has led to an increase in the energy deposited, which necessitates the taking of precautions. A study covering everything from the basic effects to the regulations in force would be so vast that it could never fit into the bounds of a presentation such as this.

Section 2.4 is not presented as a technical image-processing task, but rather as an analysis of the acquisition parameters which give rise to the contrast – a major strong point of MRI – whilst attempting to form associations between those elements and situations corresponding to clinical practice.

In section 2.5, we wanted to draw the connection between the procedures for reconstruction and the execution of the acquisition sequences by taking a number of examples from the numerous families and filiations of sequences. Another way of proceeding would have been to systematically examine the relation between contrast and sequence, or the relation between sequence and performance. These approaches, which are better perceived by users – first and foremost medical examiners, are to be found in the existing body of literature associated with the topic. Given the multitude of sequences and the varying names given to them from one manufacturer to another, we have chosen, if not to shorten our text, at least to attempt to restrict ourselves to the general principles, including in the presentation of the modes of reconstruction.

Finally, section 2.6, while it cannot cover all of the biomedical applications of MRI, focuses at greater length on the functional and metabolic aspect of MRI, such as spectroscopy and its links with imaging. Based on the same principle but in a more analytical manner, we examine the use of MRI for analyzing diffusion processes. The potential of this approach, recognized since the late 1940s, is constantly being enhanced, and its advantage for medical applications is being confirmed. It falls into the context of clinical functional imaging whilst remaining a very readily-evolving tool, particularly promising in the domain of neurosciences. The functional aspect of MRI will also be mentioned, with the example of brain activation. Finally, we have chosen to mention the association between MRI and Positron Emission Tomography (PET) to illustrate MRI's potential for evolution in the domain of multimodal imaging.

The bibliographical references are grouped by topic. We hope that they will be helpful for readers wishing to find avenues to pursue further. In view of the abundance of academic production on the subject of MRI on the one hand, and on the other the fact that we have limited the list of facts chosen – in principle for reasons of conciseness but doubtless sometimes by ignorance or omission, we hope that the indications given here will be useful and will be used.

2.2. Fundamental elements for MRI

2.2.1. *Introduction*

MRI exploits the presence, in the media being explored, of nuclei with paramagnetism. When subjected to a magnetic field, the individual magnetic moments associated with the spin of these nuclei create a magnetization which satisfies Curie's law (magnetization oriented in the direction of the field applied (the field is coined $\vec{B}$), proportional to its value, and inversely proportional to the temperature). Expressed per unit volume, the magnetization recorded at thermodynamic equilibrium can be written thus:

$$\overrightarrow{M}_{vol} = N \cdot \frac{s(s+1)\hbar^2\gamma^2}{3k_B T} \cdot \vec{B} \qquad [2.1]$$

where N is the number of nuclei per unit volume of the sample in question, s is their spin number, $\hbar$ Planck's constant divided by 2π, γ the gyromagnetic ratio of the nuclei under consideration, k_B Boltzmann's constant and T the temperature of the sample. Remember that the magnetic moment of each nucleus is the product $\vec{\mu} == \gamma\vec{s}$ where $\vec{s}$ is, in the conventional sense, the nuclear spin vector. The distribution of the nuclei according to the different levels of magnetic energy engendered by the $2s+1$ spin states (the same character is used to denote the spin number s and $\vec{s}$ the spin vector) is obtained by Boltzmann's statistic, which explains the intervention of the temperature *T*. The algebraic sum of the elementary magnetic moments, weighted by the populations of the $2s+1$ energy states, gives us the result.

The value given by equation [2.1] is expressed in amperes per meter. Therefore, the unit used for the magnetization of a given volume of matter is the ampere meter squared ($A \cdot m^2$). In a 1-tesla field, the value of protonic magnetization at thermal equilibrium of a cubic centimeter of water at ambient temperature is approximately $3.21 \times 10^{-6}\ A \cdot m^2$ Table 2.1 shows values of the gyromagnetic ratios of certain nuclei used in MRI, beginning with the hydrogen nucleus or proton, which is most commonly used because of its abundance in living tissues made up of water (concentration of around 50 mole/kg) and lipids (fats). It is also to be found (though at considerably lower concentrations – around a milli-mole/kg) in the metabolites present in organs (the brain, liver, heart and other muscles, etc.), where it is detected by magnetic resonance spectroscopy (MRS). Even if it is very abundant in a sample, an isotope with a non-null spin number (as is the case of protons in water), and therefore giving rise to nuclear magnetization, has its contribution to the magnetic susceptibility completely masked by electronic diamagnetism. Therefore, we can only access the magnetization of the nuclei indirectly (we cannot measure nuclear magnetization by conventional magnetometry) and selectively using nuclear magnetic resonance. This technique began being developed from the late 1930s by Rabi, and then independently by Bloch and Purcell in 1945. Physical chemists use quantum mechanics to describe the interactions which the magnetization may experience and the effects of those interactions. The corresponding work and subsequent applications, such as NMR spectroscopy of liquids and solids, have been developed and continue being developed on that basis. However, a vectorial description is perfectly acceptable in most situations encountered in MRI examinations.

Nucleus	Natural percentage	Spin number	$\gamma \times 10^{-6}$ (rad/s.T)	Frequency at 3 teslas (MHz)	Biomedical usage	Other domain of use
Proton (^{1}H)	99.985	½	267.5	128	Very high (MRI, MRS)	Chemistry, Biochemistry
Deuterium (^{2}D)	0.015	1	41.07	17.05	Very low	Material chemistry
Carbon 13 (^{13}C)	0.0159	½	67.26	32.1	Low * (MRS)	Organic chemistry
Sodium (^{23}Na)	100	3/2	70.76	33.8	Medium (MRI, MRS)	Chemistry
Phosphorus (^{31}P)	100	½	108.3	51.7	Medium (MRS)	Chemistry Biochemistry
Helium 3 (^{3}He)	0.000137	½	-204	97.3	Low * (MRI)	Physics

Table 2.1. *Nuclei, natural percentage, spin number, gyromagnetic moment, precession frequency, main uses (imaging/spectroscopy) (* mainly employed in a so-called "hyper-polarized" form, whereby the magnetization is strengthened by special procedures)*

2.2.2. *Vectorial description of nuclear magnetic resonance (NMR)*

The elements that have to be used, most of which are discussed in greater detail in publications in the field of nuclear magnetic resonance (NMR), are outlined below.

2.2.2.1. *Precession*

The mechanical model of precession results from the actions of a magnetic field $\vec{B}$ on an elementary magnetic moment of nuclear origin $\vec{\mu}$ – this moment is itself connected to the kinetic moment of the particle. This model gives us the basic equation:

$$\frac{d\vec{\mu}}{dt} = -\gamma \vec{B} \wedge \vec{\mu} \qquad [2.2]$$

by application of the kinetic moment theorem, noting that the resultant of the magnetic forces acting on $\vec{\mu}$ is negligible. This result shows, on the one hand, that the modulus of $\vec{\mu}$ remains constant, and on the other, that this vector rotates around the direction of $\vec{B}$ at angular velocity $\vec{\omega} = -\gamma\vec{B}$. This causes a motion of precession of $\vec{\mu}$, known as the Larmor precession, around $\vec{B}$ which we can define as the guiding static magnetic field. The magnetization $\vec{M}$ of the sample under examination (suppose that its volume is the unit volume so as to be able to conserve the notation used in equation [2.1]) can be considered to be the vector sum of the moments present. If the field $\vec{B}$ is perfectly uniform throughout the sample, then using equation [2.2] where $\vec{M}$ replaces $\vec{\mu}$, we can describe it; the motion of precession of that magnetization around the *OZ* axis is represented in Figure 2.1(a). In order to take account of the lack of uniformity of the static field, Figure 2.1(b) considers the magnetization assigned to a 3D sample element within which the directive magnetic field exhibits a very precise particular value, which is different to that governing any neighboring elements. When dealing with very small volumes, these elements, which are distinguished by their angular velocities of precession, are generally called "isochromats". This does not alter the validity of equation [2.2].

2.2.2.2. *Magnetization flipping by a pulsed RF field*

The field $\vec{B}$ determines the direction $\overrightarrow{OZ}$ in which any isochromat in precession around $\vec{B}$ tends to move before it recovers its orientation due to thermodynamic equilibrium. How, then, can we shift the orientation of an isochromat, and more generally, how can we alter the value of magnetization of a sample and render it different to that which exists at thermodynamic equilibrium? Another worthy question is: how can we switch from a state of precession characterized by an inclination in relation to $\vec{B}$ to a different state with a different inclination?

It is therefore necessary to use a second magnetic field, which evolves within a rotating frame. The analysis can be discussed without using any figures, as all the constituent parts are presented in this section. However, there is nothing to stop interested readers from graphically plotting the calculations given here, drawing inspiration from Figure 2.1(a) and (b), which will make it easier to use Figure 2.5 later on.

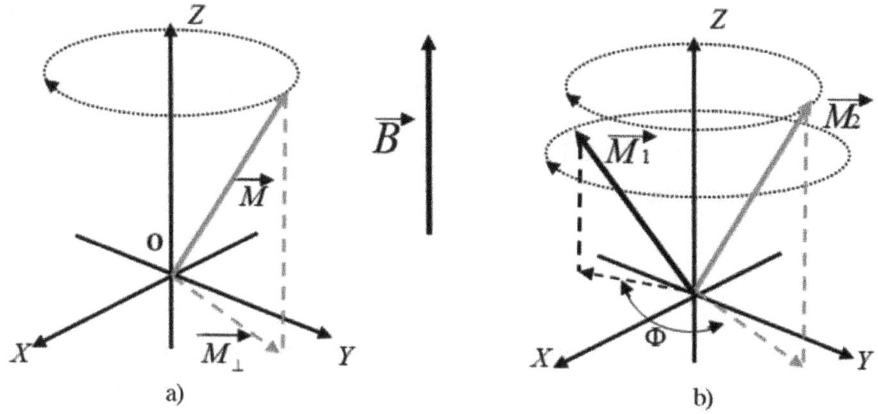

Figure 2.1. *a) Case of a magnetic field which is perfectly uniform throughout the sample: Motion, seen in the Laboratory frame ($(R) = OXYZ$) of the precession of magnetization $\vec{M}$ around the guiding static field $\vec{B}$; when γ is positive, we note that the rotation of the projection $\vec{M_\perp}$ of $\vec{M}$ takes place in the opposite direction. b) Case where the field $\vec{B}$ is not uniform, the magnetizations $\vec{M_1}$ and $\vec{M_2}$, of two different parts of the sample, are not synchronous; the angle Φ which their projections form in the plane (XOY) varies linearly over time. Note here that the description of the precession of different magnetizations can be reduced to a single common point – e.g. the origin of the frame (R)*

The rotating framework, which we shall denote as $(R') = (OxyZ)$, shares with (R) the axis OZ occupied by the guiding static field. It rotates around this shared axis at a constant angular frequency $\overrightarrow{\omega_{R'/R}}$. When we move from the fixed laboratory frame (R) to the frame (R') rotating at angular frequency $\overrightarrow{\omega_{R'/R}}$, the angular frequency composition law gives us:

$$\left.\frac{d\vec{M}}{dt}\right)_{R'} = -(\gamma\vec{B} + \overrightarrow{\omega_{R'/R}}) \wedge \vec{M} \qquad [2.3]$$

Thus, in the frame (R'), the magnetization $\vec{M}$ of the isochromat in question is animated with a motion of precession at angular frequency $-(\gamma\vec{B} + \overrightarrow{\omega_{R'/R}})$ around a fictitious field $\vec{B} + \frac{\overrightarrow{\omega_{R'/R}}}{\gamma}$ directed along OZ.

i) The isochromat remains fixed in (R') if that frame rotates in the sense of precession at the angular frequency of precession in (R); this corresponds to the condition $\vec{B} + \frac{\overrightarrow{\omega_{R'/R}}}{\gamma} = 0$. With this condition fulfilled, we then create a magnetic field $\vec{B_1}$ whose orientation is fixed in (R') which, in light of equation [2.2], acts on $\vec{M}$ in that frame in accordance with the law:

$$\left.\frac{d\vec{M}}{dt}\right)_{R'} = -\gamma\vec{B}_1 \wedge \vec{M} \qquad [2.4]$$

Equation [2.4] indicates that $\vec{M}$ is animated with a motion of precession around $\vec{B}_1$ at angular frequency $-\gamma\vec{B}_1$. Consequently, the application of the field notated as $\vec{B}_1$ for a time τ alters in (R') the orientation of $\vec{M}$ which performs a rotation of angle $\theta = \gamma B_1\tau$ around the direction defined by $\vec{B}_1$. We have generated a pulse which modifies the orientation of $\vec{M}$ with respect to OZ in the frame (R') and also in the frame (R) where the description of the magnetization motion is more complicated. In view of the usual value of the static magnetic fields generally employed, the angular frequency of $\vec{B}_1$ in the Laboratory frame (R), near or equal to $-\gamma\vec{B}$, is within the range of radio frequencies (see Table 2.1).

ii) If the condition $\vec{B} + \frac{\overrightarrow{\omega_{R'/R}}}{\gamma} = 0$ is not fulfilled, which is indeed the case for a different isochromat than the one discussed in (i), this new isochromat will be subject to an action which results in a motion of precession in (R') around the "actual" or effective field resulting from the fictitious field $\vec{B} + \frac{\overrightarrow{\omega_{R'/R}}}{\gamma}$ and the radio frequency (RF) field $\vec{B}_1$:

$$\overrightarrow{B_{eff}} = \vec{B} + \frac{\overrightarrow{\omega_{R'/R}}}{\gamma} + \vec{B}_1 \qquad [2.5]$$

We can see that the direction around which precession takes place in (R') forms an angle α with the direction OZ, such that $tg(\alpha) = \dfrac{\vec{B}_1}{\vec{B} + \dfrac{\overline{\omega_{R'/R}}}{\gamma}}$. The motion of magnetization of the isochromat in question is relatively simple in the frame (R'), and the precession around the effective field takes place with the angular frequency $|\omega| = \gamma \sqrt{\vec{B}_1^{\,2} + (\vec{B} + \dfrac{\overline{\omega_{R'/R}}}{\gamma})^2}$ (see section 2.3). This motion, though, is still complicated to describe in (R), the fixed Laboratory frame.

The RF field created from the Laboratory frame is generally linearly polarized. Its RF is within the range of frequencies of precession of the different isochromats. As it is oriented perpendicularly to the direction of the guiding static field, the two circular vibrations which make it up rotate in opposite directions; one constitutes the field $\vec{B}_1$, which we need to generate so-called RF pulses, to which we shall come back later. Although there is a slight influence of the component rotating in the opposite direction (the Bloch-Siegert effect), we shall not take this into account, given that we can directly create rotating fields using two quadrature orthogonal coils.

2.2.2.3. *Relaxation*

Once the magnetization has been altered by RF excitation, and then left to evolve freely, it gradually recovers its alignment with $\vec{B}$ by a process known as relaxation. For the sake of simplicity in our analysis, we shall suppose that the static magnetic field is perfectly uniform. Therefore, here we shall only need to consider one isochromat, of magnetization $\vec{M}$, to which we shall apply an RF pulse in two different initial situations. Let us look directly at this problem through its analytical aspect in (R').

i) The initial orientation of $\vec{M}$ is along the direction $\overrightarrow{OZ}$ of the guiding static field:

The pulse of duration τ spreads the vector $\vec{M}$ from $\overrightarrow{OZ}$, which thereby creates a component in the direction OZ (the so-called longitudinal component, which is positive if $|\gamma B_1 \tau| < 90°$ or negative if $|\gamma B_1 \tau| > 90°$) and a component perpendicular to

OZ (the "transverse" component, which can in turn be projected onto the directions x and y of the rotating frame (R') or X and Y of the fixed frame (R)).

ii) The initial orientation of $\vec{M}$ is any at all in the rotating frame and pivots by an angle $|\gamma B_1 \tau|$ around the direction defined by the effective field. The final orientation is linked to the original orientation by the amplitude $|B_1|$ and duration τ of the pulse. It is easy to verify that we generally obtain two components – one being longitudinal and the other transverse.

Once the RF excitations cease, the amplitudes of the longitudinal and transverse components change:

The longitudinal component tends to regain its thermodynamic equilibrium value M_0 in accordance with the law:

$$\frac{dM_Z(t)}{dt} = -\frac{Mz(t) - M_0}{T_1} \qquad [2.6]$$

The amplitude of the transverse magnetization tends toward zero in accordance with a law, written here in its rotating frame:

$$\frac{dM_\perp(t)}{dt} = -\frac{M_\perp(t)}{T_2} \qquad [2.7]$$

T_1 is the spin–lattice relaxation time and T_2 the spin–spin relaxation time.

These two differential equations are said to be phenomenological, as they are simply accepted and proposed by F. Bloch in order to account for his observations. They lead to exponential evolutions over time – evolutions which can generally be accounted for reasonably well. The inequality $T_2 \leq T_1$ can be verified experimentally in liquid media, and can be confirmed theoretically. This inequality and model indicate that the complete recovery of $M_z(t)$ in M_0 occurs after $M_\perp(t)$ is removed. An example of a solution to the Bloch equations where the two components M_X and M_Y replace $M_\perp$ is given in Figure 2.2.

In Figure 2.2, where the amplitudes are standardized with respect to M_0, the precession frequency has been chosen as very low, for the purposes of presentation. Note that in order to observe such low frequency, it is "sufficient" to operate in a rotating frame (R''), whose angular frequency is close to the angular frequency of precession. We have introduced relaxation times such that $T_1 > T_2$. Note that the RF pulse used to give rise to the initial conditions is applied for a length of time τ

which is generally much shorter than the recovery time of $\vec{M}$ to $\vec{M}_0$ ($\tau \ll T_2 \leq T_1$), so we can discount relaxation during these pulses. The values of the relaxation times essentially express the effect of entirely random magnetic fields, which operate at microscopic scale in the sample; these fields are modulated by molecular thermal motions, so their average values are null. These mechanisms are effective (we see short relaxation times) when the spectral power density of these fluctuating fields has an appreciable value at the resonance frequency. Therefore, it is possible to artificially shorten the relaxation times by creating additional random interactions, e.g. by introducing paramagnetic substances into the medium. This is the role played by contrast-enhancing products in MRI. When we consider magnetization resulting from different isochromats, the decrease in transverse magnetization is affected by the heterogeneity of the value of the static field throughout the sample. Consequently, the description given by equation [2.7] is valid when the guiding static field is absolutely uniform throughout the volume; this condition is rarely observed in practice.

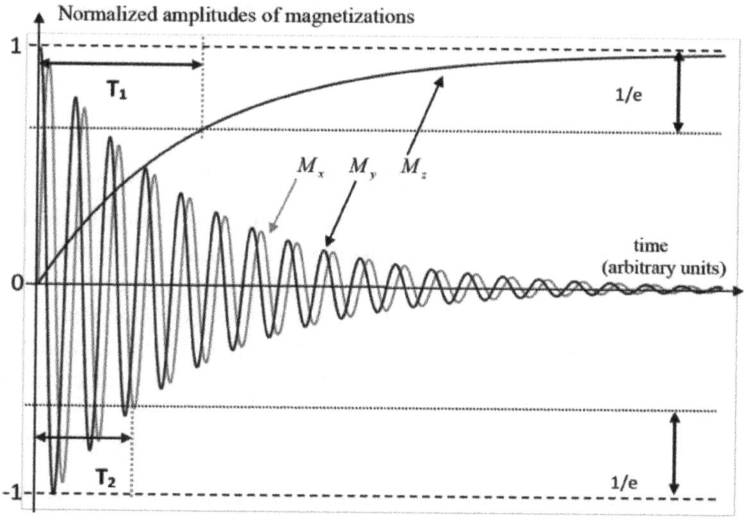

Figure 2.2. *Change in the magnetization of an isochromat following an RF pulse (not represented here). The transverse component has been projected in the perpendicular directions Ox" and Oy" in the frame (R") discussed above. Here, the initial conditions are* $M_z(0) = M_x(0) = 0, M_y(0) = 1$, *with the value of* M_z *at thermodynamic equilibrium taken to be equal to 1. In these conditions, the variations in amplitude are of the form* $1 - e^{-t/T_1}$ *for* M_z, *and* e^{-t/T_2} *for* M_y. *Hence, it is possible to determine the values of the relaxation times graphically*

In a non-uniform static field, a fairly common approach is to replace the parameter T_2 with an apparent transverse relaxation time T_{tr}, which is shorter and which tends to represent the experimental situation. It should be noted that this process is not always valid unless the spatial distribution of the static field through the sample enables us to verify the validity of the exponential decrease expected over time. The spin relaxation time of the medium, T_1, which depends on the temperature, does not depend on the non-uniformity of the field if this is not excessive. Indeed, it must be stressed that the value of the field has an effect on the value of T_1. For example, with a very strong magnetic field, the resonance frequency increases in absolute value by virtue of the law $\vec{\omega} = -\gamma \vec{B}$. In general, the spectral density of the interactions which cause spin–lattice relaxation in the medium, decreases as the frequency increases, and consequently the relaxation loses its efficiency because of lesser stimulation, and the values of T_1 increase. In pure water, from which all paramagnetic impurities such as gaseous oxygen have been removed, the relaxation times T_1 and T_2 for the protons with a value of around 2.5 seconds if the strength of the static field is approximately 2 teslas. The values observed in living tissues are shorter, and differ quite significantly depending on the nature of those tissues.

2.2.2.4. *The signal*

In view of the order of magnitude of T_1, the longitudinal component subject to spin relaxation in the medium evolves slowly over time in relation to the transverse components, which oscillate at a radio frequency. The variation of the magnetic flux induced in a circuit by these latter components creates the signal. As a circuit, we most usually use a coil, tuned to a frequency which is equal or very near to the precession frequency. We determine the electromotive force e which appears between the extremities of the circuit using the usual laws of electromagnetism: $e = \int_{coil} \vec{E}\,d\vec{l}$ where $\vec{E} = -\dfrac{\partial \vec{A}}{\partial t}$, with $\vec{A}$ being the magnetic vector potential created by the variations in transverse magnetization. With a sample of volume δV whose transverse magnetization $\vec{m}(t)\ \delta V$ during the precession, we obtain the expression of the signal induced in the coil, which serves as a sensor:

$$\delta e = -\frac{\partial}{\partial t}[\frac{\vec{B}_1}{I}.\vec{m}(t)]\delta V \qquad [2.8]$$

where $\dfrac{\vec{B}_1}{I}$ is the field per unit current at the position of δV, i.e. the field that would be created by the receiving coil if it were to function as an RF transmitter. We can immediately see that the sensitivity of the method is greater when there is greater fluctuation in the magnetization $\vec{m}(t)\ \delta V$ in relation to time. As this variation is directly connected to the angular frequency of precession, it is conditioned by the value of $\vec{B}$. Another factor favoring detection is the selective nature of the detecting coil, whose spectral response must contain the frequency band of precession. In general practice in the field of NMR, it is just as common to use a setup with a unique excitation/detection coil as to use a setup with two orthogonal coils, with one providing excitation and the second receiving the signal. The latter setup, which is very widely used in MRI, necessitates good electrical decoupling between the transmitter and receiver coils. A relatively large transmitter coil ensures the best possible uniformity of the RF field through the volume of the sample, and also enables us to use sensors which are smaller and closer to the region we have to explore. These two devices are usually referred to as RF antennas. The appropriateness of this term is debatable if we do not need to take account of propagation phenomena. However, this situation needs to be considered with strong magnetic fields in MRI: consider a 7-tesla field and a proton resonance frequency of 300 MHz, giving us a wavelength of one meter in a vacuum, but far shorter in living material since the relative dielectric permittivity of living tissues is high – similar to that of water ($\varepsilon_r \approx$ 80). Therefore, we can indeed speak of antennas, and be led to exploit propagation phenomena within the magnet to detect the signal.

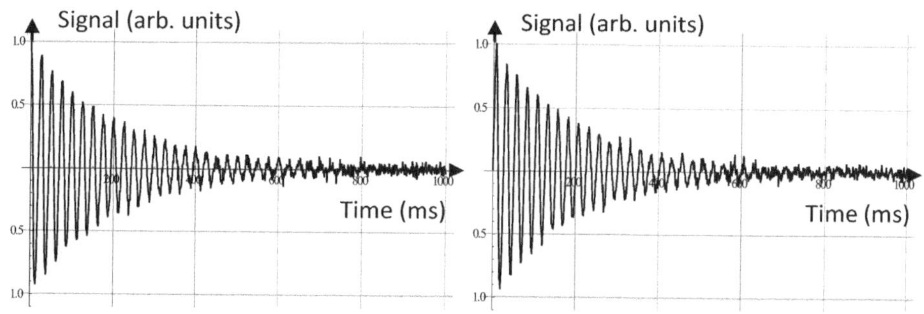

Figure 2.3. *Shape of an NMR signal observed and recorded at 1000 complex points with a sampling period of 1 ms, which enables us to use a spectral width equal to 1 kHz. Note the existence of the background noise. The signal is visualized once it has been digitized. Arbitrary vertical initial amplitude. Horizontal amplitude 1 s. Left: "Real" part; right: corresponding "imaginary" part*

Following an RF pulse, we are generally led to record a so-called free induction decay (FID) signal, whose amplitude decreases over time. This oscillating, damped signal, resulting from the contribution of all the isochromats, exhibits an average pseudo-pulsation given by the law $\vec{\omega} = \gamma\vec{B}$. This formula can be used to measure the field value in an almost punctual manner (NMR magnetometry) provided we use a sample of very small volume. Note that with weak static fields, we can observe a low-frequency signal, e.g. of around 2 kHz with the protons in the Earth's field, which is around 0.05 milli-tesla. For stronger fields leading to radio frequencies, we may use a frequency subtraction in order to bring down the frequency of the signals into the range of low frequencies (typically less than 100 kHz).

Therefore, the signals can be easily recorded, either analogically or digitally. Detection, combined with changing frequency performed on two parallel channels whose references are 90° separated from one another, enables us to receive a complex signal and offers a usable spectral range whose width is equal to the inverse of the sampling period.

The Fourier transform of this complex signal gives us its representation in the spectral domain (Figure 2.4). The denominations used in terms of absorption and terms of dispersion are a legacy from Optics. It is important to note that the signal usually corresponds fairly well to a so-called Lorentz model, due to the fact that the variation in amplitude over time is generally governed, fairly closely, by an exponentially decreasing function. In these conditions, it is easy to verify that the time constant T_{tr} – which replaces T_2 in the experimental situations – determines the linewidth at halfway height, $\Delta f_{\frac{1}{2}}$, in accordance with relation $\pi \cdot \Delta f_{\frac{1}{2}} \cdot T_{tr} = 1$. The time constant of transverse decay of the temporal signal is thus linked to the time constant T_2 by the empirical relation:

$$\frac{1}{T_{tr}} = \frac{1}{T_2} + \frac{1}{T_{in}} \qquad [2.9]$$

In equation [2.9], T_{in} represents the non-uniformity of the magnetic field $\vec{B}$ throughout the volume of the sample. This relation only truly makes sense if we can express the non-uniformity of the field (also referred to as "inhomogeneity") as a time constant, which is not always the case. However, it is useful in magnetic resonance spectroscopy (MRS) to evaluate the power of spectral resolution, in order to improve the "homogeneity" of the static field. The value of T_{in} then gives a fairly accurate representation of the lack of uniformity of the directive magnetic field leading to the angular frequencies of precession spreading out. In MRI, the resulting spectral spreading creates a convolution effect, which in turn noticeably broadens the spatial representation of objects.

Note that in the absence of inhomogeneity of the field, or we are thinking about a single isochromat, the absorption line could not be infinitely narrow because it is determined by T_2. If the signal exhibits a non-null initial phase, which is not the case in Figure 2.3, each part of the spectrum is made up of absorption terms and dispersion terms. Generally, it is possible to perform *a posteriori* phase correction, as we have done here in order to improve the presentation of the signals in Figure 2.4. When we are dealing with a spectrum with a number of lines superposed over a spectral interval, we have to take account of the frequency (fairly often, a linear dependence with frequency) to perform effective compensation.

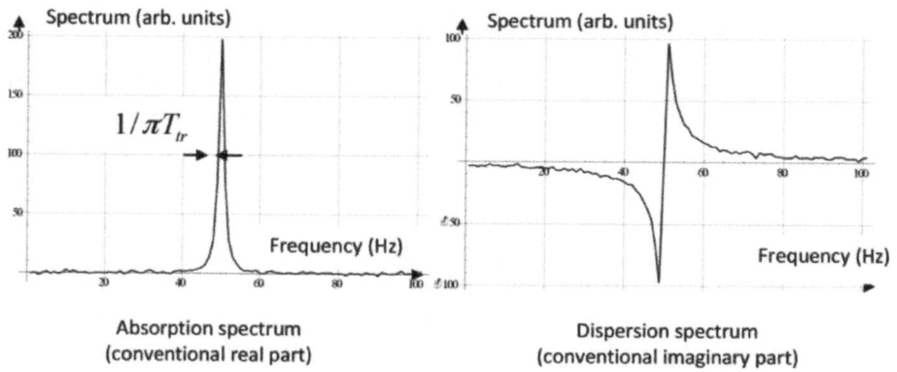

Figure 2.4. *Representation of the shape of the signal obtained by Fourier transformation of the complex time signal shown in Figure 2.3. Representation on 100 complex points so as to obtain the usable part of the spectrum occupying 100 Hz. Note that, because of the use of the discrete Fourier transform, the relative level of amplitude of the background noise is significantly less than in the temporal domain. Vertical axis: arbitrary units. Horizontal axis: 1 Hz interval between two successive points*

The signal is triggered by the application of an RF pulse, usually of very short duration (between a hundred microseconds and a few milliseconds) in comparison to the time needed to damp the signal, which is several tens of milliseconds. Viewed in the rotating frame, the pulse results in a rotation around the effective field vector. This rotation has the orientation of the vector $\overrightarrow{B_1}$ in the rotating frame at the so-called "resonance" angular frequency, which is the angular frequency of precession of the nuclei. Note that for nuclei with a positive gyromagnetic ratio, the direction of rotation around a magnetic field is reversed. However, we usually describe positive rotations in the trigonometric direction, and we shall respect this convention in further discussions of RF pulses.

2.2.3. *RF pulses and their effect on magnetizations*

The RF pulses cause transverse components of magnetization to appear, and serve to control these components with a view to recording the signal. As the generation of these pulses is combined (as we shall discuss later on) with the application of the gradients necessary to create images, the collection of measurements of the signal in digital form (acquisition) constitutes a sequence. In general, it is necessary to combine several sequences, or repeat the same elementary sequence multiple times, altering only one of its parameters each time. A particularly important aspect is choosing the RF field pulses, which are necessary at the beginning of every elementary sequence, and choosing the pulses to be applied during that sequence, which are usually different to the previous ones. Below, we discuss a few examples of the most commonly used elementary pulses.

2.2.3.1. *General principle and mode of description*

It should not be forgotten that in practice, these pulses are performed in the presence of a distribution of isochromats which, when viewed in the rotating frame, are not all subject to exactly the same effective field, because of relation [2.5]. This peculiarity needs to be examined using Figure 2.5. In this figure, we illustrate the motion of an isochromat which is not at the resonance frequency. In the frame (R') and during the duration τ of the pulse, the effective field $\overrightarrow{B_{eff}}$, the vector sum of the fictitious field $\overrightarrow{B_f}$ and the field $\overrightarrow{B_1}$, creates a rotation by an angle $\omega\tau = \gamma B_{eff}$ (traditionally notated θ), counted algebraically in the trigonometric direction defined by the direction of $\overrightarrow{B_{eff}}$. Using Figure 2.5, we can verify that an RF pulse causes multiple alterations to the magnetization of an isochromat, initially taken at rest, in the direction $\overrightarrow{OZ}$: a decrease of the longitudinal component, the creation of a transverse component (in the planes Oxy or OXY) – a component which in turn can be decomposed into components in the directions Ox and Oy of the rotating frame. If we know the initial situation of the magnetization and the characteristics of the pulse, we can determine the new components using the formula describing the rotations in a 3D space. This involves describing the magnetization in a trihedron with one of its primary axes aligned with the effective field, and then representing the result by "coming back" to the original trihedron (R'), using rotation matrices. It is possible to employ a formalism that conventionally uses 3 × 3 matrices but is rather unwieldy, or a calculation technique which combines the transverse components of the magnetizations in order to use 2 × 2 matrices (this is known as the "spinors" method).

For the examples below, we shall not go into detail about the analytical description of the pulses – a subject which poses no mathematical difficulty and lends itself well to numerical simulation. Remember that in these simulations, account must be taken of the action of the RF field on all the isochromats. The outlines of the method satisfying this condition are shown here with the example of a 90° pulse.

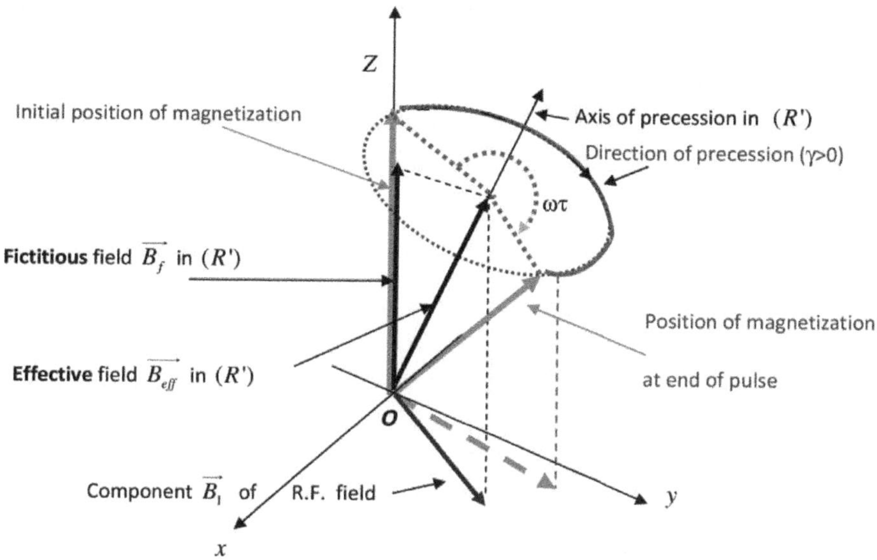

Figure 2.5. *Precession of the magnetization of an isochromat around the effective field in the frame* $(R',OxyZ)$. *The component* $\overrightarrow{B_1}$ *of the RF field is fixed in (R'). This precession occurs in the opposite direction around the effective field vector* $\overrightarrow{B_{eff}} = \overrightarrow{B_1} + \overrightarrow{B_f}$ *for nuclei with a positive gyromagnetic ratio. The absolute value of the angle of rotation around the effective field is* $|\omega|\tau$ *(here a little more than 180°).* τ *is the pulse duration*

2.2.3.2. *90° rectangular pulse on magnetization at thermodynamic equilibrium*

The RF field, in this case, is applied in the direction $0x$ of the frame (R'). This field acts on isochromats whose magnetization is at thermodynamic equilibrium (the initial situation before the pulse). These isochromats are assumed to be distributed uniformly over a wide range of the static magnetic field, on either side of the value of the field which gives exact resonance to the radio frequency determining the rotating frame (R') (Figure 2.5). For isochromats exhibiting a precession frequency exactly equal to that of the RF field applied $\overrightarrow{B_1}$, which is therefore that of the rotation of (R') around $\overrightarrow{OZ}$, we consider that the flip forms a 90° angle ($|\gamma| B_1 \tau = \pi / 2$). The pulse of duration τ, during which the amplitude of the RF field remains constant, then exactly puts the magnetization of these isochromats in the direction $+Oy$ of (R'). In (R), this angular frequency of precession is referred to as the "resonance" frequency (algebraically speaking, $2\pi f_0 = -\gamma B$). In (R'), the apparent pulsations of the isochromats with the value $-\gamma B_{fictitious}$ are notated as ω, and they are distributed on both sides of zero. Finally, the distribution of the isochromats is taken to be uniform. The reference value is M_0, in relation to which the measures of the components of magnetization are normalized on the basis of the relative deviation ω / ω_1, where $\omega_1 = \gamma B_1$, and then notated as $M_z M_x, M_y, M_\perp = (M_x^2 + M_y^2)^{1/2}$. These initial conditions of magnetizations will be retained throughout section 2.3.2 and in the examples given below.

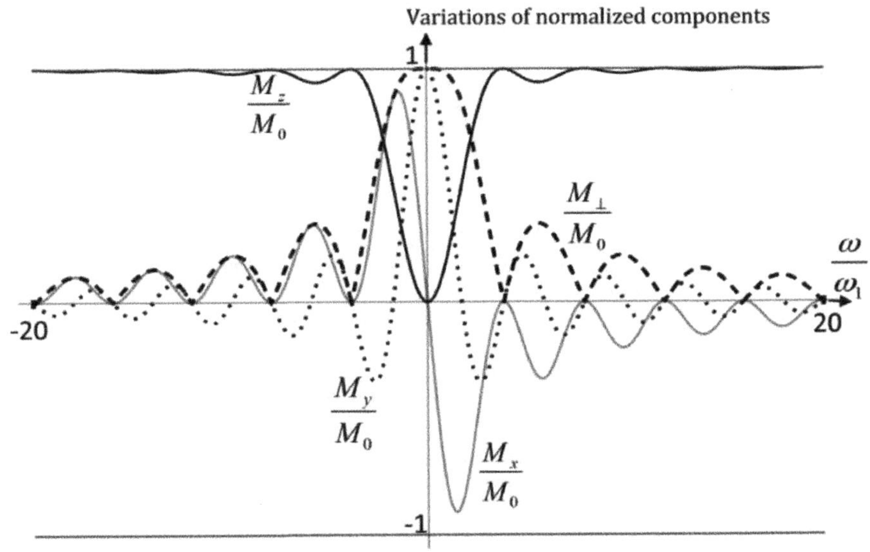

Figure 2.6. *Determination by simulation of relative values (in relation to M_0) of the components $M_z M_x, M_y, M_\perp = (M_x^2 + M_y^2)^{1/2}$ after application of a "hard" 90° pulse, with the field $\vec{B_1}$ being oriented along Ox in (R'). The abscissa value is given by the ratio of the pulsation ω around the effective field to the pulsation ω_1 observed when the fictitious field is null. The variations in this ratio range, in this case, from -20 to 20*

In Figure 2.6, we can see the effect of the pulse on isochromats whose frequencies of precession in (R) are distributed on both sides of the central frequency $2\pi f_0 = -\gamma B$. The pulse can be considered effective over a small range of variations in the ratio $\frac{\omega}{|\gamma B_1|}$, contained within the width of the central lobe of $M_\perp$. This width is determined by the duration τ of the pulse and by the values of ω and $|\gamma B_1|$, i.e. by the formula $\sqrt{\omega^2 + (\gamma B_1)^2} \cdot \tau = 4\pi$. If we want to apply a 90° flip to all the magnetizations of a sample placed in a slightly-uniform static magnetic field, the operation is therefore not feasible. We can, however, approximate this result by increasing the value of the field B_1, so that the ratio $\frac{\omega}{|\gamma B_1|}$ becomes small enough, using high-powered (or "hard") pulses. Using modulated-amplitude pulses, we can perform excitation to a better-controlled spectral width than that described immediately above. The application of these pulses needs to be longer, but they are lesser in amplitude than the previous cases, and can be qualified as "soft".

2.2.3.3. *Examples of 90° modulated pulses*

2.2.3.3.1. Gaussian pulses

In order to reduce the lateral oscillations of the curves in Figure 2.6, we can use a variation law which is Gaussian in form and necessarily truncated, to modulate the amplitude of the RF field. With regard to the curve representing the value of $M_\perp$ normalized by M_0 in Figure 2.7, note that this type of pulse is interesting in that it can be used to apply relatively selective excitation to a spectral band within which isochromats are distributed. Although the selectivity here is far from perfect – we can still alter the width affected by adjusting the parameters of the pulse – this pulse is easy to use to excite nuclei within a given spectral band, and therefore act with a certain degree of efficiency if the static field is fairly uniform. The following example shows how selectivity can be increased if we introduce phase modulation in an excitation.

2.2.3.3.2. Cardinal sine pulses

Cardinal sine (sinc) pulses use modulation of the amplitude of the RF field by a cardinal sine law. Its Fourier transform, and therefore the spectral response, is practically a rectangle, the width of which can be adjusted by altering the duration of the necessarily truncated pulse. The Bloch equations are not linear; hence, a linear correspondence between the temporal form of the RF excitation and its frequential response is only acceptable in cases of small flip angles.

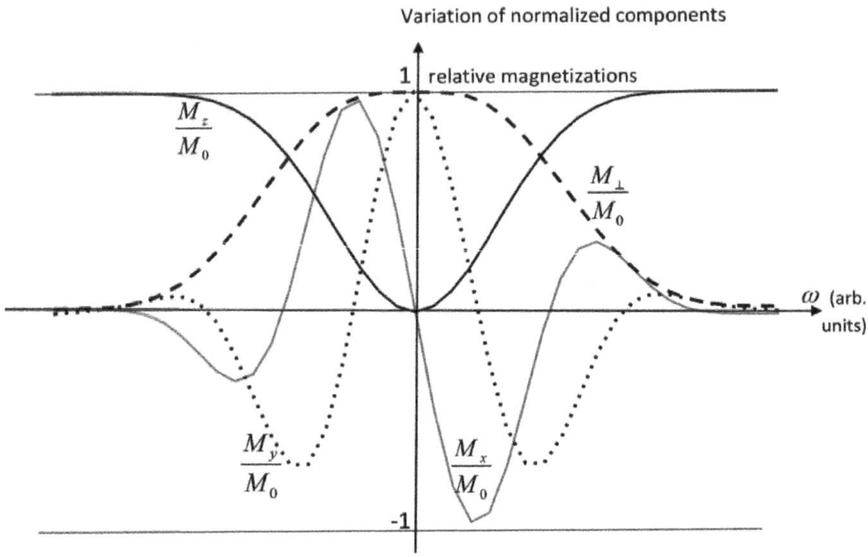

Figure 2.7. *Determination by simulation of the components* $M_z, M_x, M_y, M_\perp = (M_x^2 + M_y^2)^{1/2}$ *following a 90° Gaussian pulse oriented along* Ox *in the frame* (R') *and applied using 25 elementary rectangular pulses of equal duration, whose amplitudes are scattered according to a Gaussian distribution. These relative amplitudes range from .05 to 1. In this simulation, represented as a function of* ω*, we can consider that the effective spectral width accounts for only 10-15% of the width of the whole response. The lines plotted are assigned in the same way as for Figure 2.6*

With a 90° flip angle, the linear approximation, which is no longer valid, is sometimes accepted. In the simulations presented here, the Bloch equations are fully respected. The excitation is decomposed into a succession of small rectangular pulses matching the form of the modulation. Usually, there will be no more than three positive lobes (the central lobe and the two lateral lobes of the cardinal sine representation). This truncation causes oscillations in the response (known as the Gibbs phenomenon). However, this drawback can be overcome by "apodization", which involves multiplying the form of the pulse by the coefficients of a finite impulse response filter (e.g. Gaussian or Hanning (etc.) filters).

The pulse used in Figure 2.8, including five positive lobes and ten zeros with the beginning and ends of the pulse, is described by 64 elementary small pulses, of the same duration and whose amplitudes are modulated by the model imposed. We obtain an almost rectangular profile for $M_\perp$, with relatively slight oscillations. On the other hand, we do observe oscillations of the components M_x and M_y within the determined profile. Thus, the RF pulses serve to "manipulate" the magnetizations, as we shall see in the next section.

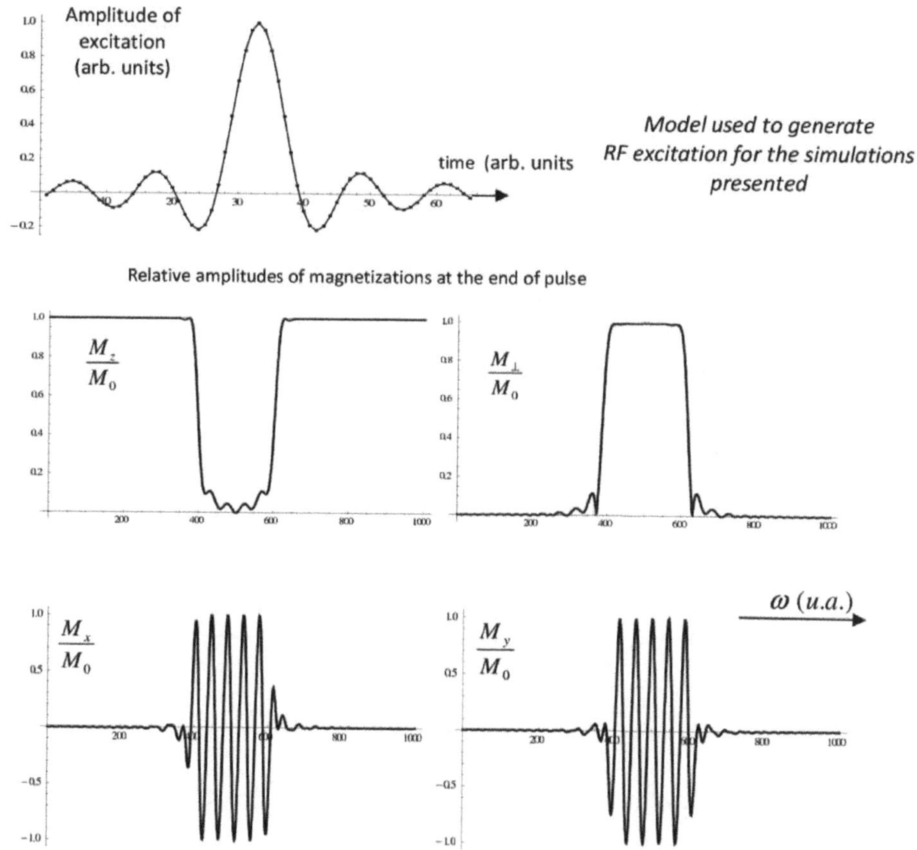

Figure 2.8. *1000-point simulations of the normalized components of* M_z, M_x, M_y, and $M_\perp$ *in relation to* M_0 *with changing frequency (here in arbitrary units) in response to a 90° sinc pulse oriented in the direction* Ox *of the rotating frame* (R')

2.2.3.4. *"Refocusing" pulses*

The above examples are just some of a hugely wide variety of pulses, which cannot possibly all be discussed here. They illustrate variations in amplitudes with ω in the transverse components M_x and M_y at the end of the RF excitation. These frequency-dependent oscillations indicate that the different transverse magnetizations within the central band are not in phase. This fact also indicates that the phase of the transverse magnetizations has an important part to play in NMR detection, because this phenomenon will influence the signal received by induction. As the components are out of phase because of the pulse, the signal is weakened, and it is then helpful to compensate for this deterioration by correcting the phase of the signal during the experiment itself, in real time. The corresponding operation is called "refocusing" of the magnetizations. It may take place after the pulse with the application of a new pulse, or may indeed be done by the pulse itself. We shall now briefly examine each of these two cases.

2.2.3.4.1. Self-refocusing using a 270° Gaussian pulse

From the examples given in Figures 2.6–2.8, we have seen that the application of the 90° pulse to isochromats which are exactly at the radio frequency causes their transverse magnetization to be oriented in the direction $+Oy$ of the frame (R'), which can be used as a reference point for the phase (Φ). Refocusing should be viewed as the gathering together of the transverse magnetizations in any single given direction in the plane xOy. This process occurs directly in the direction $-Oy$ with a single Gaussian pulse of +270°, which is therefore said to be "self-refocusing" (Figure 2.9).

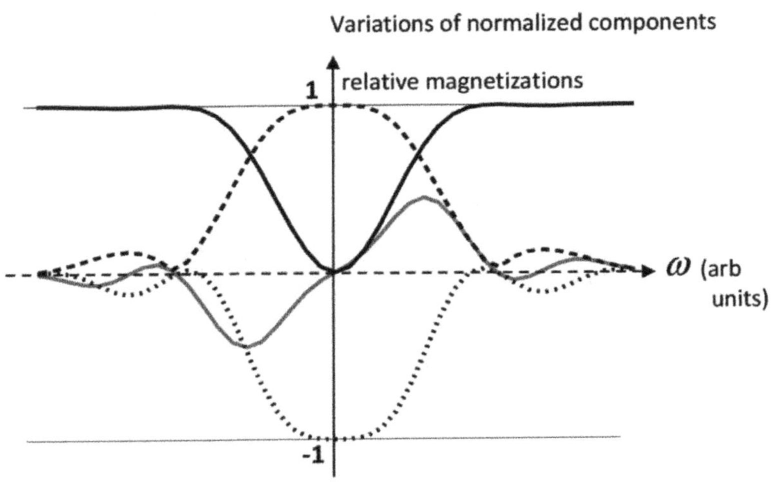

Figure 2.9. *Self-refocusing of the transverse magnetization of the component M_y in the frame (R') in response to a +270° Gaussian pulse in the direction Ox. This result is gained with a simulation identical to that used in Figure 2.7, where only the required value of the flip angle was modified. The lines plotted for the trends are the same as in Figures 2.6 and 2.7*

2.2.3.4.2 Refocusing using a 180° "hard" pulse

For an isochromat whose angular frequency of precession in (R) is shifted by a value ω with respect to the RF, the angle Φ mentioned in the previous section, formed during the application of the 90° pulse illustrated in Figure 2.6, is, according to analysis with the nonlinear Bloch equations, equal to the product of ω by 1.034 times half of the duration of the pulse. This property is specific to the use of a cardinal sine modulation.

At the end of this pulse, that angle will continue to evolve at the angular frequency ω, so that after an additional time period equal to 0.517 times the pulse duration, the value of Φ will have doubled (for instance) – see Figure 2.10. If, at the end of the initial excitation pulse, we apply, in direction Ox, a 180° rectangular pulse covering the whole of the spectrum (a hard pulse, as mentioned in section 2.3.2, and therefore of a very short duration), this corrects the phase of the components M_y for all of the isochromats affected by the initial excitation after 0.517 times the duration of the original pulse. This phase correction takes place in the direction $-Oy$, as shown by the simulation in Figure 2.10. Here, it is also important to note that the 180° hard pulse in direction Ox has transformed M_y into $-M_y$ without affecting M_x. Finally, remember that the descriptions and their associated simulations do not take account of the phenomena of relaxation, for which the timescale is much longer than that for pulses.

2.2.3.4.3. Refocusing under the influence of a gradient

The distribution of the isochromats chosen in Figures 2.6–2.10 corresponds to a uniform distribution of the nuclear spins according to the field value. This distribution can be idealized if we use a highly uniform magnetic field, on which we superpose a field varying in a perfectly linear manner in one direction in space. This second field thus creates a uniform gradient in this direction; this gradient causes a variation in the isochromats' angular frequency of precession which is proportional to the distance. The same is true for the variation of the dephasings Φ created by a pulse and observed at a certain time after the pulse. It is then possible to refocus these dephasings if the sense of the gradient can be modified immediately after the initial pulse. For the simulations, this involves changing the sign of the angular frequency ω at the end of the initial pulse. After a short waiting period, equal to 0.517 times the duration of the initial pulse, we find the result shown in Figure 2.10, except that the refocusing of the component M_y takes place in $+Oy$ instead of $-Oy$ (Figure 2.11). A graphic analysis of the process – which is not difficult to perform and which readers can do for themselves – will illustrate this peculiarity.

The simulations shown in Figures 2.10 and 2.11 use the same truncated sinc function, with three positive lobes and a very moderate Gaussian weighting function to limit the Gibbs oscillations. In these representations, the positive sense of ω is that which exists before the gradient is reversed. The example just given is a particular form of gradient echo, to which we shall return in section 2.4.2.

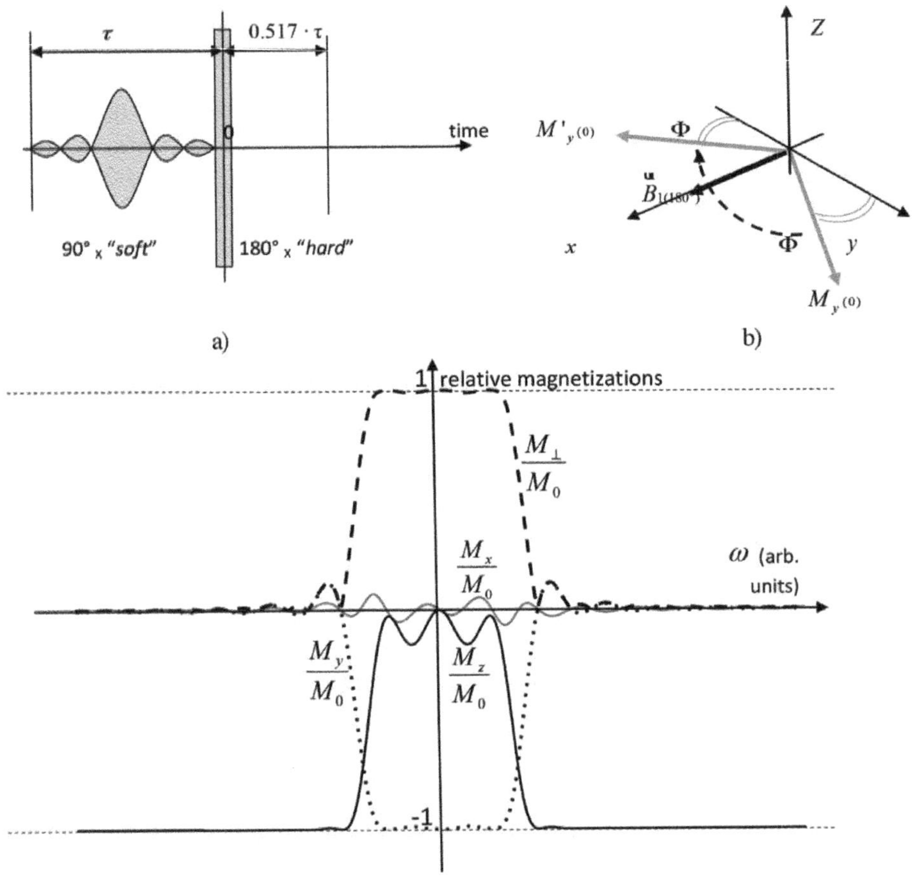

Figure 2.10. *Top: sequence formed of a 90° cardinal sine pulse of duration* τ *, applied in the direction* Ox*, followed by a "hard" 180° pulse in the direction* Ox*, centered at date zero (plot marked a)). The action of the compound pulse represented here (plot marked b)) transforms the component of magnetization* $M_y(0)$ *of any isochromat in* $M'_y(0)$ *by a complete 180° rotation around* Ox*. Bottom: representation by simulation of the components of the magnetizations at date* $0.517 \cdot \tau$*. These components are mainly aligned with* **–Oy** *at date* $0.517 \cdot \tau$*, given that* Φ *is proportional to* ω*. (Simulation where* γ *is positive. The duration of the "hard" pulse is assumed to be negligible). The lines plotted for the trends are the same as in Figures 2.6, 2.7 and 2.9*

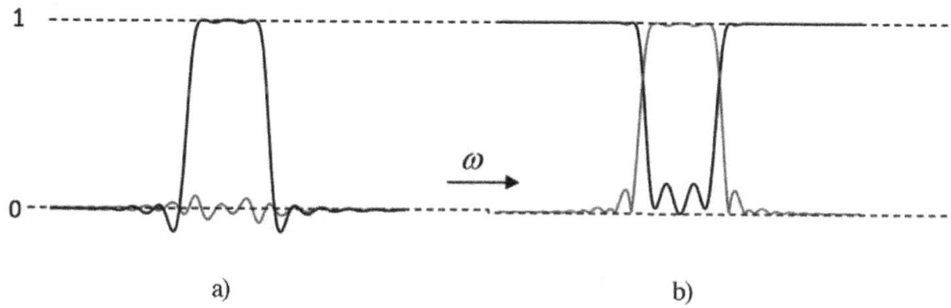

Figure 2.11. *Representations of normalized components a) M_x (gray curve) and M_y (black curve); b) $M_\perp$ (gray curve) and M_z (black curve) after a 90° cardinal sine pulse along Ox, followed immediately by the reversal of the sign of the pulsations ω, caused by changing the sign of the gradient and leading to the refocusing of the component M_y after 0.517 times the duration of the pulse*

General remark on the effects of pulses

Remember that the examples discussed here do not take account of the spin–lattice (T1) and spin–spin (T2) relaxation. This hypothesis is acceptable because the durations of the RF excitations are short in comparison to the relaxation times. Another remark which is verified by the brief experiment just shown in Figures 2.5–2.10: we can establish a qualitative summary of the effects caused by the RF pulses (Table 2.2). When an RF pulse is applied to a set of identical nuclei, Table 2.2 shows that we must expect three possible results if the initial magnetization is longitudinal, and also three possible results with transverse initial magnetization. The problem which may then arise in creating transverse magnetizations is that of overlap of the signals which these magnetizations may generate. In Figures 2.9–2.11, we saw that the presence of the residual of the component M_x, null when $\omega = 0$, which contributes in the same way as M_y to the generation of the signal in the affected spectral band; however, as this component is 90° different in phase, it is preferable to make it as small as possible.

2.2.4. *Elementary pulse sequences using the refocusing technique*

We have seen (in Figure 2.10) a particular example of the combination of two pulses forming a refocusing sequence. This sequence has the advantage of re-phasing the different isochromats after a short delay ($0.517 \cdot \tau$ seconds), placed after a pulse with modulated amplitude and phase, acting in the presence of a non-uniform field. Thus, we have a significant signal $0.517 \cdot \tau$ seconds after the second pulse. It is also possible to delay the action of the hard pulse, which would also delay the refocusing time when most of the signal is reformed. This signal is a spin echo. We have also seen that changing the sign of the gradient of a static magnetic field can be an effective refocusing technique. Hence, we can expect to be able to create an echo which is "delayed" with respect to the initial excitation simply by adjusting the polarity of a gradient. The properties of this second sequence are then different to those of the former one, and establish the distinction between the spin-echo and gradient-echo sequences.

Nature of initial magnetization	Nature of magnetization obtained	Degeneration
Longitudinal	Longitudinal	1
Longitudinal	Transverse	2
Transverse	Longitudinal	2
Transverse	Transverse	4

Table 2.2. *Magnetizations produced by a pulse applied to a given type of magnetization (longitudinal along OZ or transverse in the frame xOy in (R'), which is superimposed on the plane XOY in (R)). The degeneration is used to make a distinction between the magnetizations M_x and M_y. Hence, if we start with a transverse magnetization with two different possible orientations, a pulse may cause two new possibilities for transverse magnetizations, which is why the degeneration is 4 in the bottom row*

2.2.4.1. *The spin-echo (SE) sequence*

2.2.4.1.1. *Sequence with a hard 180° refocusing pulse*

The operation is illustrated in Figure 2.12 with a simulation. Once the soft RF pulse is applied along Ox, followed by free evolution of the magnetizations for a time-period $T_E/2-\tau/2$, we apply a short rectangular pulse which affects the whole of the sample. The temporal origin is placed at the center of the first pulse; the echo signal observed T_E seconds later is at its maximum.

2.2.4.1.2. Sequence with a "soft" refocusing pulse

The diagram of RF excitations given in Figure 2.12 (top) needs to be changed if we use a second modulated pulse with a view to refocusing after a period of free evolution of the isochromats (Figure 2.13). The first pulse may correspond to a flip angle of less than 90° at the center of the excitation band. On the other hand, the second pulse, modulated in exactly the same way as the first, with the same duration, should cause a 180° rotation at the center of the central band, which requires an amplitude calibrated to a higher value than that of the excitation pulse. Note that in the rotating frame (R') it may, like any pulse, be directed along one of the four referential directions: Ox, $-Ox$, Oy or $-Oy$.

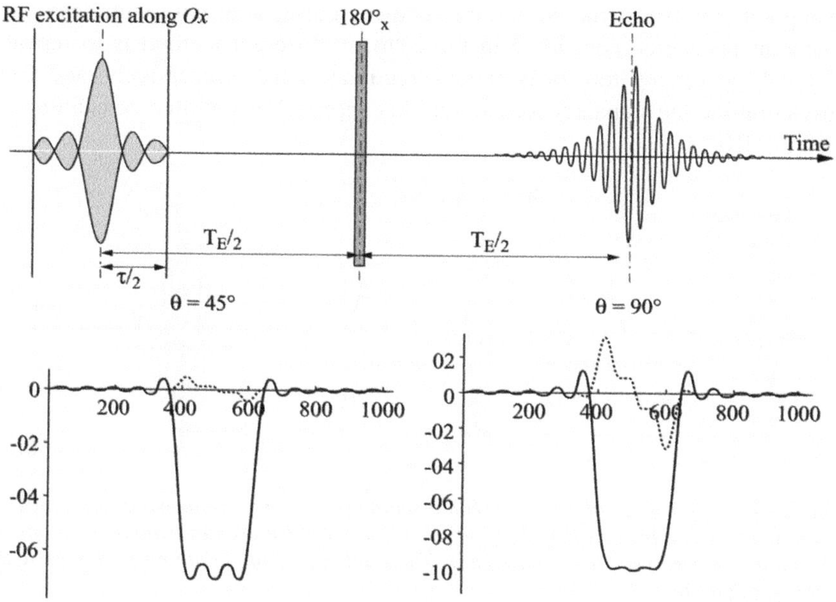

Figure 2.12. *Simulations for the spin-echo sequence with a hard refocusing pulse. The RF pulses are directed along Ox. Top: diagram of the sequence and of the echo formed. Bottom: 1000-point computation on the basis of ω. At the time of the echo, M_y (in black) is directed along $-Oy$, as indicated by the significant negative values obtained for its normalized value, which becomes -1 when $\theta = 90°$. By comparing the relative amplitudes of M_x (in gray) and M_y, we can see that the non-linearity of the Bloch equations, which is not taken into account in the sequence (τ instead of 1.024 τ), has relatively little effect if the flip angle θ of the pulse is only 45°, whereas the fluctuations of M_x reach up to $\pm 30\%$ of the value of M_y when $\theta = 90°$*

The very short rectangular pulse shown in Figure 2.12, which is quite deserving of the label "hard", causes the switch from M_y to $-M_y$ by an action which is uniform throughout the spectrum of the isochromats. The same is not true of the 180° modulated pulse. A complete study, which we shall not dwell on here, indicates that the refocusing soft pulse, as predicted by Table 2.2, creates two terms in addition to the term $-M_y$. As these terms may interfere with the usable signal, they are eliminated by placing a spreading gradient on both sides of the refocusing pulse. Figure 2.13 does not represent a simulation; merely an illustration.

Note that in the process shown in Figure 2.13, the same spreading gradient needs to be applied both before and after the refocusing pulse, so as not to ultimately affect the components which contribute to the echo. A different method is to repeat the experiment, occupying four orthogonal directions of (R') successively, and then to linearly combine the signals recorded, but this approach is too time-consuming to be applied in MRI.

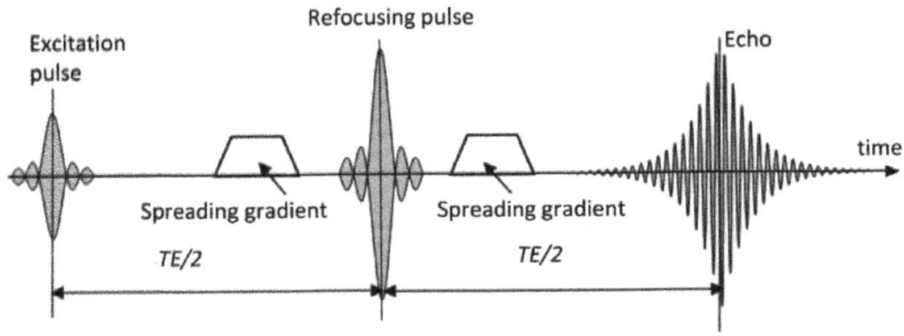

Figure 2.13. *Description of the spin-echo sequence using a refocusing soft pulse with orientation Ox in the frame (R'). The initial excitation pulse creates transverse components which could interfere with the echo signal. The application of spreading gradients tends to eliminate these terms*

To examine the mechanism by which spin echoes are formed, we can monitor the evolution of the phase of an isochromat with pulsation ω during the course of the sequences. We are only interested in isochromats which belong, in the spectral domain, to the center of the echo. The diagram in Figure 2.14 uses a set of results illustrated in Figures 2.8, 2.10, 2.12 and 2.13. Under the influence of the 180°x pulse, all the isochromats which will be rephased (belonging to the central plateau of the distribution My at the time of refocusing) have had their component M_y changed to $-M_y$. This explains why point P_1 becomes point P_2 at the end of the refocusing pulse. Remember also that the condition $\Phi \approx \omega \cdot \frac{\tau}{2}$ is better realized when the flip angle is much smaller than 90° and therefore that ω determines the slopes of the variations in phase of the isochromat. Regardless of the associated values of ω and Φ, the paths of the isochromats belonging to the central plateau of the distribution My at the time of refocusing do indeed pass through the same point on the time axis, giving the time of the echo.

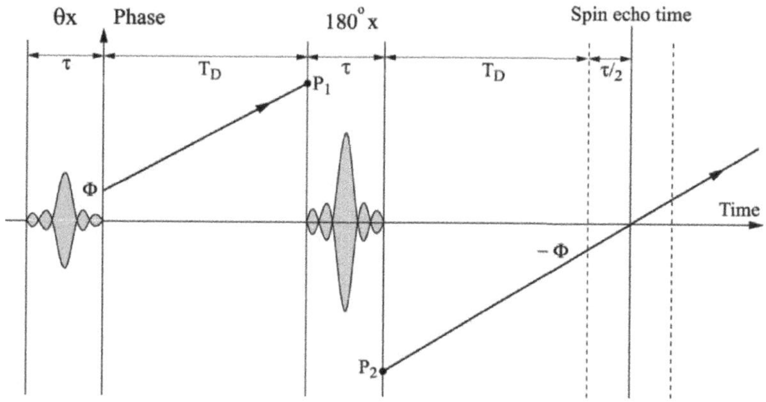

Figure 2.14. *Representation of the phase evolution of an isochromat effectively re-phased by the spin-echo sequence: time is counted from the middle of the first RF pulse (θx). A solid line represents the linear increase of phase angle with respect to O_y before the refocusing pulse (180°x), and then with respect to the focusing direction $-O_y$ after the refocusing pulse. Here, the echo time is equal to $T_E = 2 \cdot (\tau + T_D)$*

Note that this sequence – similar to that using hard refocusing pulses – is able to compensate for the lack of homogeneity of the static field, because the variations in phase which these sequences cause during the trajectory up to P_1 are compensated by those which take place after P_2. The amplitudes of the echoes are therefore, in principle and if we do not take account of diffusion (see section 2.6.3), governed by the relaxation time T_2. This behavior is not observed with the gradient-echo sequence discussed above. The amplitude of the echo will, in all but exceptional cases, be determined by transverse time constants. When the samples are large in volume and heterogeneous, these time constants usually present different values when observed in different positions. The ideal case would be where only one of the transverse time constant, defined as T_2^*, is involved – that is, if the spatial distribution of the field permits it.

2.2.4.2. *Gradient-echo sequence*

2.2.4.2.1. Gradient echo in a uniform field under the influence of a uniform refocusing gradient

Refocusing under the influence of a gradient, the principle of which we have already encountered (see section 2.3.4.3) can be envisaged to perform delayed refocusing, shown in Figure 2.15.

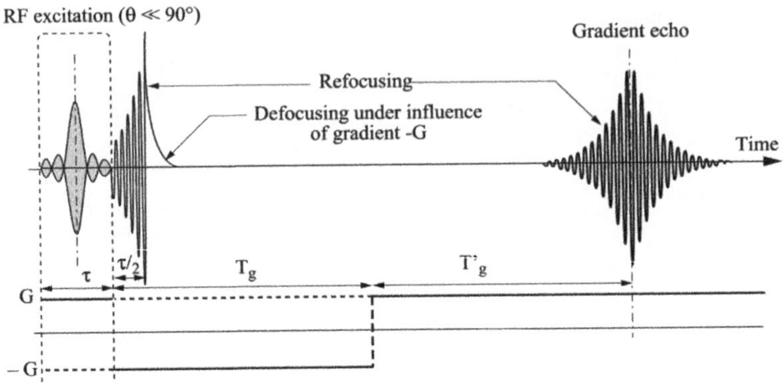

Figure 2.15. *Idealized experiment for creating a delayed "gradient echo" in the case of a small excitation angle ($\theta \ll 90°$), such that the first refocusing takes place $\tau/2$ seconds after the end of the sinc modulated RF pulse (linear model acceptable). The changes of sign of the gradient are idealized by steps. One possible variant is not to alter the sense of the gradient immediately after the modulated pulse and to create only one gradient switch. Simulation results are shown in Figure 2.16*

The gradient, whose sign has been changed after 1.034 times $\tau/2$, half the duration of the initial pulse if it is at 90°, or around $\tau/2$ if the pulse is weaker (linear model acceptable) remains negative for the duration T_g (Figure 2.15). Then it changes sign. The gradient echo forms after a new time interval $T_g^{'}$ counted after the gradient step. In this ideal experiment, the magnetic field is perfectly uniform, as is the gradient superposed upon it. When θ is much less than $90°$, a first refocusing takes place at time $3\tau/2$. This is followed by a dephasing gradient for a duration $T_g - \tau/2 = T_g^{'}$ after the sign inversion of the gradient. It is also possible not to modify the sign of the gradient immediately after the $\theta \ll 90°$ pulse, but instead modify it only after a given duration $T_g^{''}$ (this case is not represented in Figure 2.15, and the reader may plot it if they so desire). The echo obtained is then such that $T_g + \tau/2 = T_g^{''}$.

The simulation of the relative components M_x and M_y at the time of the echo is shown in Figure 2.16, for both of the situations discussed above. As the duration of evolution $T_g - \tau/2$ is fixed (case (a)) or $T_g + \tau/2$ (case (b)), we look for the values of

$T_g^{'}$ or $T_g^{''}$ respectively, which give us the smallest relative oscillations in amplitude of M_x.

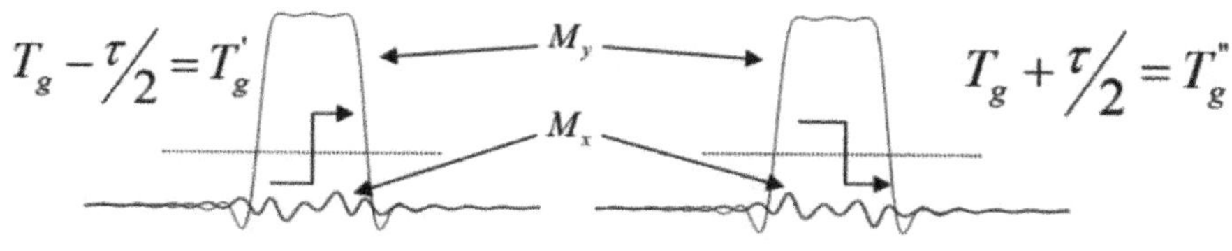

Figure 2.16. *Results of simulations of the components M_x and M_y at the refocusing time with two "delayed" gradient echo sequences in the case of very good spatial uniformity of the static magnetic field and of the gradient. The steps superposed in the figure indicate the sense of the changes in gradient with the switch between T_g and $T_g^{'}$ (a, corresponding to Figure 2.15) and T_g and $T_g^{''}$ (b, not represented in Figure 2.15)*

2.2.4.2.2. Application to slice selection in MRI

Gradient echo plays a major role in the selection of a volume in MRI. Here, we shall consider the example of slice selection. The gradient used is an additional gradient whose value is large with respect to the heterogeneities of the static field. Thus, we can place ourselves in the situation described in the previous section, considering that because of the relatively short duration of the pulse with respect to the durations of the subsequent evolutions, we can accept that the static field is sufficiently uniform. Thus, by applying a gradient with absolute value G during the RF pulse of spectral width Δf, we obtain a slice whose thickness is Δu, perpendicular to the direction u of the gradient in the sample, such that:

$$\Delta u = \frac{2\pi \cdot \Delta f}{\gamma G} \qquad [2.10]$$

Given that all gradients are vectors, it is therefore possible, by compounding two or three gradients in the main directions of the laboratory frame (R) in which u is defined, to determine any given slice within the sample. Note that the fact of selecting slices to capture 2D images does not, in any way, take away from the 3D, volumetric nature of MRI, because if we want to have a signal, it is necessary for the slice to have a certain thickness.

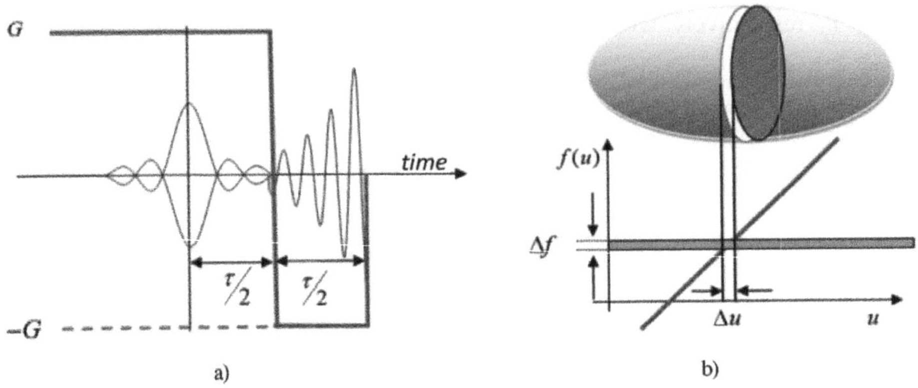

Figure 2.17. *Selective pulse for slice selection: a) Refocusing of magnetizations selectively excited in a slice by idealized reversal of the gradient* G *of direction u. b) Elementary representation of the creation of a slice: the oblique line indicates the variation* $f(u)$ *of the frequency with increasing distance under the influence of the gradient in direction u. The frequency band* Δf *of the pulse therefore determines the thickness* Δu *of the slice*

2.2.4.2.3. Gradient echo obtained from the selection of a volume

Once again, we shall consider the case of a slice which results from the use of a so-called "selection gradient". The dimensions of the slice are supposed to be such that within it, the heterogeneities of the static field are modest in comparison to the variations imposed by the selection gradient. This does not mean that this lack of inhomogeneity is to be ignored, but rather that the refocusing, after changing of the sign of the so-called "slice gradient", is sufficiently good for us to consider that at that time, the isochromats in the slice are re-phased. Figure 2.18 gives a very concise illustration, using the magnetization phase diagram – used previously in Figure 2.14 but with spin echo – of how a gradient echo forms under the influence of a bipolar gradient, independently of the selection gradient, even if it occupies the same direction.

Figure 2.18, which refers to Figures 2.11, 2.16 and 2.17, shows how the rephasings of the different isochromats are dispersed with respect to the central time of the echo. With a similar echo time, the amplitude of that echo is therefore lesser than that of the spin echo obtained for the same slice. The difference lies in the fact that the action of the 180° pulse rephases the isochromats, although changing the sign of the gradient $\pm G_D$ cannot correct the sense of the phase variations connected to the difference of ω from zero. The attenuation of the amplitude of the echo is therefore governed by one (or possibly several) time constants, which may be replaced by T_2^* if the spatial distribution of the static field is sufficient.

Note that the refocusing results in a reformation of the signal amplitude; the phase of the signal will depend on the adjustable phase difference between the transmitter and the receiver. In the case of Figure 2.17, this difference is apparently 180° and, for the purposes of explanation, it has been chosen as 0° in Figure 2.18.

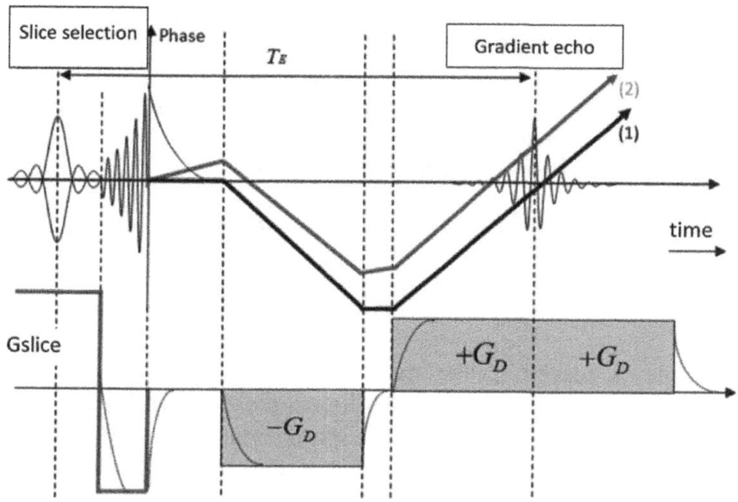

Figure 2.18. *Formation of a gradient echo under the influence of bipolar gradient $\pm G_D$ different from the gradient which is used to establish the slice (see Figure 2.17). The trajectories of the isochromats whose pulsation ω is near to the median pulsation of the RF excitation are practically identical to trajectory (1), which corresponds to a significant collective rephasing at time T_E. The other isochromats are rephased on both sides of the center of the echo. (A single situation of this type is represented by trajectory (2)). Consequently, the isochromat does not contribute to the central amplitude of the echo. The drags of the gradient switching are indicated by curved lines, though their complicated influence on the trajectories is not taken into account*

Another practical remark relates to the switching of the gradients, which is not quite as simple as in the clear-cut boundaries illustrated in the diagrams. At the time of the application and of the cutoff of the pulse, there is a significant drag effect in spite of all the power and the extensive bandwidth which may be enjoyed by the gradient controls. The essential point, then, is to account for the surface beneath the curve representing the variations of the gradients. The surface beneath the curve before switching must be equal to the surface beneath the curve of variation of the gradient after switching, until the point at which the gradient echo peak occurs. This is the reason for the curved line plots superposed on the ideal representation of the gradients in Figure 2.18.

2.2.5. *Spatial discrimination of signals using gradients: fundamental principle of MRI*

As the width of the resonance line (see Figure 2.4) is subject to the spatial inhomogeneity of the static field, it is possible to account for the spatial distribution of the isochromats if we increase this lack of uniformity of the static field. This approach was suggested by Robert Gabillard in the early 1950s. Hence, in order to independently examine two regions of the same sample, Paul Lauterbur had the idea of differentiating them by the application of a field gradient. This spreads out the magnetic resonance signals in the direction of the gradient. Herein lies the fundamental principle of one-dimensional (1D) MRI.

2.2.5.1. *One-dimensional model: 1D image or profile of magnetization*

Consider a sample with a known and not-overly-complicated shape, e.g. a cylindrical test tube containing water, whose protons will serve as probes.

This sample is placed in the main static magnetic field, upon which we superpose a gradient field, e.g. along the direction *OZ* in the fixed frame (R) (case A) or *OX* (case B); see Figure 2.19. The experiment consists of applying an RF pulse which flips the magnetizations, and then digitizing (here at 128 points) the temporal signal recorded in the presence of the gradient without particular selection of an element of volume. The profiles are calculated by discrete Fourier transform of the signals available to us in complex form during their detection. In view of the very special symmetry of the sample, all the profiles obtained with a horizontal gradient such as G_X are identical regardless of their orientation in the plane perpendicular to G_Z. This experiment is a simulation, with no background noise, but it would be possible to introduce such noise. The real part of the Fourier transform of the complex signal gives the profile of the nuclear magnetizations (protons) contained in the sample. These are 1D images representing the projection of the sample in the direction of the gradient in question.

2.2.5.2. *Generalization to several dimensions: from spin space to phase space*

For a 2D model, all of the above indications are such that we can record a signal that is proportional to the amount of magnetization contained in a slice whose thickness is Δu (Figure 2.17). This signal would be distributed along the direction of a so-called reading gradient G_v, because it is maintained throughout the duration of the recording in v, which is necessarily orthogonal to u. The resonance frequency of the nuclei contained in the slice with abscissa u varies linearly with v. Then, frequential encoding of the space is written as:

$$f(v) = f_0 + \frac{1}{2\pi} \cdot \gamma \cdot G_v \cdot v \qquad [2.11]$$

where f_0 is the radio frequency in the absence of gradient G_v .

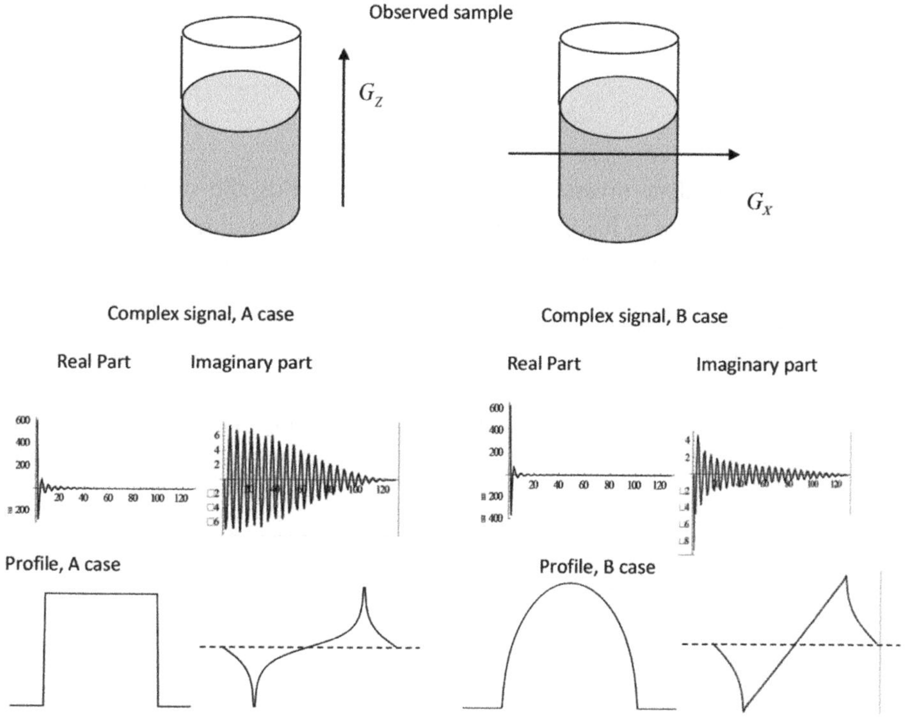

Figure 2.19. *Principle of 1D imaging with a simulation: obtaining profiles or 1D images, of the same sample (liquid water, shown here in gray) using two different orientations of a gradient of the static magnetic field (the so-called "reading gradient"). Note the considerable difference in level between the real and imaginary parts of the temporal signal observed in the presence of the gradients – these signals are represented using identical arbitrary units. The zero level of the imaginary part of the profiles, which has been shifted for the purpose of presentation, is represented by a dotted line*

For the sake of simplicity in our analysis, we shall not take account of transverse relaxation. This phenomenon is very slow in comparison to the variations imposed on the signal; hence, we shall not take account of the corresponding convolution of spectral response. This hypothesis is generally fairly well borne out, because the 1D profiles are in fact highly spread out with respect to the width of the resonance line. The expression of the temporal signal recorded therefore takes the form:

$$s_0(t) = \int_{\mathrm{v}} \rho(\mathrm{v}) \cdot e^{-j[2\pi(f_0 + \frac{1}{2\pi} \cdot \gamma \cdot G_{\mathrm{v}} \cdot \mathrm{v})]t} d\mathrm{v} \qquad [2.12]$$

which, after subtraction of the frequency f_0, gives us:

$$s(t) = \int \rho(\mathrm{v}) \cdot e^{-j \cdot \gamma \cdot G_{\mathrm{v}} \cdot \mathrm{v} \cdot t} d\mathrm{v} \qquad [2.13]$$

The time-dependent inverse Fourier transform yields the distribution $\rho(\mathrm{v})$, which is the 1D profile in the direction v. Note that the encoding of the direction v by the frequency is also the encoding of the generalized phase of the signal.

We can repeat the above approach for a 3D model, with the aim of determining a density of magnetization $\rho(X,Y,Z)$ dependent on the coordinates of the lab frame. In order to do so, we would successively, over a period of time, apply three gradients – G_X, G_Y and G_Z – with mutually orthogonal directions, for the respective durations τ_X, τ_Y and τ_Z. Equation [2.13] is then generalized, taking the form:

$$s(\tau_X, \tau_Y, \tau_Z) = \int_X dX \int_Y dY \int_Z dZ \cdot \rho(X,Y,Z) \cdot e^{-j \cdot \gamma \cdot (G_X \cdot X \cdot \tau_X + G_Y \cdot Y \cdot \tau_Y + G_Z \cdot Z \cdot \tau_Z)} \qquad [2.14]$$

A 3D Fourier transform with respect to the variables τ_X, τ_Y and τ_Z in $s(\tau_X, \tau_Y, \tau_Z)$ in turn leads to the determination of $\rho(X,Y,Z)$.

Instead of equation [2.14], the following formulation is preferable:

$$s(\tau_X, \tau_Y, \tau_Z) = \int_X dX \int_Y dY \int_Z dZ \cdot \rho(X,Y,Z) \cdot e^{-j \cdot 2\pi . (\frac{\gamma G_X \tau_X}{2\pi} . X + \frac{\gamma G_Y \tau_Y \cdot Y}{2\pi} \cdot \tau_Y + \frac{\gamma G_Z \tau_Z}{2\pi} . Z)} \qquad [2.15]$$

in order to reveal the reciprocal variables of *X*, *Y* and *Z* which are $k_X = \dfrac{\gamma G_X \tau_X}{2\pi}$, $k_Y = \dfrac{\gamma G_Y \tau_Y}{2\pi}$ and $k_Z = \dfrac{\gamma G_Z \tau_Z}{2\pi}$, which must be measured in m^{-1}.

Note that the generalized term of phase $(\gamma G_X \tau_X X + \gamma G_Y \tau_Y Y + \gamma G_Z \tau_Z Z)$ takes the form of the scalar product $\vec{k} \cdot \vec{r} = k_X X + k_Y Y + k_z Z$, with the vector $\vec{k}$ here defining a 3D space reciprocal to the lab one. We can therefore summarize this result using the pair of Fourier transforms:

$$s(\vec{k}) = \iiint \rho(\vec{r}) \cdot e^{-j2\pi\vec{k}\cdot\vec{r}} \cdot d\vec{r} \qquad [2.16]$$

$$\rho(\vec{r}) = \iiint s(\vec{k}) \cdot e^{j2\pi\vec{k}\cdot\vec{r}} \cdot d\vec{k} \qquad [2.17]$$

In practice, the use of digital techniques leads us to sample the signal by successive measurements in order to cover the reciprocal space. For instance, the motion of the measuring point can be created by quantified variations of the gradients, and for each set of values G_X, G_Y, G_Z we determine the amplitude of the signal, to then reconstitute the distribution of the magnetizations in three dimensions. As we shall see later on, it is useful to combine the taking of the measurements in both spaces (direct and reciprocal) knowing that the recorded signals are representative of the reciprocal space.

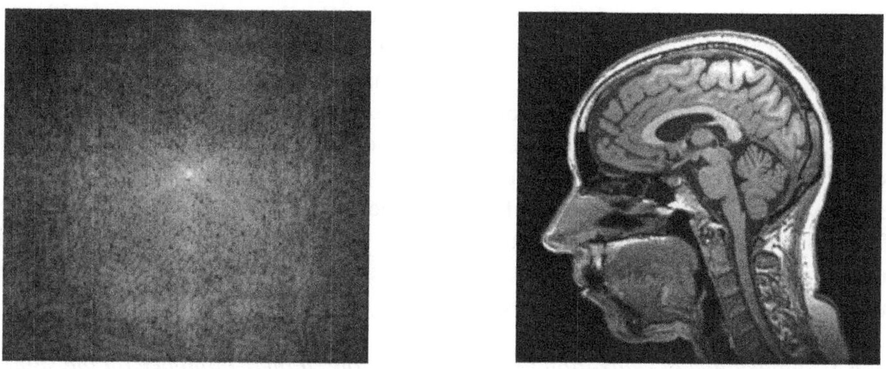

a) b)

Figure 2.20. *Example of a 2D image: a) example of the real part of the reciprocal k-space. b) example of an image of a volunteer, sagittal view, from a 1.5-tesla clinical installation, reconstructed in 256 × 256 pixels by a 2D Fourier transform and presented as absolute values. Image provided by D. Sappey-Marinier, Cermep, Creatis*

The Cartesian model weighs in favor of using the Fourier transform to switch between the reciprocal and image spaces. Because of its digital formulation, this operation is now very commonly used, and it provides MRI with a simple technique for immediate reconstruction. The image shown in Figure 2.20 is part of a series of images, all taken in the sagittal plane and created by a gradient echo sequence at 1.5 teslas. The founding principle of this process is discussed in section 2.5 of this chapter. Note that this image gives us no information about the density of protons, without which the contrast between the different tissues would be far less clear. In fact, the strengthening of the signals from the white matter and lipid-rich regions, and the weakening of those from the cerebrospinal fluid, involve no post-acquisition digital processing at all. It relies on parameters specific to the protons inhabiting those tissues which determine the appearance of the image. We shall briefly discuss these parameters in the next section.

2.2.6. *Multi-parameter aspect of MRI*

What is represented by the image we obtain? Indeed, the parameters influencing the signal are many, and are of great interest for different purposes depending on the type of nuclei being observed (hydrogen (proton), phosphorus, sodium, etc.) – which will be discussed section 2.6 – or depending on the domain and the type of application intended. From anatomical or functional MRI to industrial non-invasive controls, the choice of nuclei detecting is not strictly limited to protons, although they account for the vast majority of uses. In all these domains, we are dealing with "static" observations (anatomy, compartmentalization, etc.) or dynamic observations, which do not all share the same timescales (motions, transport of matter, physiological reactions, chemical reactions, etc.). To directly or indirectly exploit the physical properties characterizing these various situations entails a certain variety of the parameters influencing the image formation. Anything which affects the value of the signal at the moment the image is captured can provide information. Even the discontinuities in magnetic susceptibility creating artifacts in the image are potential sources of information.

2.2.6.1. *Importance of relaxation times and principle of their influence on the signal*

To begin with, the density of nuclei resonating may be a parameter for the image, on the condition that its distribution is sufficiently well differentiated. This is not the case, usually, in medical MRI, where protons are widely used as probes. With a few exceptions, the relatively uniform distribution of water in human tissues precludes its use as a discriminating element for MRI. Instead, we use the differences between the relaxation times of the nuclei depending on the medium they occupy to create contrast, or else artificially and specifically modify the relaxation times between the different compartments. The principle of weighting the images using the relaxation time T_1 is based on the fact that this parameter governs the return of the magnetization values to thermodynamic equilibrium. We can therefore use the repetition frequency, which defines the repetition time T_R, of the RF excitations to enhance the image of those regions where T_1 is shortest. Weighting the image contrast using the relaxation time T_2 exploits the fact that in a spin-echo sequence, the nuclei with the highest T_2 values retain a measurable transverse magnetization for the longest. For a given medium, the relaxation times T_1 and T_2 have different values, but are nonetheless correlated, in that the tissues with the highest T_1 values may have the highest T_2 values. We shall see later on that the attenuation of the gradient echo caused by a low local value of the transverse relaxation time can also be exploited. This so-called "T_2^* weighting" coefficient tends to be used when exploiting "super-paramagnetic" nanoparticles whose role is to destroy the uniformity of the static field in a near-punctual manner. The local variation in magnetic susceptibility causes a disappearance of the signal, resulting from the random dephasing experienced by the magnetizations in the zone of influence of these particles. Thus, truth to tell, this is not a method of weighting by a random relaxation time, but rather visualization by adjusting the amplitude of the signal. This modification may be done by a number of complex means, and it is in this context that the BOLD (Blood Oxygenation Level Dependent) method – which is used in cerebral activation imaging, and will be discussed in section 2.6.4 – fits in. It is common to use MRI in a dynamic regime to measure the volume of blood passing through a given volume in an organ per unit time, mainly for brain imaging. This technique, known as perfusion imaging and illustrated in Figure 2.21, uses the presence of a contrast agent in the blood moving through the capillaries to provide information about numerous values – in particular: the average time taken by the bolus of contrast agent to travel through the capillary network, the proportion of the cerebral volume occupied by those capillaries, the cerebral blood flow for a reference mass (generally 100 g), etc. The injection of the contrast agent, usually based on Gadolinium III (see section 2.4.5.1.1), creates a difference in magnetic susceptibility between the capillaries and the surrounding tissue. The passage of the bolus therefore results in a transitory drop in the MRI signal. By modeling the signal thus obtained, it is then possible to work back to the values relating to the perfusion of the organ, which represent useful indications for diagnosis.

The method for measuring tissue perfusion using MRI, when it was initially developed, used contrast agents (see section 2.4.5) for labeling. However, we can simply use the fact that the relaxation times T_1 of the protons in blood are fairly long (around 0.5 seconds). Arterial Spin Labeling (ASL) is made possible in the brain because we can easily "label" protons in the blood with a 180° RF pulse as they enter into the carotids. Also, the spin–lattice relaxation time of the protons in arterial blood is sufficiently high in comparison to the transfer time of blood in the tissue. As with the previous example, the images are then obtained by way of a very fast imaging technique (see sections 2.5.2.6 and 2.5.3.4) for dynamic monitoring of the consecutive variations of the signal through the organ being visualized. The attenuation of the signal in comparison to that of a reference observation (without a 180° pulse) is modeled using the Bloch equations in the presence of exchange in a two-compartment model, i.e. where the nuclei in turn are to be found in the blood and the surrounding tissues (see the remark about the description of relaxation in the presence of exchange processes, in section 2.4.5). This procedure, which is not, as yet, widely used in diagnostics, exhibits a signal-to-noise ratio (SNR) which can be improved by raising the value of the static field. However, today, 3-tesla systems accommodate the first clinical versions of ASL sequences.

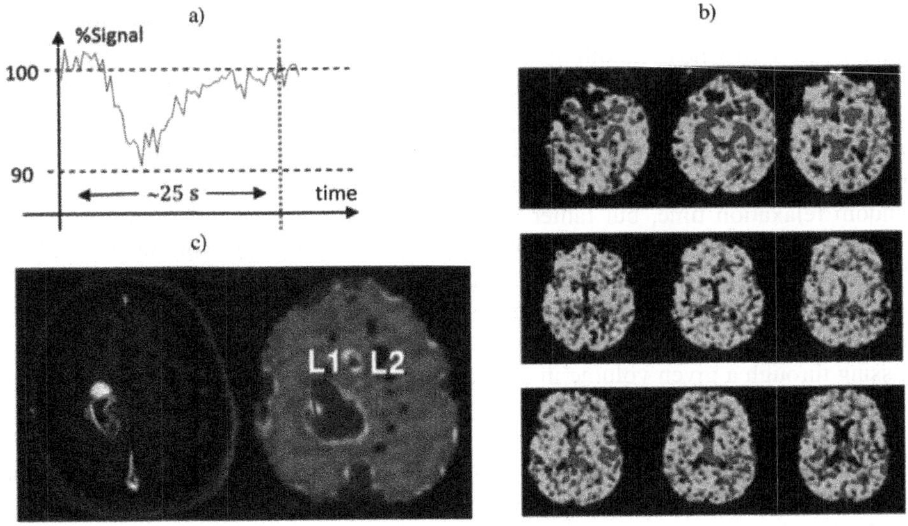

Figure 2.21. *Perfusion MRI: a) Shape of the variation in the MRI signal with the first passage of the bolus of contrast agent in the capillaries of a given volume of tissue. b) Extract from a series of 24 cerebral slices (3D image) obtained in 1 minute and mapping the blood flow in each voxel, which in this case is called the rCBF (regional Cerebral Blood Flow), its highest values being shown in brownish red. c) left, image weighted by the parameter T_1 obtained with a "gradient echo" sequence (see section 2.5.2.4) indicating the probable existence of a heavily irrigated region revealing the development of a brain tumor; the color chart of perfusion rate tends to confirm the existence of increased vascularization. Images provided by F. Cotton, Creatis, CHU Lyon. For a color version of this figure, see www.iste.co.uk/fanet/medimagnet.zip*

The phenomena of translational diffusion of water molecules in tissues can be used to perform diffusion weighting. This incoherent displacement modifies the apparent relaxation time T_2, which can therefore be used as an indicator of the microscopic tissue dynamics. The principle will be summarized in section 2.6.2. Matter transport thus involves the velocity of coherent displacements. Encoding the flow velocity using gradients enables us to make the switch from the concept of weighting of the signal by a physical parameter (with the velocity of motion being the source of variations in the imaged signal), to that of imaging that parameter, directly or indirectly.

2.2.6.2. *Fluid velocity as a fourth dimension to the image*

From the very dawn of MRI, there has been great interest taken in the variety presented by the image signal returned by the blood vessels. The observations which showed the flows up were also able to characterize them (identification of the regime, measuring of velocities and of their distribution in the laminar flows). Consequently, one of the earliest medical applications was the acquisition of vessel images, known as angiography. The diverse strategies for acquisition and reconstruction gradually led to the visualization of increasingly fine-grained blood networks. To avoid going into detail about these medical and also industrial uses, we can simply limit this mention to the influence of the velocity of a set of nuclei on the phase of their resonance signal. As the velocity is capable of modifying the phase of the signal in the presence of gradients, it can be encoded and then provide an extra dimension for the image.

Figure 2.22 illustrates the principle of encoding the phase of isochromats by their velocity when they are subjected to a bipolar gradient. This gradient is applied, in each sense, for exactly the same duration (or more accurately, yields the representation of the same surface). In this simple model, the velocity of the isochromats V_U in a direction U in the fixed frame (R) is constant and the phase acquired after application of two bipolar gradient pulses occupying the same direction and the reciprocal velocity value, are given by:

$$\varphi_{V_U} = \gamma \cdot G_U \cdot V_U \cdot T_g \cdot \Delta t \qquad [2.18]$$

$$k_{V_U} = \frac{\gamma \cdot G_U \cdot T_g \cdot \Delta t}{2\pi} \qquad [2.19]$$

We do indeed obtain encoding of the phase of the isochromats in question by their velocity. This encoding can be superposed on the spatial encodings, and thus associated with an additional reciprocal coordinate. The bipolarity of the gradient, idealized in the calculations but adjustable in the experiments, makes it possible to remove the spatial contribution (the initial position along U) in the phase acquired during the course of the motion. The finally expression of the phase retains only the term dependent on the velocity at the moment the signal was recorded. Note that if we reverse the order of application of the gradient pulses, the sign of equation [2.18] is switched. The phase acquired is then added to the other contributions relating to the spatial encoding.

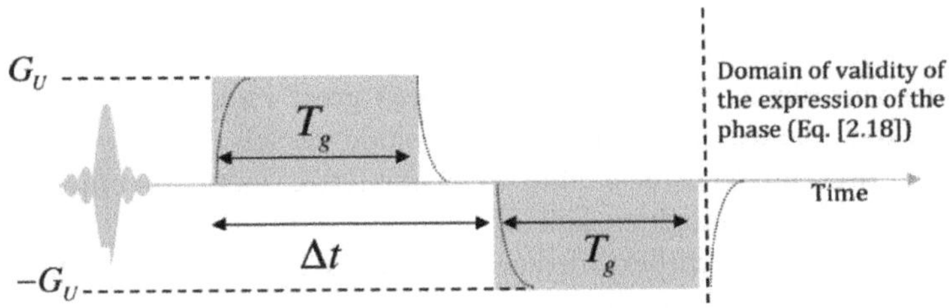

Figure 2.22. *The principle of encoding the phase of the magnetization by the velocity of linear motion in the direction of a gradient G_U*

An extra dimension – that of the velocities in the direction U – can thus be added to the three spatial dimensions. *A priori* and in principle, there is nothing to stop us creating a 6D image if we wish to explore all three components of the velocity, except for the prohibitive duration required for the recording. In practice, it is easier to use the velocity as a third component in the image of a slice of the sample. Used relatively little in biomedical MRI, velocity field imaging is a time-consuming operation. However, it has proved to be pertinent in the context of applications in non-invasive control.

2.2.6.3. *Chemical shift imaging*

Another example of an extra dimension being added to the image, but in a different register, is the difference in resonance frequency of nuclei of the same nature but immersed in different intramolecular environments because they belong to different molecules. The screening effect caused by the binding electrons depends on the chemical function of these nuclei. Consequently, in a given static magnetic field, these nuclei do not all have exactly the same value of the resonance frequency, depending on whether they belong to one molecule or another. This property, which is at the root of NMR spectrometry, comes into play in proton MRI because the lipidic protons present resonance frequencies shifted in terms of average relative value of -3.5 ppm (parts per million) in comparison to that of water. When we look at cerebral metabolites, the range of resonance spreading is around 10 ppm, giving us a range of frequencies of 1280 Hz in a 3-tesla field, and the difference in resonance frequency between the protons in water and lipids is therefore 448 Hz. Consequently if we choose the phase reference to be the signal of metabolite given at the frequency f_0 expressed in MHz, then after a time-period *t*, the phase acquired by a metabolite with chemical shift δ_C expressed in ppm is:

$$\Delta\varphi = 2\pi \cdot \delta_C \cdot f_0 \cdot t \qquad [2.20]$$

Adding a new factor creating a linear phase variation over time involves creating a new dimension in the image, associated with the values which that factor may take in the sample being observed, the factor of chemical shift being used for chemical shift imaging or indeed spectroscopic imaging. The proposed technique therefore involves assigning a spectrum to each pixel (or voxel) of the image. This representation is possible, in practice, when the number of pixels (or voxels) is low. Once this task is accomplished in the computer's memory, we can reconstitute as many images as there are metabolites present (generally each metabolite can be represented by a characteristic peak). The level of gray or color coding in the image is then modulated by the intensity of the peak; we separately obtain the spatial distributions of the components of the sample. Superposing an image of a metabolic value on the anatomical image is another means of presentation – see Figure 2.23.

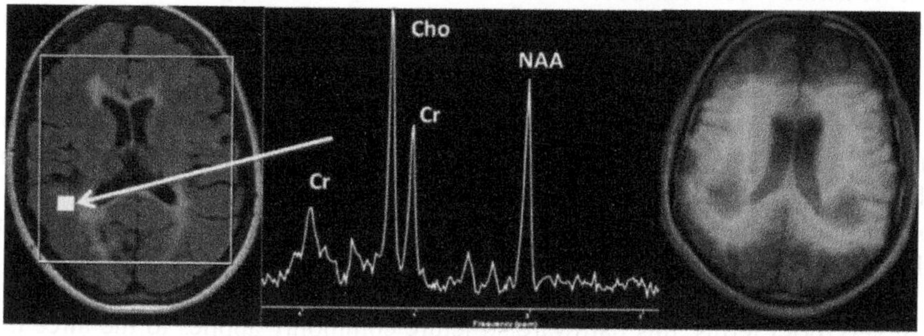

a) b) c)

Figure 2.23. *Images of the brain of a patient suffering from multiple sclerosis. a) Anatomy of an axial slice showing major peri-ventricular lesions. b) Proton spectrum from a voxel (pixel × slice thickness) of volume ~1 × 1 × 1.5 cm^3: left to right, mainly Creatine, Choline, Creatine, N-acetyl-aspartate. The metabolic indicator is the ratio of the Choline peak to the N-acetyl-aspartate peak; its distribution within the selected parallelepipedic volume (white contour box in (a)) is shown in (c). Its color coding expresses the local increase in Choline concentration and concomitant decrease in N-acetyl-aspartate, which is characteristic of the disease. Image provided by D. Sappey-Marinier, Cermep, Creatis. For a color version of this figure, see www.iste.co.uk/fanet/medimagnet.zip*

2.3. Instrumentation

2.3.1. *Introduction*

Magnetic resonance systems are similar to one another, in that we need to have a static magnetic field, an RF field and a measuring instrument downstream of the sensors. This is, of course, common to magnetic resonance spectroscopy (MRS) and imaging (MRI) which are the two primary uses of nuclear magnetic resonance (NMR). As the spectrum of applications extends into each domain with possible overlaps, we should first highlight the common points, which are also very numerous. We shall group them into the following three broad categories:

The static magnetic field part includes a fixed-field magnet. The range of values currently covered extends from around a milli-tesla to over 20 teslas. For reasons of cost, only one value of the static field is usually assigned to a given device. This set is supplemented by systems to correct the spatial distribution of the static field and by field gradient systems, whose electric supplies are independent in terms of their nature, size and power.

The RF magnetic part includes at least one powerful RF source for pulse generation. Its power requirement varies considerably depending on the intended use of the hardware – between a few hundred watts in spectroscopy and tens of kilowatts in imaging. The array of sensors necessary for detecting the signal is another central element of this part, but working at low power level.

Electronic equipment is used to control the different operations: observations, conditioning and recording of the signal, followed by processing. This subset includes digital functions which, as NMR techniques have evolved, like in most experimental domains, have overtaken analog elements which cannot be fully replaced during the application of the RF field to excite resonances.

In very brief terms, this hardware structure commonly applies to the three main applications which we ought to mention. We shall leave aside the extreme-case NMR experiments – e.g. in pulsed and very high intensity static magnetic fields.

In nuclear magnetic resonance spectrometry, devoted to the structural and functional analysis in Chemistry, Biochemistry, Biology, Physical Chemistry and Solid-State Physics/Chemistry, the conventional proton resonance frequencies are generally multiples of 100 MHz with a coefficient ranging from 1 to 10. The devices may be equipped for the observation of all the stable isotopes presenting a non-null spin number (multi-nucleus devices). Their access volumes are around a cubic centimeter.

Biomedical research requires the functions of the spectrometer and imager to be combined; but on a relatively limited number of nuclei – essentially those which have a biological aspect (hydrogen (proton), phosphorus, carbon 13, sodium) or hold additional interest (hyper-polarized helium or xenon nuclei). The magnets must have sufficiently large usable volumes to be able to deal with animal subjects (rodents) up to frequencies of around 600 MHz.

Full-body imaging for human beings requires large instruments for clinical practice or clinical research. As the access volumes are wide, the values of the fields used are usually limited to 1.5 or 3 teslas in a clinical environment, extending to 7 teslas for research, which corresponds to a frequency of 300 MHz for the proton. The imaging techniques use mainly the hydrogen nucleus and, to a lesser extent, the sodium nucleus. These devices may therefore be equipped for *in vivo* proton spectroscopy, phosphorus and sometimes carbon-13 spectroscopy.

There is a certain variety of magnetic systems which use permanent magnets. These systems can be used to perform observations for very diverse purposes, including biomedical, with field values rarely above 1 tesla. Finally, it should be pointed out that the non-invasive control applications in the broadest sense, both in research and industry, are progressively using the all of the techniques developed for the aforementioned domains of application, evolving toward specific instruments and methods.

2.3.2. *Recording the signal*

2.3.2.1. *Resonators, coils and antennas*

At present, the signal is acquired mostly by recording the electromotive force caused by the motion of precession of the transverse magnetizations in a conductor circuit covering all or part of the sample. This circuit must present a mode of electrical resonance at the frequency equal or very close to the frequency of precession of the macroscopic magnetization M_0. The physical device used could be a simple tunable coil, which can easily give rise to an elementary analytical model as a "resonator" serving as a sensor. With a transverse magnetization $\overrightarrow{M_\perp}(t)\,\delta V$ assigned to a volume δV of the sample, the instantaneous electromotive force created at the terminals of the coil is given by the circulation, along the conductor constituting that coil, with the variable electric field caused by precession. In view of equation [2.8], the result can be written in the following form:

$$e = -\frac{\partial}{\partial t}[\overrightarrow{M_\perp}(t)\cdot\frac{\overrightarrow{B_1}}{I}(t)]\,\delta V \qquad [2.21]$$

In this expression, $\frac{\overrightarrow{B_1}}{I}(t)$ is the RF field created at the position of δV when a unit current is passing through the coil. This relation indicates that if an antenna creates a high value of field B_1 per unit current, it will therefore present good sensitivity.

The electromotive force from the whole of the sample, the sum of the elementary electromotive forces, is transmitted to the receiving system, considered as a pure resistance, generally having a 50 Ω value. If we wish to optimize the transfer of power to the system (power matching) we need to transform the impedance experienced by the coil into a resistance equal to 50 Ω using a circuit, the very basic principle of which is given in Figure 2.24. Simple calculations indicate that the maximum variation in the resistive part Re of the impedance of the probe is located at a higher frequency than that chosen for its operation at 50 Ω. This spectral distance is lesser when the coefficient of quality of the self-inductance coil L, with equivalent resistance r, is high. This arrangement enables us to use a negative impedance (that of the capacitor C2) in order to bring the value of the imaginary part of the probe's impedance down to zero. For this reason, we use the term "resonator" to speak of the whole sensor or probe. Note that the coupling performed here by the capacitor C2 can be replaced by inductive coupling between the set (L,r,C1) and the input to the measuring system.

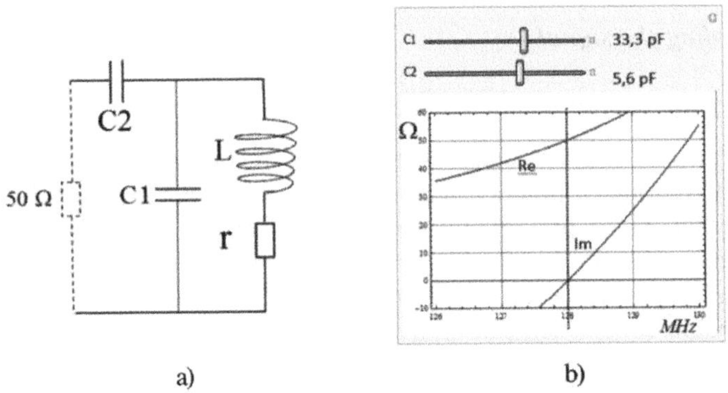

Figure 2.24. *Optimization of power transfer at 128 MHz to a 50 Ω charge R_O. a) Diagram of setup; b) Simulation using the "Manipulate" function in Mathematica™. With a 33.3 pF capacitor C1, the real part of the sensor's impedance is 50 Ω. Adjusting C2 at 5.6 pF enables us to null out the imaginary component. Representation between 126 and 130 MHz, in the presence of a sample (equivalent resistance r = 1Ω, self-inductance L = 40 nH)*

The devices used to transmit the RF field operate on the same principle and could therefore also serve as sensors. In MRI, these transmission coils are distinct from the sensors used, because they are intended to cover the whole of the subject's body, which is why their dimensions are large, giving them the name of "transmitting antennas". The term is also used for the sensors, which are generally smaller and thus become receiving antennas.

2.3.2.1.1. Traditional structures

We shall look at two examples with completely different geometric structures, knowing that between these two extremes, any geometry is, *a priori*, possible. This means that resonators are a collection of accessories specific for the regions or organs examined in MRI, and are therefore particularly varied.

Note that in the diagrams below, the coordinates used are not necessarily those of the lab frame (R), although they may coincide with that frame.

Surface coil

The structure, which is simple, can be applied to the region of interest (the eye, the heart, etc.). It can be simulated by a circular conductor, although this planar form does not always correspond very closely to the shape of the instrument.

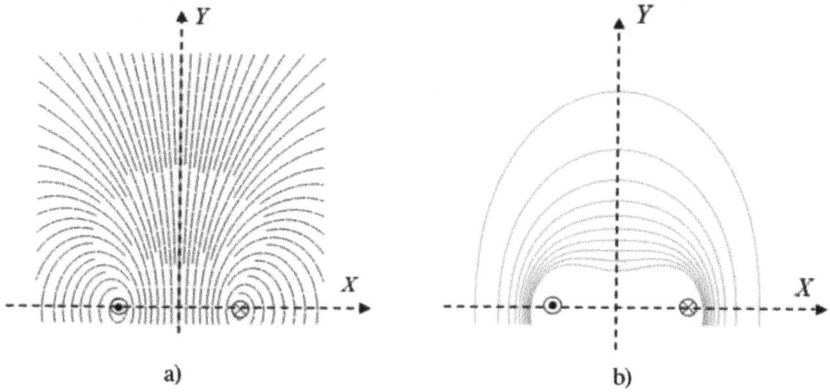

Figure 2.25. *In a median plane of a circular coil with a current passing through it: a) magnetic field lines; b) contour plot of the magnetic field of revolution around* OY. *In b), the value of the field varies tenfold between the outermost and innermost curves*

On the other hand, the circular loop model is interesting because of its simplicity, as indicated by the representation of the magnetic field given in Figure 2.25. The model's cylindrical symmetry helps limit this representation to a single median plane perpendicular to that of the coil. If we accept the hypothesis that the regime is quasi-stationary, the chart of the field B_1, polarized linearly at each point in space, and whose value governs the local sensitivity (relation [2.21]), shows that this field is not spatially uniform in terms of value or direction. The first consequence of this is that the spatial response of this sensor, relative to a small element of volume of the sample, becomes increasingly slight as that volume moves further from the conductor plane, which corresponds to use as a "surface coil". A second remark: if that coil is used for receiving, the phase of the signal received from an element of a given volume will depend on where the element is placed. The RF excitation, delivered by an antenna of significantly large dimensions, may be considered to be uniform, i.e. of an RF field with the same direction throughout the volume observed. It presents different relative orientations with respect to the detecting field B_1 depending on the position in question. Consequently, the overall response will be affected by these phase differences. A third remark relates to the use of the surface coil both for spin excitation and signal recording, which complicates the distribution of the sensor's spatial sensitivity. However, this arrangement is very rarely employed in MRI.

Bulky resonators

Conversely, bulky resonators – which surround the subject – offer RF fields whose spatial uniformity is very acceptable. As it is impossible for us to discuss all models of resonators here, we shall focus on the so-called "birdcage" structure, because it is one of the most interesting and most widely-used systems. It is able to deliver an RF field with good uniformity, which can be circularly polarized. A simplified "low-pass" model is shown in Figure 2.26.

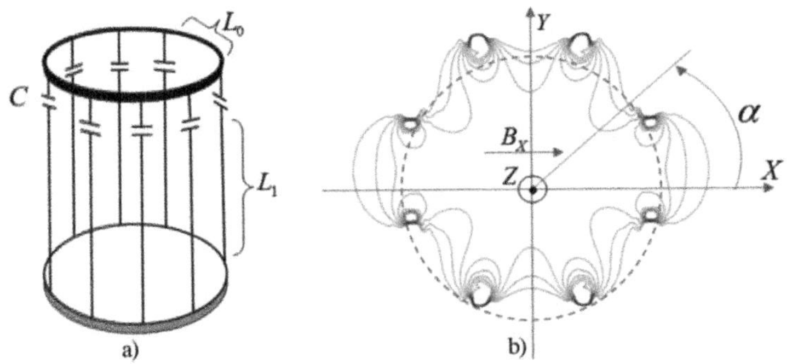

Figure 2.26. *Birdcage resonator (low-pass) formed of eight cells. a) Structure of conductive elements. The capacitors C all have the same value, as do the self-inductances L_0 on the one hand, and the self-inductances L_1 on the other. b) Contour plot of the component B_X of the field with a $\sin(\alpha)$ distribution of the currents in the conductors (L_1, C). Case of a very long cylinder operating in the fundamental mode of pulsation $\omega_1 = [\frac{4\sin^2 \pi/8}{C(L_1 + 4L_0 \sin^2 \pi/8)}]^{1/2}$. In this case, the relative variation of B_X between the center of the coil and the first contour is around 3%, and the component B_Y can be discounted*

The principle uses the fundamental mode of resonance of a row of discrete conductive elements, closed in on itself. It is therefore based on the fact that continuous distribution of currents, over the surface of an infinitely long straight cylinder, parallel to the generator and distributed in accordance with a trigonometric law, gives a perfectly uniform magnetic field inside that cylinder. The "high-pass" type is obtained by introducing the capacitors into the arcs instead of the parallel stems. The circular conductor forming the base is considered at the reference potential. In MRI, this structure of resonators is often used for brain examination. In axial position in the tunnel of a superconducting magnet (the OZ axis in Figure 2.26 is identical to the OZ axis in the fixed frame (R)), a birdcage resonator of large diameter finally constitutes an excellent antenna for RF transmission. In the literature, readers can find descriptions of other possible structures, which are also appropriate for the observation of large volumes.

2.3.2.1.2. Receiving antenna arrays

Let us stick with the domain of reception, which is increasingly done using antennas organized in arrays, with the basic device for these antennas being the surface coil. This can be represented diagrammatically as a simple flat circuit, circular or polygonal in shape. The drawback posed by the non-uniformity of their RF field is only apparent, given that we are perfectly well aware of their "spatial sensitivity profile". This property, once it has been mapped in terms of amplitude and phase, can be taken into account when exploiting the signals to create the final image. Also, the fact of using relatively small coils ensures good sensitivity in a volume whose distance to the plane of the antenna is similar to the dimension of it. Consequently, arrays of surface antennas of around 15 cm in diameter can provide effective observations of the whole of the human body. The most significant problem is the magnetic coupling observed between the coils making up the sensor array (Figure 2.27).

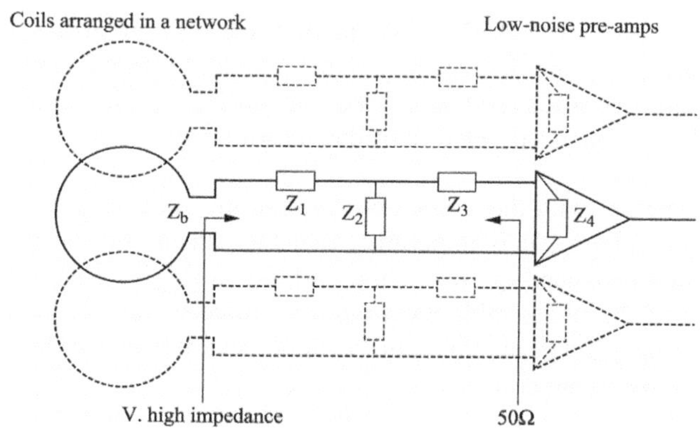

Figure 2.27. *Partial representation using an array of three circular-coil elements. So as to reduce the current in each coil, the impedance experienced by each coil on the pre-amplifier side must be as high as possible*

An initial solution to this problem is to have the immediately neighboring coils partially overlap, so as to cancel out their mutual induction coefficient. To supplement this decoupling, it is then helpful to connect these sensors to high-level impedances in order to limit the current in each coil and thereby its interactions with its neighbors. Thus, we do not need to match the power input to each low-noise pre-amplifier whose power gain is able to provide a weak signal with sufficient level whilst preserving the initial SNR due to its low noise figure (less than 0.5 dB). There is no matching on each pre-amp whose input impedance can be considered to be purely reactive. Thus, instead of having a circuit which improves power transfer from the source coil, with impedance $Z_b = r_b + jX_b$, to the input impedance Z_4 of the pre-amp, it is then necessary to "isolate" the receiving coils from the pre-amps associated with them. A T-setup using the impedances Z_1, Z_2, Z_3 enables us to perform the impedance transformation which satisfies this condition for all channels. Indeed, with each antenna there is a corresponding pre-amp, and usually its own detection system. The number of array elements depends on the intended application: imaging of the brain, heart, abdomen or spine; usually greater than 10, this number is more commonly 32 and can reach up to 128 in certain devices. For a given setup, the switch from one type of array configuration to another, or even to a single-coil configuration, is automatically handled by the control unit. In clinical practice, we can therefore move the region of interest where the observation takes place without having to move the patient in the machine. It is possible, for instance, to simply combine two neighboring coils so that they behave like a single coil, or in such a way that they produce a circularly polarized detecting RF field, which has the advantage of enhancing the sensitivity in part of the region of observation of the two coils. In general, we need to take account of the phase shifts between different elementary antennas, so the term "phased-array" is used. Another advantage, which we shall discuss later on (see section 2.5.3.4), is that the use of arrays of coils very significantly speeds up the process of image acquisition. For this reason, we must process the data recorded and use them according to the knowledge of the "amplitude–phase" sensitivity profile of the sensor array.

Note that, whether with the single coil, used in the "old-fashioned" way, or with the phased-array antenna, it is imperative that the sensors not be receptive during any RF excitation produced by the excitation body-coil. This condition necessitates the attenuation of the undesirable electromotive force which is directly induced in the sensors during excitation; this can be done by momentary detuning of some capacitors in the receiving coil.

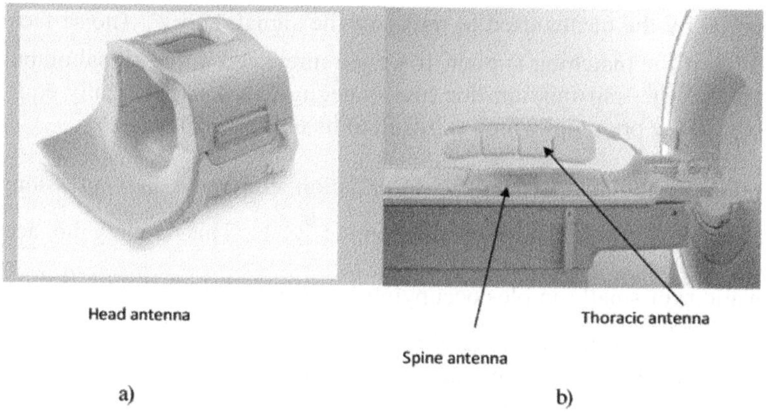

Figure 2.28. *Resonators used as receiving antennas employed in clinical MRI. a) Enveloping "birdcage" coil for the head; b) "array" of surface coils for thoracic observation, located above the patient, and spine coil beneath him/her to examine the spine. Documents provided by Philips Healthcare France*

2.3.2.2. *Integration into a measuring chain*

2.3.2.2.1. Signal-to-noise ratio (SNR), and sensitivity criterion

Using the results from section 2.2 and relation [2.21], the expression of the SNR in the temporal domain, at the input to the processing process, can also be written thus:

$$SNR_t = \frac{M_t.\omega.V_E.(B_1/i)_{av}}{\sqrt{4.k_B.T.r.\Delta f}} \cdot \frac{1}{F_N} \qquad [2.22]$$

In this expression, we have considered a sample of volume V_E within which the transverse magnetization (resulting from preparations specific to the observation sequence) is equal to M_t per unit volume. The term $(B_1/i)_{av}$ characterizes the average value of the amplitude of the RF field which the recording coil is capable of creating in the volume V_E. The equivalent resistance r of the coil, which is greater than its ohmic resistance, also takes account of the other energy losses (magnetic and dielectric losses through the sample, radiations). We estimate the quadratic noise voltage using Johnson's thermodynamic model; hence, T here is the temperature of the coil, Δf is the spectral band required by the observation and k_B is Boltzmann's constant. The coefficient $\frac{1}{F_N}$ relating to the noise corresponds to the losses caused by the means used to transport the signal, and F_N (noise factor) may become high if the matching is poor. If we are interested in the signal output by the pre-amp and if the transmission line loss is negligible, it is practically equal to the noise factor of the pre-amp, which is linked to its noise coefficient.

Attention should be paid to the formulation employed in expression [2.22], which takes an average value of the term (B_1/i). This is only an acceptable approximation for small samples occupying a position within which (B_1/i) changes very little in terms of modulus and direction. In general, the non-uniformity of the RF field is considerably greater than that of the static field, in terms of relative value. Thus, if the sample properly fills the detection coil (which tends to increase the SNR), we will be faced with regions in which the direction of the field B_1 is not the same. The local expression of the sensitivity is therefore complicated by the existence of phase differences between the signals coming from these regions. Without going into further detail, we can note that a detection coil must indeed be characterized by a chart of its sensitivity where that represented value appears as a complex number, because of the existence of phase.

Finally, particular mention must be made of the equivalent resistance of the detection coil, which depends on the working frequency. This remark, relating to energy losses, brings us back to the topic of the operation of the two types of coils – transmitting and receiving. If we do not take account of the radiation, which comes into play at high working frequencies, i.e. with a very powerful static field, the two main types of losses are caused by the coil's own resistance and the resistances corresponding to the conductive properties of the subject being examined, whose presence loads the receiving coil. We can show that the power which is dissipated into the subject on its transmission is of the form:

$$P_{dis} \sim \sigma \cdot f^2 . ((B_1/i)_{av})^2 \cdot V_E^2 \qquad [2.23]$$

where σ is the conductivity of the medium (the subject) and f the working frequency. This indicates that the corresponding contribution, which affects the value of the equivalent resistance r, is proportional to $f^2.((B_1/i)_{av})^2$ although the contribution of the ohmic losses is proportional to $\sqrt{f}$ because of the skin effect. In biological media, this second contribution, which is significant at low frequency and therefore weak fields (e.g. in the old 0.15-tesla systems), becomes negligible when using a strong field (beyond 1.5 teslas), in comparison to the major contribution of the charging of the coil by a conductive subject. Therefore, it is the subject being examined who is the main source of noise. With a recording coil, the resistance will essentially be of the type $r \approx C_F \cdot \omega^2 \cdot ((B_1/i)_{av})^2$, where C_F is a factor dependent on the shape of the sample. Expression [2.22] then becomes $\frac{M_t.V_E}{\sqrt{4.k_B.T.C_F.\Delta f}} \cdot \frac{1}{F_B}$, and the dependence on frequency – which, for a given nucleus, only appears through M_t – indicates that the SNR is approximately proportional to the working frequency dictated by the magnet. We can therefore conclude from this brief examination of the SNR in the temporal domain that with MRI installations whose working frequencies are greater than 60 MHz (1.5 teslas) and are intended for observation of conductive subjects like the human body, the sensitivity is practically proportional to the resonance frequency of the nuclei observed.

2.3.2.2.2. Signal acquisition: from analog to digital

The fact of having to use a large number of receiving channels functioning in parallel, as suggested by Figure 2.27, does not pose too many problems, provided we reduce the analog part to an absolute minimum, which is costly and tricky to regulate, in favor of digital techniques. The example chosen here, drawn from numerous strategies which are constantly evolving, satisfies this condition. The digitization is performed directly on the RF signal available on output from the low-noise pre-amp which delivers a real signal with high average frequency denoted F_0, which corresponds to the electromotive force induced in the coil. Its frequential distribution, of width ΔF, is governed by the amplitude of the gradient applied during the recording (the reading gradient) multiplied by the sample size in the direction of that gradient. Consider, for example, a frequency of precession F_0 of 130 MHz (proton in a 3.054-tesla magnetic field) and a spectral width ΔF of 1.7 MHz (reading gradient of $100\ \mathrm{mT/m}$ in a $40\ \mathrm{cm}$ field of view). This signal is assumed to be of the form $\cos[2\pi(F_0 + \delta F) + \varphi]$, and it exhibits two frequency bands of width ΔF, which are symmetrical with respect to the zero frequency. The amplitude of this analog signal is adjusted by amplification and attenuation depending on the dynamic range of the converter and the signal itself undergoes anti-aliasing filtration before its analog-to-digital conversion. In the example shown in Figure 2.29, inspired by reference to the work of M. Décorps, the sampling frequency *Fe* taken as *50 MHz* is lesser than the frequency of the signal. This arrangement, because of the resulting under-sampling, leads to overlap between two spectra with period *Fe* for which each pattern (or alias) occupies a certain frequency band. It is important to avoid overlapping between these patterns, which form symmetrical pairs around directions parallel to the axis of the amplitudes. These spectra are then formed of discrete values, the number of which is generally far higher than that imposed by the Nyquist–Shannon law if we operate with respect to the frequency range ΔF. If we consider that the duration of the signal capture equals

5 milliseconds, then for this example we get 250,000 samples at a width of *1.7* MHz for each alias. We exploit this intermediary over-sampling in order to improve SNR due to quantification. The over-sampling is followed by decimation, e.g. by a factor of 1,000. The use of a bandpass filter isolates an alias whose median frequency is Fm (in Figure 2.29, we have chosen $Fm = 20$ MHz).

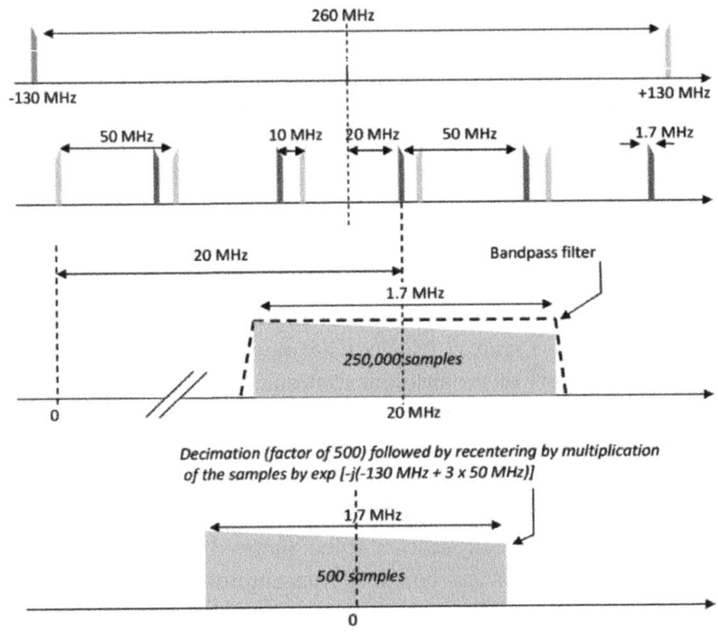

Figure 2.29. *a) The analog signal whose spectral width is centered on 130 MHz occupies a range ΔF of 1.7 MHz . Therefore, it is represented by two bands of that width, symmetrical with respect to zero and situated 260 MHz apart. b) Partial spectrum corresponding to under-sampling at 50 MHz . c) Selection of the band situated at $Fk = -(130 + 3 \times 5)$ MHz. If the duration of acquisition is 5 ms, that band contains 250,000 samples. d) Decimation by a factor of 500, and multiplication of the content of the selected band by exp-j(- Fk t) performing a frequency subtraction, leads to a profile of 500 samples occupying a range of 1.7 MHz centered on the zero frequency*

Frequency shift obtained by multiplying the discrete spectral distribution in this pattern by $\exp(-j \cdot Fm \cdot t)$ finally leaves us with a single pattern of 500 complex samples centered on the frequency zero. For this operation, we use multiplication by two digital signals, differing in phase by $\pi/2$ and provided by a digital oscillator which is, itself, synchronized with all the sources of frequency used in the course of the process, including with the excitation frequency.

These indications constitute an example which can be qualified as "hybrid", because it corresponds overall to the different solutions proposed by manufacturers. Indeed, there are detection techniques which lower the analog signal to an intermediary frequency, followed by (possibly digital) detection, or which use detections performed directly on the signal with a very high sampling frequency. The use of an intermediary frequency is obviously an option which is justified in multi-nucleus NMR spectroscopy, but it is not necessary in MRI – particularly if we limit ourselves to using the proton. In the case of sodium-23 imaging – which is currently being developed, using a lower frequency but which is still 79 MHz at 7 teslas –, detection by initial under-sampling as illustrated by Figure 2.29 or detection by direct over-sampling can be envisaged. It remains to be seen which technique is the best suited to the dynamic range of the recorded signals and its impact on the architecture of the digital part.

2.3.2.2.3. Dynamic range and impact on system architecture

We can briefly show how the architecture of the "digital detection" setup can evolve under the influence of the measuring techniques (mainly the signal sampling methods) when certain conditions are imposed by the dynamic range of the data and by the modes of data transmission towards the image reconstruction process.

In the example given in Figure 2.29, over-sampling at 50 MHz with respect to the spectral band of 1.7 MHz occupied by the resonances causes an increase in the dynamic range, which is expressed in the form:

$$\Delta n = \frac{1}{2}\log_2\left(\frac{Fe}{2\Delta F}\right) \qquad [2.24]$$

with Fe=50 MHz and $\Delta F = 1.7$ MHz, we obtain $\Delta n \approx 1.95$ bits.

The converter supposed to function, for instance, at 16 bits (dynamic range DR_{conv} = 96 dB) has its dynamic range increased to 108 dB. This result depends on the image required (here we make use of a very broad spectral band ($\Delta F = 1.7$ MHz) because we have used a large reading gradient applied to a very extensive field of view: 40 cm). If we focus the observation on a field two times less wide, using a gradient half as great, the value of ΔF is divided by four, and the increase in the dynamic range becomes $\Delta n \approx 2.94$ bits, which gives us practically 19 bits for the acquisition and a new dynamic range of 114 dB. This indeed corresponds to the fact that the SNR in MRI can be all the stronger when the reading gradient is moderate – provided the uniformity of the static field and spatial linearity of the gradients are appropriate. Of course, the range thus determined by the analog-to-digital conversion (with or without over-sampling) must correspond to the dynamic of the signal, which is governed by the average value of the signal.

Obviously, this is the signal in the reciprocal space. We can show that with a 2D image formed of N_1 by N_2 pixels, we can express the dynamic range specific to the signal by way of the following expression:

$$\mathrm{DR}_{signal} = 20 \cdot \log_{10}[2 \cdot \sqrt{N_1 N_2} \cdot \overline{(S / B)}_{pixel}] \qquad [2.25]$$

where $\overline{(S / B)}_{pixel}$ is the average SNR for each pixel, evaluated in the time domain, using the results of section 2.3.2.2.1.

In turn, this value will depend on the conditions of use of the imager, and primarily on the sensitivity of the antenna in relation to the volume of the sample concerned (here the slice whose thickness is fixed by the value of the slice gradient and the spectrum of the excitation pulse). Based on the distribution of the RF field in the volume of the slice selected, the value to be assigned to M_t in the permanent regime of excitation by a given RF pulse is accessible and we can use relation [2.25] provided we know all the parameters involved in the expression of the resistance r, or know its value. One problem arises when the value of $\mathrm{DR}_{\mathrm{signal}}$ is greater than that of $\mathrm{DR}_{\mathrm{conv}}$, because this situation leads to a loss of dynamic range in the image, and consequently causes reduced performances in terms of sensitivity. These issues lead to distortions, additional noise, or a loss of maximum resolution in comparison to that which could legitimately have been expected from the imaging technique. The problem can be circumvented by a variety of approaches described in the literature, including, in particular, dynamic raising by over-sampling. The problem arises, notably, when we wish to obtain high-resolution images with a strong field.

If the necessary dynamic is reached ($\mathrm{DR}_{\mathrm{conv}} \geq \mathrm{DR}_{\mathrm{signal}}$), a second problem concerns the compatibility of the transmitting system after recording of the data and temporary storage of those data for processing. When transmitting analog data, we encounter problems relating to the attenuation along the coaxial transmitting cables. These problems may be greatly accentuated by imperfect matching of these cables to different elements: wall passages, inputs to amplifiers, multiplexers, etc. These difficulties, which are compounded by the problems of interaction between cables in the same bundle, cannot be resolved by using fiber-optic transfer, as the attenuation with fiber-optic technology is usually too high with the bandwidths that are then required. However, if we are transmitting data which have already been digitized, this difficulty is prevented, and it is possible to use fiber-optic.

The resulting architecture, therefore, may be remarkably simplified – particularly if each antenna has an immediate corresponding sampling device, and the digitized data are then evacuated by an optical fiber. The diagram in Figure 2.27 corresponds to the link between the antenna array and the detection equipment in a conventional setup. This setup brings together the sensors and all the processing devices, and an important part is given over to the analog functions and the digitization of the data. In quite the opposite manner, the solution envisaged in Figure 2.30 is to equip each antenna with its own detection and digitization device, and then use only one optical fiber to transmit the data from the whole measuring network.

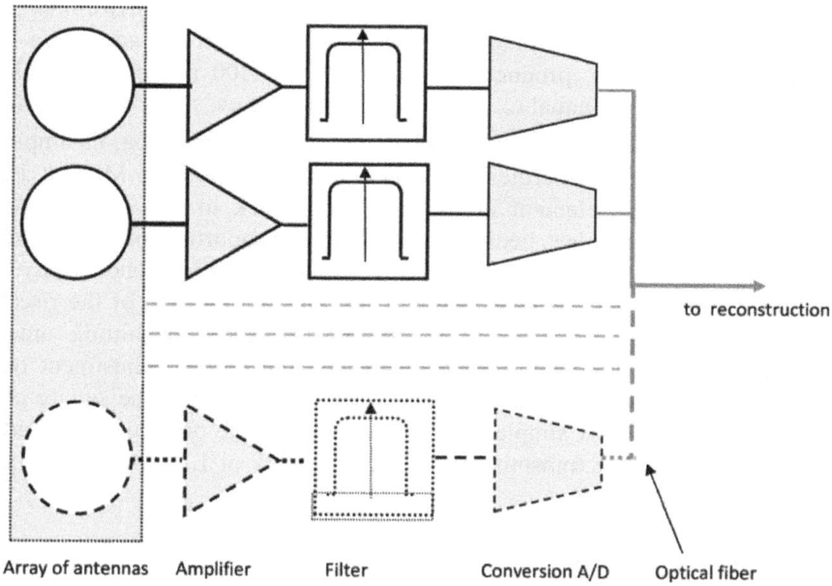

Figure 2.30. *Digital architecture using individual receivers on each antenna of a network, and fiber-optic transmission of the digitized data on each channel (dStream architecture from Philips Healthcare)*

2.3.3. *Magnetic systems*

2.3.3.1. *Radiofrequency unit*

RF devices (generators, power amplifiers, antennas, etc.) form a particular magnetic system, which we shall not discuss in detail. From what we saw earlier, the expected properties of the RF field whose role is to manipulate the magnetization during an acquisition sequence are (i) to present a sufficient amplitude to flip the magnetization for short durations in comparison to the relaxation times; and (ii) to have a sufficiently uniform spatial distribution throughout the sample.

The transmitting signal, at a frequency very close to the average frequency of precession of the nuclei distributed through the sample, is created at a low level by a synthesizer, and then is modulated in terms of amplitude (or in certain applications, its frequency is modulated). It is significantly amplified before being addressed to the transmitting coil. The transmitting and receiving coils must present as little magnetic coupling as possible. This is not entirely guaranteed by simple orthogonality of their respective axes of symmetry. During the transmission it is therefore necessary to neutralize the recording circuits to avoid saturating the receiver – e.g. by de-tuning the sensor. It is not overly complicated to evaluate the RF field amplitude corresponding to a given flip angle in a fixed time-period. For instance, producing a 180° pulse in 500 microseconds ideally requires an amplitude B_1 equal to $\pi / \gamma\tau$, where $\tau = 500$ µs, so $B_1 \approx 24$ microteslas in the case of the proton. If the RF field has rectilinear polarization, its amplitude therefore needs to be 48 microteslas. Although it is helpful, in MRI, it is not sufficient to know this element in order to work back to the value of power required to create the pulse, because a very large proportion of this power is dissipated through the sample, which has a certain degree of conductivity. The simulation is more or less approximate and complicated because of the rise time (and cutoff time) of the RF at the terminals of the transmitting antenna. Experimentally, it is relatively easy to determine the correct adjustment of the desired pulses and the power requirement. This varies, as it is the square of the value of the static field. A simple way of working out the order of magnitude of the power needed for the transmission is to take 15 kW at 1.5 teslas, which gives us around 60 kW at 3 teslas.

2.3.3.2. *Static magnetic field*

For MRI, which almost exclusively uses proton resonance, it has become usual to speak of installations by the value of their static magnetic field. At the beginning of the 1980s, the earliest installations usually functioned at 0.15 tesla with so-called “resistive” magnets, i.e. comprising huge copper coils, cooled by water circulation. These magnets were soon replaced by cryogenic devices whose dimensions were appropriate for the size of a human body, operating at 0.35 and then 0.5 tesla, before the dawn of systems using 1.5 and later 3 teslas. Their development was stimulated by the dawn of the market for MRI. The efficiency of using such systems resulted in the replacement of the liquid nitrogen thermal screen by super-insulators and the addition of a helium recovery system/cryogenerator. Helium’s boiling point of 4.2 K makes it possible to propagate currents dozens of amperes strong in the windings of dozens of kilometers of wire superconductors: niobium-titanium alloys with a copper sheath. The leakage field is then channeled by counter-current coils, and this solution, vindicated by the use of high values of magnetic fields, has become commonly used in the domain of NMR. Systems operating with a lower field (1 tesla) use permanent magnets often based on alloys of iron, neodymium and boron. The temporal stability of the field generated by these magnets depends primarily on the quality of their thermal insulation, which therefore ought to be ensured as fully as possible.

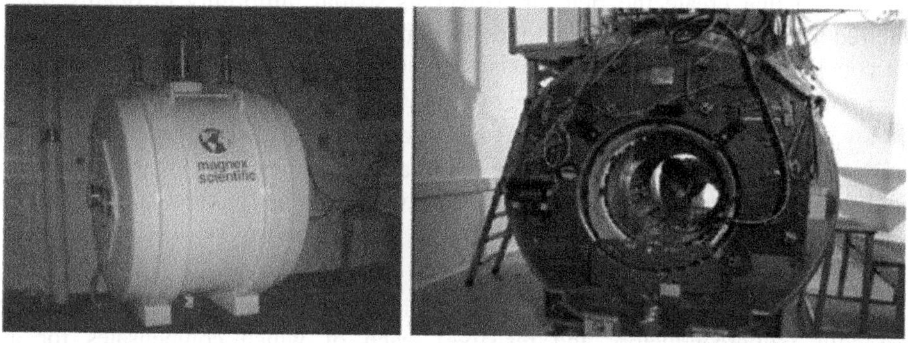

a) b)

Figure 2.31. *Cryogenic magnets: a) 4.7-tesla Magnex magnet, with access diameter of 15 cm for pre-clinical research; and b) 70 cm 3-tesla "full body" magnet from General Electric for clinical imaging. Documents obtained from the manufacturers*

The free space (or free bore) where the subject enters (the tunnel or the air gap in the magnet) has a relative permeability which is practically equal to that of a vacuum. Because of the properties of cylindrical symmetry around the direction OZ defined by the main orientation of the field, that field can easily be modeled. The advantage of modeling the field is that it serves as a guide for compensation of the lack of uniformity, and for the realization of constant gradients, additional fields whose spatial linearity needs to be guaranteed within the volume of interest placed at the center of the accessible space. The spherical harmonic functions (or, if readers prefer, the polynomials and associated Legendre polynomials) prove useful as solutions to the Laplace equation $\Delta\vec{B} = 0$, satisfied in a free space located within the currents. The three components of the static field, B_X, B_Y, B_Z, taken in the laboratory frame, are separately solutions to the Laplace equation. In order to ensure the spatial uniformity of the field in that space, so as to render the component B_Z as uniform as possible, we need to render B_X and B_Y negligible or null. The expression of the component B_Z can be developed on a set of Legendre functions whose general form is $r^{\ell} P_{\ell,m}(\theta,\varphi)$, with P being a polynomial form dependent on the angles θ and φ, respectively co-latitude and longitude in the frame (R). The integers ℓ, m are such that $m = -\ell, -\ell+1, \ldots, +\ell$. In the development of entire powers of r, the distance from the origin of the frame (R), the higher the order ℓ and the degree $|m|$, the smaller the coefficients assigned to the Legendre functions, indicating the value of the field at a point (r, θ, φ). As the expressions of these coefficients also involve Legendre functions, the result depends on the properties of symmetry of the currents distribution that create the field or perfect its spatial uniformity (a quality often referred to as "homogeneity"). It is therefore important to choose an arrangement of the conductors along the common axis OZ, so that with the exception of the term $\ell = 0$ (which gives the nominal value of the field), the first-order coefficients, which are involved in the development, only appear at sufficiently high values of ℓ, such as 6 or 8. It is just as important for the coils carrying the currents creating the main field to conserve a mean plane orthogonal to OZ so as not to create notable terms with degrees $|m|$ different to zero.

In order to correct the non-uniformity (or inhomogeneity) of the static field, we use coils (superconductors and resistive), each of which compensates for an undesirable term of given order ℓ and degree m, by creating a term of the opposite sign. These terms are identified by the field map measured before correction. The orthogonality of the Legendre functions thus enables us to adjust the currents in the compensation coils almost independently. Here we shall limit ourselves to the very first terms, because they are particularly important:

– The term $\ell = 0$ represents *one* uniform component of the field. It also corresponds to the correction coil, noted Z_0, which is used to modify the nominal value of the field in a range that is generally small, but that it is very useful to have at our disposal.

– The three terms $\ell = 1; m = -1, m = 0, m = 1$ express dependencies in $r Sin\theta Cos\varphi$, $r Cos\theta$ and in $r Sin\theta Sin\varphi$. Expressed in Cartesian coordinates, this corresponds to linear variations of the field in the vicinity of the origin, along *Z*, *X* and *Y* – that is, the gradients of the static field. They can be created by three coils called *Z*, *X* and *Y* and of course yield three gradients in the directions indicated.

– The terms $\ell = 2; m = \pm 2, m = \pm 1, m = 0$, of which there are five, correspond to the five 2nd-order correction coils $2Z^2 - X^2 - Y^2$, *ZX*, *ZY*, $X^2 - Y^2$ and *XY*, which are well known in spectroscopy. The symmetry of the currents with respect to the plane XOY leads to a development of the magnetic field containing only even-ordered terms. In the case of the Helmholtz coil, the component B_Z is corrected by elimination of the second-order terms. The distance between the planes of the two circular coils traversed by identical currents is theoretically 1.0002 times the radius Rc. In the case of a Maxwell coil, the antisymmetry with respect to the plane XOY conserves only odd-order terms. The realization of a gradient of the component B_Z in the sense of the axis Z with elimination of the 3rd-order terms imposes a theoretical distance of 1.7325 Rc between the planes of the coils (Figure 2.32).

Hence, when manufacturing magnets for MRI, one tries to eliminate the even-order terms at least up to $\ell = 6$. The uniformity obtained with full-body imaging magnets, where the constraints are considerably lesser than with a physico-chemical NMR device, does not mean we cannot perform proton spectroscopy in voxels of a few cubic centimeters (see section 2.6.3) with a linewidth at half-height of 15 Hz at 64 MHz (1.5-tesla field).

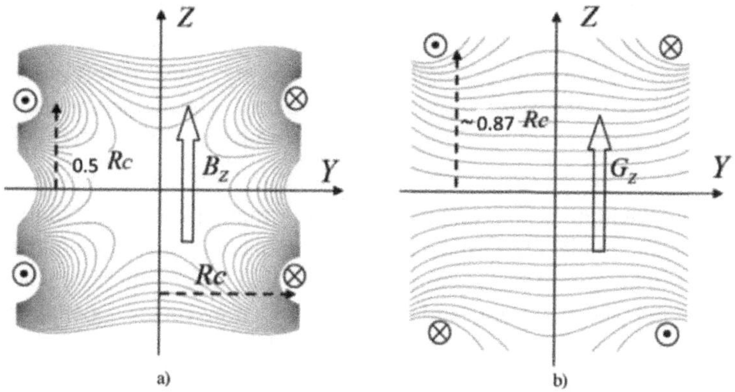

Figure 2.32. *Models of magnetic fields created by circular currents of radius Rc. a) Helmholtz coil: the relative variation in the value of the component B_Z between two level curves is 1.3%. b) Gradient G_Z obtained with a Maxwell coil: the difference in value of B_Z between two successive level curves is constant*

2.3.3.3. *Magnetic field gradients*

2.3.3.3.1. Structure of gradient devices

The first-order terms, $\ell = 1$, indicate $2\ell + 1 = 3$ possibilities which correspond to the three gradients of the component B_Z. These gradients are oriented along the main directions of the frame (R). The same is true for any components B_X, B_Y which obey the same laws as B_Z. Consequently, the set of the gradients of the three components B_Z, B_X, B_Y is represented by the following 2nd-order tensor T_G:

$$T_G = \begin{pmatrix} G_{XX} & G_{XY} & G_{XZ} \\ G_{YX} & G_{YY} & G_{YZ} \\ G_{ZX} & G_{ZY} & G_{ZZ} \end{pmatrix} \qquad [2.26]$$

The three gradients which come into play in MRI, G_X, G_Y, G_Z, are the terms in the last column in expression [2.26], relating to the component B_Z. The property

$\overrightarrow{rot}\vec{B} = 0$, which is valid in the free volume of the magnet and more specifically within the gradient coils, means that the gradient of a component along a direction orthogonal to it causes the appearance of a gradient equal to the component of the field which has that direction, along the direction of the first component. Consequently, the use of the gradients G_X, G_Y of the component B_Z, directed respectively along OX and OY, causes the appearance of gradients directed along OZ in the components B_X, B_Y (the so-called Maxwell terms). Their presence influences the spatial structure of the gradients field. This effect, which is sufficiently slight to typically be neglected in imaging, must nevertheless be taken into account when using very high gradient values. The device which creates a gradient G_Z corresponds to antisymmetrical distributions of the currents in relation to the plane XOY. As we can see in Figure 2.32(b), a coil with gradient G_Z creates a magnetic field which, when superposed on the main field, is an additional field whose value in the plane OXY must be zero. It must vary linearly along OZ in the vicinity of O. The cylindrical symmetry of the coil means there are no terms of degree m in the development and no spatially transverse dependence of the associated field. In the case of the transverse gradients G_X and G_Y, the coils G_X must present antisymmetry in relation to YOZ and in relation to ZOX respectively. The two systems may be similar, but shifted by a 90° rotation around OZ. We use the property $\overrightarrow{rot}\vec{B} = 0$, in fact creating gradients along OZ of the component B_X, which gives us G_X, and with the component B_Y, which yields G_Y. The realization can be done with coils which stick closely to a cylindrical form introduced into the bore of a cryogenic magnet. More generally, regardless of the geometry of the air gap in the magnet, we can create transverse gradients with layers of current. The principle of this approach is illustrated in Figure 2.33.

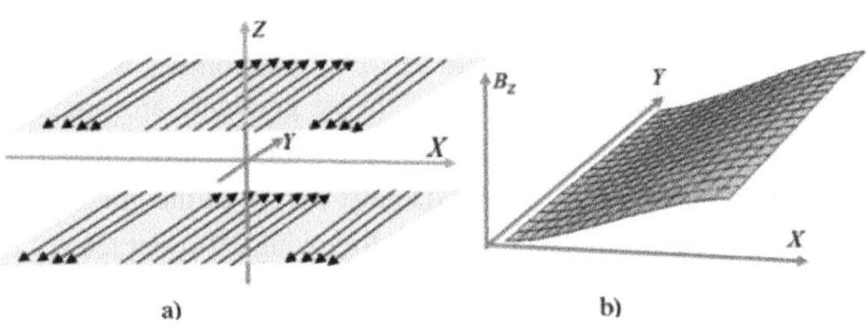

Figure 2.33. *Principle of the creation of a gradient* $G_X = \dfrac{dB_Z}{dX}$ *using layers of current: a) idealized distribution of the currents along rectilinear conductors which form a closed system far from the origin of the system of axes shown. b) The result corresponding to the calculation of* B_Z *in the plane* XOY*. It is also possible to use conductors arranged on the surface of cylindrical shapes in order to obtain the transverse gradients*

2.3.3.3.2. Dynamic functioning of gradients

During the excitation/acquisition sequences, the gradients are applied in pulsed mode with durations of application of a few milliseconds, while the flip durations remain around a millisecond. The idealized rectangular schemas used in section 2.2 are of value only in principle, not in practice. Power supply devices have to efficiently control the waveform gradients and provide sufficient energy for fast switching. Indeed, the brevity of the flips means that substantial electrical power is required for systems with very high output bandwidths. One constraint stems from the fact that the conductors forming the gradient coils must be sufficiently far away for the useable volume where the subject is placed so that the conditions of spatial linearity of the pulsed gradient fields are satisfied. This leads to a rise in the values of the coefficients of self-inductance of the gradient coils, and consequently slows down the installation of the current in those conductors. Moreover, bringing the currents of the gradients closer to the metal structure of the magnet strengthens the eddy currents, which in turn disrupt the effectiveness of these gradients.

The magnetic energy stored by the coil of a gradient G_U $(U = X, Y, Z)$ can be evaluated, e.g. in the case of Z:

$$E_{GZ} = \frac{1}{2\mu_0} \int_{V_{GZ}} (G_Z \cdot Z)^2 dZdXdY \qquad [2.27]$$

where μ_0 is the permeability of air, taken to be equal to that of a vacuum, and V_{GZ} is the volume covered by the coil.

The calculation shows that this magnetic energy is proportional to the product of the square of the value of the gradient by the power 5 of the dimension of coil, schematically assimilated to a cube. The power required to install the gradient G_Z is proportional to the ratio of E_{GZ} by the "rise time" of the gradient of value 0 to the value occupied by G_Z. This value will in fact depend on the type of switching envisaged, which requires a certain range of power availability on the part of the supply. We can fix the order of magnitude of that range when we know the value of the gradient created for a given current *I*. This current depends on the coil and also on its geometry. Its order of magnitude is generally several tens of amperes for a gradient of 1 milli-tesla per meter ($1 mTm^{-1}$). Let us take $25 A/(mTm^{-1})$ and suppose that the highest gradient is $15\ mTm^{-1}$, the value obtained starting from zero in a rise time t_r of 1 ms with a self-inductance L of the coil equal to $250\ \mu\text{H}$. We can show that the power necessary to install this gradient is ($\frac{L.I^2}{t_r}$), which gives us a value a little over 35 kW. This result gives a certain idea of the amount of power that must be made available to feed the gradients, with this electrical power being required by the switching operation alone. While the gradient is maintained, the energy expenditure is comparatively slight, as the resistance of the conductors is only a fraction of an ohm. The eddy currents induced in the conductive frame of the magnet during the switching unfortunately create a distribution which decreases the gradient applied, introducing undesirable harmonics which persist beyond the switching time. One of the effective remedies is to create active screening, by surrounding the gradient coil with a set of conductors acting in the opposite sense to that of the gradient currents. From outside this setup, the effects of switching are hardly – if at all – observed, and inside, the desired form for the resulting gradient is preserved, at the cost of reduced amplitude and cramped space.

In order to represent the gradient pulses, it is useful to approximate the real form of these pulses. In practice, the ratios between the durations of applications of the gradients used to describe the sequences must therefore be replaced by ratios of areas beneath the representative curve of those gradients. Indeed, in order to take account of the temporal fluctuation of a gradient $\overrightarrow{G(t)}$ during its application, we can look at the phase variation undergone by a transverse magnetization placed at the coordinate point $\vec{r}$, between two times t_1 and t_2, generally given by relation [2.28]:

$$\varphi(t_1) - \varphi(t_2) = \int_{t_1}^{t_2} \gamma \overrightarrow{G(t)} \cdot \vec{r} dt \qquad [2.28]$$

This relation alone illustrates both the importance of being able to control the gradient pulses and the difficulty of that task, which becomes necessary in order to avoid artifacts appearing in the images.

2.3.4. *A typical MRI installation in a clinical environment*

As we have just seen, the exploitation of the physical principles of NMR to create an image of the human body requires heavy machinery and complex techniques.

The installation of an MRI system in a clinical environment will be subject to both technical and practical constraints, with a view to optimizing the quality of the examination, on the one hand, and the safety of the patient and the people working in that environment, on the other.

2.3.4.1. *General view of an installation*

In the diagram in Figure 2.34, we see three "levels", which correspond to the three types of interventions practiced in a laboratory or on a clinical site.

Geographically, the installations of the technical premises are isolated from the examination room, which itself is independent of the place where the results are exploited. This organization is echoed in research laboratories using advanced nuclear magnetic resonance devices.

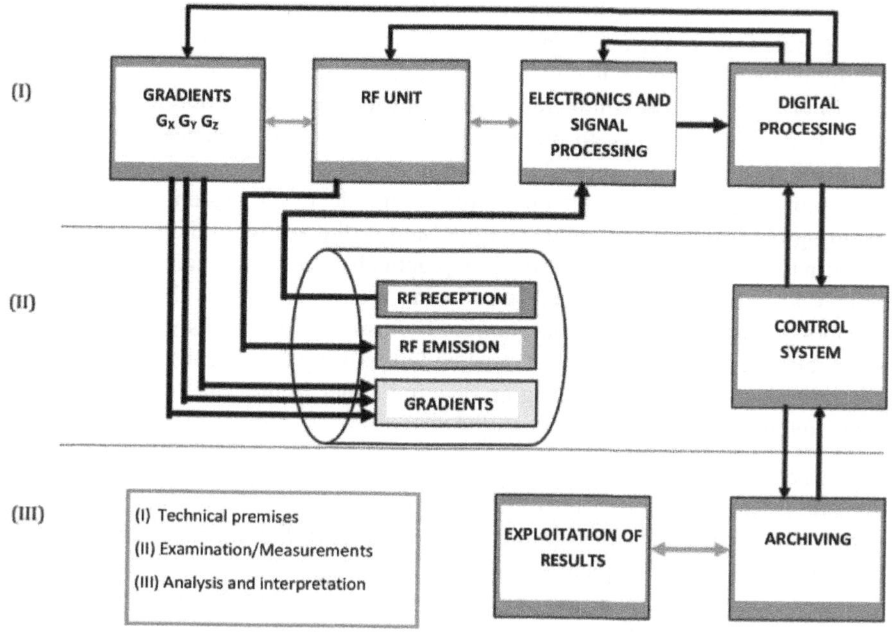

Figure 2.34. *Diagram of the functional structure of an MRI installation appropriate for a research site or a clinical site, independently of the part specific to the preparation and handling of the subjects, and to the applications*

2.3.4.2. *Implantation of a clinical MRI installation*

Depending on its use, which may be devoted to clinical practice or include a greater or lesser amount of clinical research, the site of the installation or its degree of integration in the imaging department, the architectural makeup of an MRI unit may vary slightly from one site to another. In general, it will include the following installations, which can be grouped into three sectors:

1) Rooms for the patient's use, such as the waiting room, cloakrooms, toilets, patient interview room and a care room, equipped with medical fluids and a bed, for preparation (drip feeding of a vein for the injection of the contrast agent), or recovery of the patient after anesthesia.

2) Rooms devoted to the whole of the MRI system, including the examination room, the control room and the technical booth.

3) Rooms for the personnel, both paramedical and medical, including the reception and secretaries' offices, the reading room and the medical office, and the image processing room, for research, which is attached to the computer room for the storage and archiving of the data.

4) At the heart of the MRI unit is the MRI system, all of the elements of which are distributed in three adjoining rooms. The central room is the exam room, containing the magnet and all the amagnetic materials associated with the acquisition. This room is completely sealed, for technical and safety reasons. 1) It is built within a Faraday cage to prevent any external RF disturbances; 2) it must be kept at a constant temperature (23°C) and humidity rate (60%) by a dual-flow air-conditioning system for proper operation of the electronic systems and gradients, preventing the formation of arc discharges; and 3) the access to the magnet must be secured and under constant control. For practical reasons and safety reasons, it is desirable to keep the 0.5 mT field lines inside the same room as the magnet, either by adapting the dimensions of that room or by insulating the walls with iron sheaths. Finally, outside access to this room must be left to bring the magnet in, at the time of the building's construction, leaving an opening either in the wall or the roof, which is later covered over, insulated and sealed.

5) Adjacent to, and often on the side of this room, is the technical booth, which houses all the technical elements of the system, often divided into three cabinets: the magnet control cabinet, the gradients cabinet, and the cabinet containing the electronics for the RF sequence and the computing equipment. This space also contains the electrical inputs and cold water inlet; this cold water is used to cool the "coldhead" of the magnet (cryogenerator) and the gradient cabinets.

This room, which also needs to be kept at the constant temperature of 18°C, is connected to the magnet room through a "filter plate", which channels all the connections between the electrical cabinets and the magnet. During acquisition, other devices may be needed to monitor the patient's physiological signs, or to automatically inject contrast products. As they are placed in proximity to the magnet, these devices need to be amagnetic, and must be protected against the disruption that the magnetic fields could cause to their electronics. In the case of an exam under anesthesia, an amagnetic respirator and medical fluids (oxygen, nitrogen, medical air, medical vacuum) will also be needed, in proximity to the magnet. Finally, with the gradual integration of functional MRI into neuroradiology, the room must also be equipped with systems for visual stimulation (projection of images onto a screen or onto special glasses), audio stimulation (headphones adapted to the magnetic field) and systems for recording the subjects' manual or vocal responses. While of course, these electronic devices must be amagnetic, it must also be ensured that they do not induce a parasitic signal which would case artifacts to appear in the image.

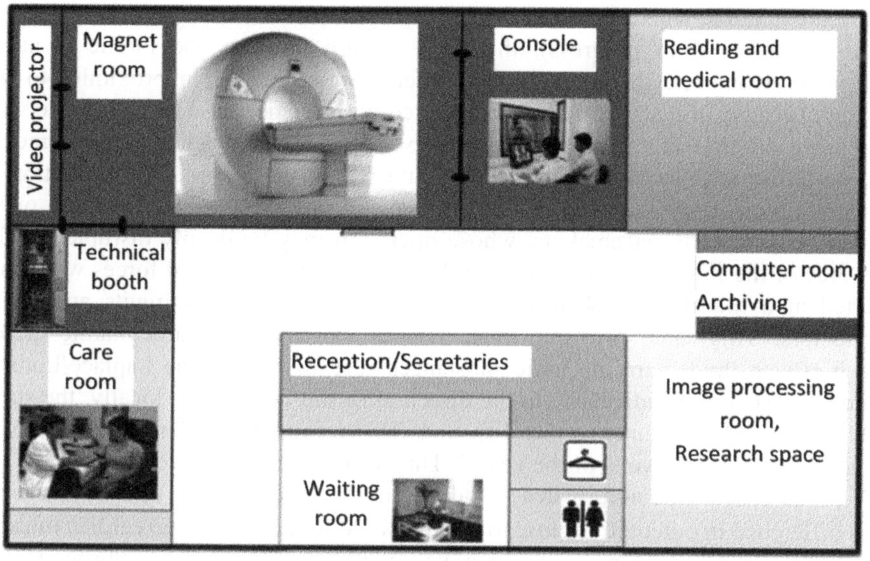

Figure 2.35. *Diagram of a clinical MRI installation and its environment*

2.3.5. *Operation and safety*

The presence of magnetic fields significantly stronger than the Earth's magnetic field, covering relatively large areas of space in places frequented by the public imposes very strong constraints in terms, firstly, of the patient's safety; secondly, of the safety of those working near to the installation; and finally of the protection of the machinery itself. These constraints have become severe because of the emission of an RF field and the gradient switching operations.

2.3.5.1. *Exposure to static magnetic fields*

The presence of the intense magnetic field produced by the superconductor magnet in the persistent mode necessitates very stringent safety precautions. This magnetic field could be a danger to the patient (or to any other person present in the exam room). The elementary precautions relate to the protection of people and goods against the effects of the field leakage which exists in the vicinity of a magnet. It must be remembered that, under the influence of the field, all ferromagnetic items acquire an induced magnetic moment, on which the action of the field results in a force proportional to the local gradient (a force which pulls the object towards the magnet, in regions where the field is most intense) and a torque which tends to align the object's magnetic moment in the direction of the field. This may result in twisting, which will have a mechanically destructive effect. Note that inside the magnet, where the field is perfectly uniform (there is no gradient at all if no examination is in progress), the force is null whereas the torque is at its maximum. However, the force becomes very strong indeed in proximity to the magnet, because of a static gradient associated with the decrease in field value with increasing distance from the magnet. The existence of these actions simply means that no ferromagnetic objects must be brought within that perimeter. In addition to this precaution, there is the *caveat* of not keeping any electronic devices on one's person – particularly pacemakers, whose operation may be deeply disrupted in the presence of the field. What is happening here is the intervention of forces which are exerted at a microscopic scale on the charge carriers in the circuits and semi-conductors. This same principle gives rise to the magnetohydrodynamic effect, which affects fluids carrying ionic substances such as blood. The Laplace/Lorentz force is exerted perpendicularly to the direction of the flow when, locally, the static field is oriented differently to that direction. This results in a difference in potential between the opposite walls of the vessel. This difference is estimated at a few dozen mV in the case of the aorta when blood is running perpendicular to a 3-tesla field. This difference in potential is close to the threshold to depolarize the cardiac muscle. However, to our knowledge, no problem has ever been encountered in this area.

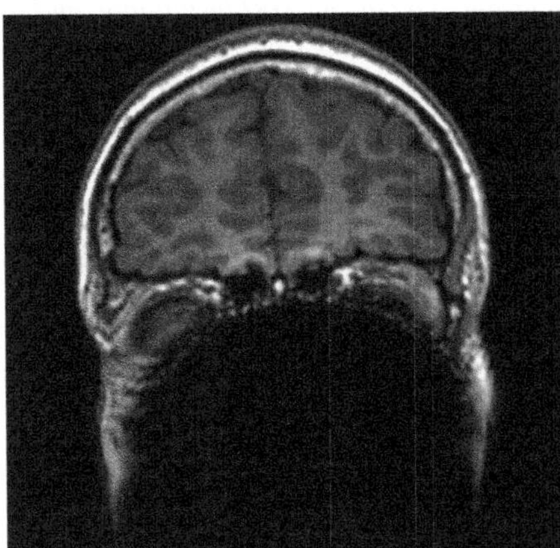

Figure 2.36. *The operation of the imager may be unexpectedly impeded, as shown by this MRI image of a patient with dental braces, with materials able to obscure the image around them. Other magnetic elements may be used in the composition of numerous products, such as makeup, hair gels, clothes, tattoos, etc., and cause problems in the image. Image provided by D. Sappey-Marinier, Cermep, Creatis*

Thus, the magnet room is a space to which access must be very closely controlled, and continuously guarded by competent personnel. Hence, anyone entering this environment must either be accompanied or receive prior training.

Systematic questions should be put to the patient to verify that he presents no contraindication with respect to strong magnetic fields. If it is necessary to bring a device into the exam room, it must be verified that it is truly compatible with operating in proximity to the magnet. It is good practice to explicitly quote the value of the field by labeling on the floor, and to constantly reinforce users' awareness of its presence during their practical training. Public access to the MRI machine is therefore generally prohibited within the volume wherein the field is greater than 0.5 millitesla– a low value, which is only ten times that of the Earth's field.

Although in the long term, no particular problems are observed in people regularly exposed to static magnetic fields, legislation (usually specific to each country) tends to be in place to regulate these situations, for precautionary reasons. However, it is not the aim of this exposé to give a detailed discussion of these legislative precautions, and so as not to get lost in the jungle of the very diverse regulations, we shall content ourselves with mentioning the recommendations found on the following sites:

http://europa.eu/legislation_summaries/public_health/health_determinants_environ ment/c11150_en.htm http://www.greenfacts.org/en/static-fields/index.htm

2.3.5.2. *Effects of RF fields*

Exposure to an RF field can cause heating of the tissue after a prolonged accumulated period of RF excitation. In view of the importance of this issue, it is helpful to give a brief analysis of the mechanism of "energy deposition" in living tissues subjected to RF excitation. Earlier we pointed out how the conductivity of the subjects played a part in the losses of RF energy. It is certainly true that these losses are by far the highest during the application of RF pulses and that they may therefore become harmful when they are repeated in the course of a complete imaging sequence. The normal regulations aim to prevent a heating effect resulting in a 1°C increase in local temperature. Clinical devices are designed to evaluate and therefore prevent this risk, based on the patient's physiological data – primarily his weight – and on the RF power emitted during a given sequence of a given duration.

We then define the specific absorption rate (SAR) as being the RF power absorbed per unit tissue mass during an observation. This observation corresponds to the application of a complete sequence of duration T_S, during which the total duration of exposure to RF excitations is T_{RF}.

By way of example, we can consider a cylindrical sample whose base surface has the radius r, of height h, with an axis along the direction of the field $\vec{B_1}$ (rectilinear polarization). The instantaneous power dissipated per unit volume in that element is therefore $p = \sigma \cdot E^2$, where $\vec{E}$, taken to be approximately uniform, is the electrical field associated with the field $\overrightarrow{B_1}$, such that $\overrightarrow{rot(\vec{E})} = -\frac{\partial \overrightarrow{B_1}}{\partial t}$. In an AC regime, the power absorbed by the sample is therefore $p_t = \frac{1}{2} \cdot \sigma \cdot E^2 \cdot S \cdot h$. If we take the circulation of $\vec{E}$ around the cylinder such that $2\pi \cdot r \cdot E = \pi \cdot r^2 \cdot \frac{\partial B_1}{\partial t}$ in terms of absolute value, we obtain a total power per unit mass of the sample equal to $\frac{p_t}{\rho . S . h} = \frac{1}{2} \cdot \sigma \frac{\pi^2 \cdot r^2 \cdot f^2 \cdot B_1^2}{\rho}$, where ρ is the density of the sample, σ its conductivity and f the frequency of the B_1 field. In order to obtain the expression of the SAR corresponding to the sequence in question, we need to take account of the total observation time and the time for which the RF field is being applied, so that finally:

$$SAR = \frac{1}{2} \cdot \sigma \frac{\pi^2 \cdot r^2 \cdot f^2 \cdot B_1^2}{\rho} \cdot \frac{T_{RF}}{T_S} \qquad [2.29]$$

The increase in temperature of this volume of a sample subjected to RF excitation will be approximately $\Delta T \approx \frac{1}{c_{th}} \cdot SAR \cdot T_S$, where c_{th} is the thermal capacity of the element under consideration, which is itself proportional to its volume.

In humans, the standardized indications relating to exposure to RF fields are highly variable depending on where that exposure takes place (the head, the torso, the extremities, the whole body), and of course from country to country. We refer the reader to the documentation relating to the dangers posed by electromagnetic fields (see the Websites mentioned above).

The use of powerful RF waves can lead to other risks, such as localized burns in the presence of conductors for monitoring (such as ECG wires) that are incorrectly positioned over the patient inside the magnet. As a general rule, MRI-compatible monitoring systems have connectors made of low-conductive material (such as carbon), as short as possible and which do not form loops, which can cause "hot points".

2.3.5.3. *Gradient switching*

Electromagnetic induction phenomena also come into play during gradient switching. These phenomena are heightened when the three gradients are applied simultaneously during the selection of an "oblique" slice perpendicular to the direction of the resulting gradient. The important thing is not to locally surpass too high a value of current density induced by rapid application of the gradient. This can occur with an observation field occupying regions far from the center of the gradient coils, which is also the center of the magnet. We can imagine a subject as a cylindrical form with radius r, a section of which receives flux variation $\pi r^2 \cdot \frac{dB}{dt} = 2\pi \cdot r \cdot E$ with the value of current density on the disc with radius r is $j = \sigma \cdot E = \frac{1}{2} \cdot \sigma \cdot r \cdot \frac{dB}{dt}$. If we place ourselves at a distance from the center such that the gradient creates an additional field on 5 mT at that site (around 33 cm from the center for a 15 mTm^{-1} gradient) the term $\frac{dB}{dt}$ is 5×10^{-3} times the inverse of the time taken for the gradient to arise. With a conductivity σ of approximately 0.4 S/m (muscle) and a radius r of 10 cm, we obtain the following current densities: $j \approx 0.1 A/m^2$ with a switching time of 1 ms, so $1\ A/m^2$ if the gradient supply is sufficiently powerful to perform the switching in $100\ \mu s$. In the second case, the current density induced is too high with respect to the different thresholds for electro-physiological stimulations, and corresponds to too high a "slew rate" (speed of variation of the gradient). Although this model is very approximate, we can also highlight the influence of the transverse dimensions of the subject; these dimensions are represented here by the parameter r.

Another problem encountered with static field gradients relates to the noise created by electromagnetic forces which act on the conductors forming the gradient coils, by tending to deform their support rather violently during the current application. This causes a very loud noise, which is commonly accepted to reach a maximum level of $100\ dBa$ (pain threshold at $120\ dBa$). This may require anti-noise devices to be worn.

2.3.5.4. *Testing conditions*

If the patient is not claustrophobic and presents no contraindications, his cooperation – and particularly his ability to stay still during the exam – will have a positive effect on the quality of the images. The precautions taken make the installation comfortable and calm, and aid in the success of the examination. The receiving antenna appropriate for the anatomical region being explored is placed around or over the patient. The region of observation is automatically placed at the center of the magnet by moving the bed. While the patient is being moved into the magnet, the operator stays in contact with him, before leaving him alone in the exam room. Subsequent contact, which takes place *via* a sound link between the control room and the inside of the magnet, is supplemented by video-surveillance. In the case of a patient in a particular situation (sedation, anesthesia, cardiac trouble, etc.), his physiological parameters (ECG, oxygen tension, etc.) are constantly monitored from the control room. If necessary, a contrast agent is administered by an automatic, amagnetic injector installed near the magnet, with a catheter leading from the syringe to the patient's arm vein. The triggering and control of the injection are done remotely, from the acquisition console. All precautions are taken in order to be able to respond instantly to the slightest problem.

2.4. Image properties

2.4.1. *Introduction*

The recent development in terms of instrumentation, with access to intense magnetic fields and greater use of digital processes in signal recording, has significantly marked the progress of the technique of Nuclear Magnetic Resonance for Imaging and Spectroscopy. From this point of view, MRI is very interesting, in that this approach has a great capacity to create fairly numerous types of contrast using essentially only one probe molecule: the water present in living tissues. This is certainly the most noteworthy characteristic, and therefore a place is reserved in this section for a discussion of the contrast. We saw earlier that the background noise affected the sensitivity at the time of detection. It also has the same penalizing influence on the contrast. First, we need to establish the link between image formation and the instrumental and experimental conditions. For this purpose, it is not necessary to go into detail about the sequences used, or about the various strategies combining observation sequences and modes of reconstruction. The correspondence between the measuring space (k space) and the image space is the crucial element. Having demonstrated that these spaces are linked by the Fourier transform, we shall not deviate from this easy-to-use concept. For this purpose, the notions of sampling may be used, in addition to the notions discussed in the previous section. Thus, we shall begin by examining the notion of field of view, followed by questions relating to spatial resolution before discussing contrasts.

2.4.2. *Field of view*

The field of view may be thought of as the region where it is possible to perform observations inside the magnet. It defines an available useable space, which is subdivided into smaller volumes of interest. As the devices creating the gradient fields are placed inside the magnet, the spatial properties of those devices first need to be taken into account.

2.4.2.1. *Field of view, gradients and RF coverage*

The resonance frequency per unit magnetic field of protons is 42.57 MHz per tesla, so a gradient of 1 milli-tesla per meter, used for readings, stretches the frequencies at a rate of 42.57 kHz per meter. The value of the gradients, and the domain of linearity of the magnetic fields which create them, determine the ranges of frequency and the spatial range which can be explored in each of the three directions of application of those gradients. Care is taken not to exceed the bounds of the frame defined by the region where the gradients are constant (Figure 2.37). In a two-dimensional model, with the subscript 1 being associated with the direction of reading with the gradient G_1 and the subscript 2 assigned to the direction of phase encoding G_2 in an orthogonal direction, we obtain an image formed of $N_1 \times N_2$ pixels by a Fourier transform. When N_1 and N_2 are greater than 1 (i.e. so that $N_p - 1 \sim N_p$, $p = 1,2$) the sampling rate is identical in both directions in the reciprocal space if we impose the following condition:

$$G_1 \times t_1 = N_2 \times \Delta G_2 \times t_2$$

where ΔG_2 is the step of the phase-encoding gradient, with t_1 being the duration of measurement of the signal in the presence of G_1 and t_2 that of the application of G_2. It is therefore useful to bear these data in mind in order to avoid subsidence or stretching of the image and not spill over into the region where the gradients are no longer considered constant.

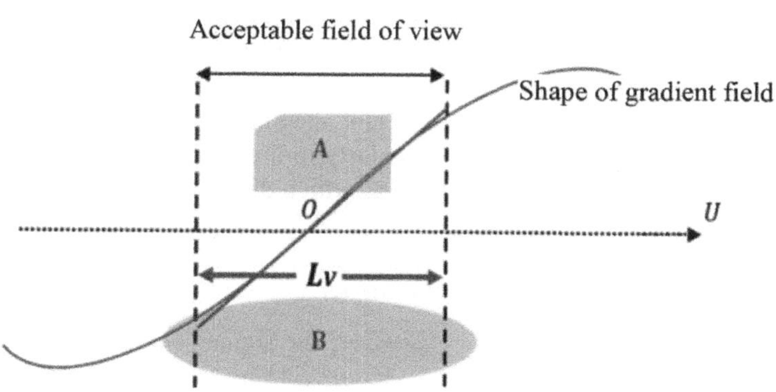

Figure 2.37. *Definition of a 1D field of view L_v delimited by a gradient (here represented with a slight tolerance over linearity) oriented in the sense $\overrightarrow{OU}$. A magnetic field presenting a value increasing linearly along OU in its usable part here defines the Reading Gradient. The nonlinearity, which does not affect object **A**, deforms the extremities of the image of object **B**. This nonlinearity comes into play when the observation moves away from the central region of the gradient coils*

A gradient created in a direction OV perpendicular to OU, used to encode the phase, may not necessarily share the same properties of linearity. Therefore, the acceptable field of view in that direction may not extend as far as in the direction OU. Note, finally, that the spectral bandwidth will be imposed by the highest value of the gradient usable in the direction corresponding to the largest dimension of the sample.

For simplicity's sake, let us stick with the 1D model from Figure 2.37, with the length L_v being determined by the properties of linearity of the gradient field (case of object **A**), we consider the spatial extents occupied by the RF fields LRF_{exc} and LRF_{det}, for excitation and detection respectively, along the direction of the gradient. The sample has the length L_{sam}. Of the relationships which may exist between these various lengths, we can look at three cases in particular:

$LRF_{exc} \approx LRF_{det} > L_v \geq L_{sam}$: situation corresponding to conventional imaging,

$LRF_{det} > L_v \geq L_{sam} > LRF_{exc}$: only part of the sample is subjected to RF excitation; the response comes mainly from that part, but with spatial selectivity of excitation, weak signals will be received on the boundaries of the field defined by L_v.

$LRF_{exc} > L_v \geq L_{sam} > LRF_{det}$: this third case could correspond to conventional imaging using a single receiving antenna to observe a limited region of the sample. Here, there is a risk of recording signals from the peripheral regions, which we may not necessarily want to measure. There is a second option, with the use of phased-array receiving coils, each of which has a nearby region of accentuated sensitivity, which diminishes as we move towards the neighboring coil. This arrangement expands the overall field of view – particularly in the case of fast imaging. Finally, we can always adjust the width of the filter, which delimits the frequential range of the signal.

2.4.2.2. *Field of view and sampling of the signal*

Once the condition $L_v \geq L_{sam}$ is satisfied, the signals coming from the sample occupy a range $\Delta F_{GU} = \frac{\gamma G_U}{2\pi} \cdot L_{sam}$, which is usually centered on the frequency zero using the complex description of the recorded signal. The Nyquist–Shannon condition, which helps prevent aliasing of the periodic patterns in the frequential domain (in this case 1D profiles) following sampling at the frequency Fe, necessitates that $Fe > \Delta F_{GU}$.

Owing to this, we need to choose the sampling rate so that there is no overlap in $n\times\frac{1}{\delta k_U}$ or in $n\times\frac{1}{\delta k_U}$.

This inequality becomes an equality in the extreme case where $L^{(U)}_{\text{samMax}}=\frac{2\pi Fe}{\gamma G_U}$, the maximum dimension of the sample in direction U. Consequently, the additional constraint $L_{\text{sam}}<L^{(U)}_{\text{samMax}}$ needs to be imposed. Expressed in the reciprocal space, where the general term $k=\frac{\gamma Gt}{2\pi}$ gives a sampling rate of $\delta k_U=\frac{\gamma\cdot G_U}{2\pi\cdot Fe}$, the non-aliasing condition becomes $\frac{1}{\delta k_U}\geq L^{(U)}_{\text{samMax}}$.

For a 2D image, this condition corresponds to the direction of the reading gradient; it is transposed directly onto the direction of phase-encoding in the direction V in the form $\frac{1}{\delta k_V}\geq L^{(V)}_{\text{samMax}}$, where $L^{(V)}_{\text{samMax}}$ represents the limit of the field of view in that direction and, using the same type of notation for the phase-encoding gradient applied for the duration t_Φ, we would have $\delta k_V=\frac{\gamma\cdot\Delta G_V\cdot t_\Phi}{2\pi}$ (Figure 2.38).

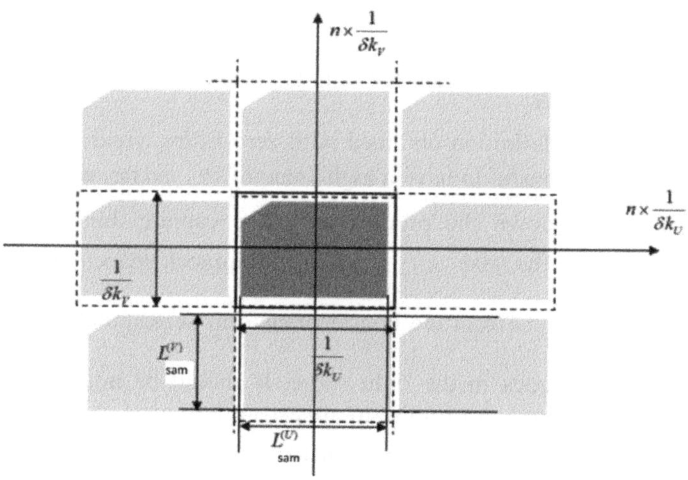

Figure 2.38. *2D imaging using a discrete Fourier transform: periodization of the image because of the sampling performed in the reciprocal space and limitation of the field of view. The phase-encoding gradient is directed vertically (V) and the reading gradient horizontally (U)*

2.4.3. *Spatial resolution*

2.4.3.1. *Limit of spatial resolution and digital resolution*

The limit of spatial resolution corresponds to the smallest possible observable detail. Thus, it is the smallest distance which corresponds to two distinct and independent image signals. It can be evaluated in the image by taking the length of the side of a pixel in a given direction as a unit of length. To simplify the process, it is preferable to limit ourselves to a single dimension, e.g. U . Knowing the field of view $U_{\max} = \dfrac{1}{\delta k_U}$ and the definition of the 1D image, which is the total number of pixels N_{tp} covering the length $U_{\max}$ in the direction U , it is possible to measure distances using the side of the pixels and the scaling factor given by the ratio $\dfrac{U_{\max}}{N_{\text{tp}}}$. This element does not correspond to the limit of spatial resolution defined above.

– The first reason is that the reconstruction of the image by the Fourier transform may have been performed after we have added a series of zero values to the list of measured values. In this case, $N_{\text{tp}} = N_{\text{p}} + N_{\text{zero}}$ where N_{p} is the number of points of measurement carried out in the reciprocal space and N_{zero} the number of zeros added to the list of transformed values. This technique is interesting, as it improves image definition (see section 2.5.3.2) and gives an appreciably better visual perception, but it is not a way of improving the power of spatial resolution.

– The ratio $\delta U = \dfrac{U_{\max}}{N_{\text{p}}}$ defines the "digital resolution" of the image. For the same reasons that the definition obtained with zero-filling yields an advantageous qualitative aspect, it is useful to have a high value of N_{p}, as far as possible. Indeed, increasing N_{p} slows down the process of image capture, but we can use the technique to increase the size $k_{U\max}$ of the reciprocal space, and the relation $\delta U = \dfrac{1}{k_{U\max}}$ which results from the Fourier transform correspondence between the two reciprocal spaces goes in the right sense. It should be noted that increasing $k_{U\max}$ causes a larger amount of noise to appear in the image, and that conversely, choosing extended pixels causes so-called "partial volume" effects which correspond to the fact that a pixel contains overlapping signals from two different tissues.

When we look at the so-called "point spread function (PSF)" (see section 2.4.3.2, immediately below), we can see that the limit obtained still remains greater than the "digital resolution", which should be chosen to be sufficiently low, as illustrated by Figure 2.39.

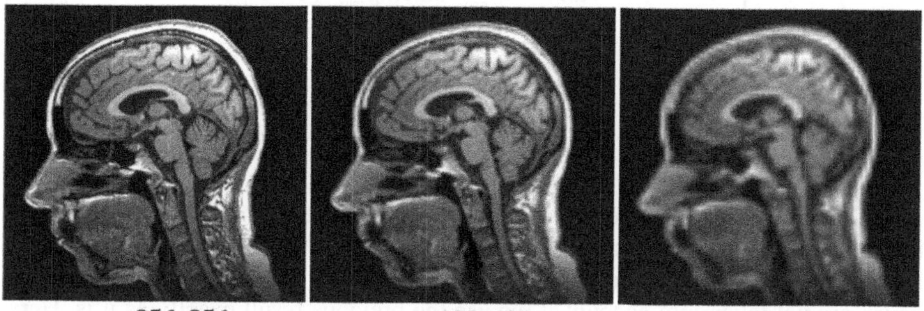

Figure 2.39. *Influence of definition (digital resolution) on detail perception. Initial image obtained in 256 × 256 points (see Figure 2.20) and then reformed in 128 × 128 and 64 × 64 points after reconstruction*

2.4.3.2. *Point spread function (PSF)*

The spread function of a point belonging to the object conditions the distribution, in the image, of the information attached to that point. This information corresponds to the value attributed to that point in the object – e.g. the value of the magnetization in a very small volume of matter containing resonating nuclei assigned to that point. Staying with the 1D example, let us take a point of unit value placed at the abscissa U_0. Using relations [2.16] and [2.17], $s(k_U) = \int_{-\infty}^{+\infty} \rho(U) \cdot e^{-j2\pi k_U U} dU \Longleftrightarrow \rho(U) = \int_{-\infty}^{+\infty} s(k_U) \cdot e^{j2\pi k_U U} dk_U$, thus giving the expression of the signal in the reciprocal space, so that for the point positioned at U_0:

$$s(k_U) = \int_{-\infty}^{+\infty} \delta(U\text{-}U_0) \cdot e^{-j2\pi \cdot k_U \cdot U} dU \qquad [2.30]$$

The result of this sum is simply $e^{-j2\pi \cdot k_U \cdot U_0}$, whose inverse Fourier transform with respect to k_U gives the image of the point through a truncation window of width $k_{U\max}$, which is the "PSF" of that point, determined using the following calculation:

$$\rho(U) = \int_{-\frac{k_{U\max}}{2}}^{+\frac{k_{U\max}}{2}} e^{j2\pi \cdot k_U \cdot (U - U_0)} dk_U = k_{U\max} \cdot \mathrm{sinc}(k_{U\max} \cdot (U - U_0)) \qquad [2.31]$$

where $\mathrm{sinc}(k_{U\max} \cdot (U - U_0)) = \dfrac{\sin(\pi \cdot k_{U\max} \cdot (U - U_0))}{\pi \cdot k_{U\max} \cdot (U - U_0)}$.

The expression obtained in equation [2.31] is the 1D PSF. It is, *a priori*, infinitely wide, and therefore obviously larger than the field of view shown in an image. This leads to phenomena of aliasing related to the sampling – phenomena which are not taken into account if we use the previous result exactly as it is. Here, we shall limit ourselves to looking at the undulations, or the Gibbs phenomena, caused in the image by the truncation of the reciprocal space.

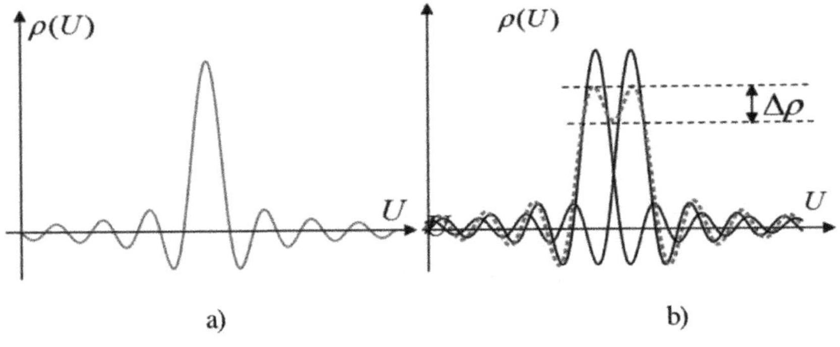

Figure 2.40. *1D model of PSF: a) representation of the PSF according to formula [2.31]; b) overlay of the responses from two neighboring points – real responses of the same amplitude – with the acceptation that the signals from these two points are properly in phase. The criterion of separation of these two points depends on the possibility of distinguishing the difference in level $\Delta\rho$ in relation to the value of the maxima of the resulting curve (dotted line): particularly in the presence of noise, which is not taken into account in this example*

In two dimensions, the PSF will depend firstly on the shape of the *k*-space used. If it is a rectangle (or more usually, a square) and particularly when the image is reconstructed using an inverse Fourier transform, the PSF will simply be given by the product of two functions $\rho(U)$ and $\rho(V)$, similar in type to formula [2.31]. An example is given in Figure 2.41(a) for a 24 × 24 image matrix. The choice of this low digital definition, which is significantly lower than that used in conventional imaging, is a situation encountered in metabolic imaging, which necessitates the use of fairly voluminous voxels (i.e. pixels multiplied by the thickness of the slice) for reasons of sensitivity.

If the coverage of the k space is circular in shape, in this domain we need to use the reciprocal values of the cylindrical coordinates (r,θ), so these coordinates would be (k_r, k_θ), such that, when k_0 determines the extent of the disc explored in the k space, the PSF becomes:

$$PSF(r,\theta) = \int_0^{2\pi}\int_0^{k_0} k_r \cdot e^{j2\pi k_r r.Cos(k_\theta - \theta)} dk_r dk_\theta \quad [2.32]$$

In view of the cylindrical symmetry depending only on r, the PSF at the center of the image is denoted $PSF(0)$, and the calculation yields:

$$PSF(0) = 2\pi k_0^2 \frac{J_1(2\pi k_0 r)}{2\pi k_0 r} \quad [2.33]$$

where J_1 is the 1st-order Bessel function of the first species. The corresponding 2D distribution is shown in Figure 2.41b.

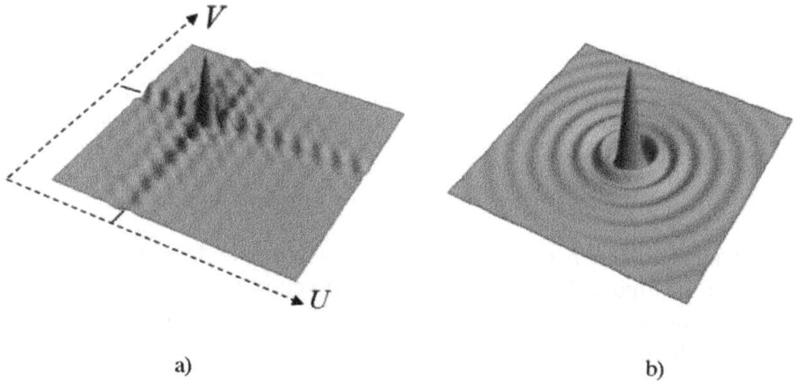

Figure 2.41. *3D representation of point spread functions. a) Truncation effect on the real image of a point with numerical coordinates (U=8, V=15) with a 24 × 24 Cartesian representation; simulation entirely analytic. b) 3D view of $PSF(0)$ (case of cylindrical symmetry) simulated at 256 × 256 computation points*

2.4.3.3. *Apodization of truncation effects*

In order to reduce the Gibbs phenomena, which result in oscillations of the signal around the position of the represented point, we know that, at the price of a loss in power of spatial resolution, it is necessary to soften the sharp boundaries which border the window used in the reciprocal space. Progressively reducing the "transparency" of that window is an operation wherein, *a priori*, the number of possibilities is limitless. The weighting windows used for generating the finite impulse response digital filters indeed correspond to the usage of this procedure. More usually, we can use a Gaussian or Lorentzian weighting window. This latter option is taken naturally but very moderately by way of the transverse relaxation which occurs during the recording of the temporal signal. Figure 2.42 illustrates the use of the Gaussian window, which gives only average performances but is easy to use. It could be replaced by an entirely different type of weighting window; on this point, we invite readers to refer to the existing literature on the subject.

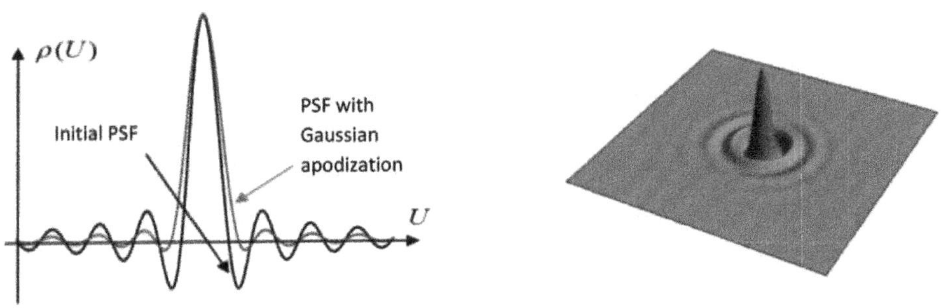

Figure 2.42. *Reduction of Gibbs phenomena by a Gaussian weighting function and subsequent loss of spatial resolution: the apodization stretches the point spread function (PSF) as shown by the 1D example (left), whilst also reducing its undulations. This property is also seen on the right with the 2D model of* $PSF(0)$ *used for the previous figure*

We still need to touch on how this strategy is applied, with reference to the example given in Figure 2.43. The Gibbs phenomenon only affects the fronts with sharp variation in amplitude of a signal, and this behavior has consequences for the presentation of the images.

In the example chosen, we have used the Fourier formula (relations [2.30] and [2.31]) to establish the expressions of the samples in the reciprocal space with a square-shaped uniform distribution of magnetization. Then, the square shape is simply relocated (with or without undulation effects) by using 2D Fourier transforms.

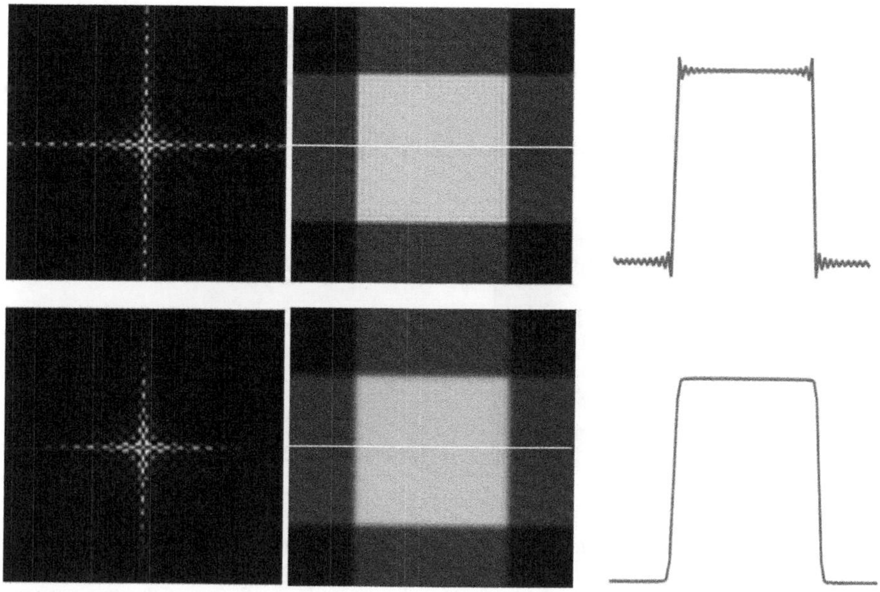

Figure 2.43. *Reduction of Gibbs phenomenon by apodization in the simulation of the image of an object with a square cross-section. Top, left to right: image of the samples calculated in the reciprocal space; image obtained by reconstruction using a 2D Fourier transform; profile corresponding to the plot shown in the image. Bottom: the elements are similar to the previous ones, but in this case, a Gaussian apodization is used. Note the slight fuzz which appears on the boundary of the square because of the limitation of the spatial frequencies by the apodization. 86 × 86-point matrix, images in absolute mode*

2.4.3.4. *Influence of noise on power of spatial resolution*

The above simulations can be repeated, taking account of the level of noise which appears in the images.

In particular whilst, for a given width of the reciprocal space, we can consider that it is possible to separate two neighboring points in the sense of the PSF, it is not certain that this maximum resolution will be maintained if the noise level is high.

To begin with, we can use an apodization which attenuates the Gibbs oscillations – we know that this stretches the spatial response whilst also reducing noise. Therefore, we need to strike a balance between the limitation of the noise and the loss of power of spatial resolution. Examples using a 1D model (a and b) and a 2D model (c and d) are given in Figure 2.44, where we can see the effect of the noise on the response of two neighboring PSFs, stretched identically by the use of a Gaussian apodization.

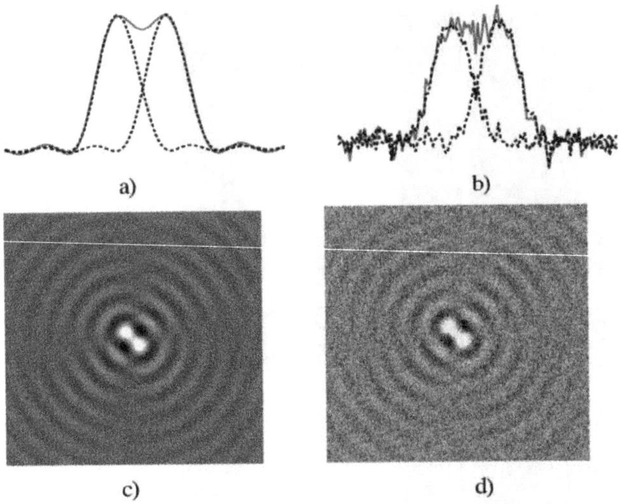

Figure 2.44. *a) Spread functions for two neighboring points, expanded by Gaussian apodization, limiting the undulations: we can consider that these two PSFs correspond to the boundary of separation in the absence of noise, because the summing of the two responses reveals an appreciable gap between them. b) When the noise level reaches the height of the gap, it becomes less and less easy to distinguish the two PSFs, and the power of spatial resolution drops noticeably. c) 2D representation when the noise level is again sufficiently slight to be able to separate the two PSFs. d) With a higher noise level, there is no longer as clear a separation between the two PSFs, although the shape of the response suggests that one exists. In addition, here, the noise causes an overall rise in the level of gray, which fades the representation and attenuates the undulations. The representations of the spread functions c) and d) are without apodization, and simulated in 128 × 128 matrices*

The reduction in power of spatial resolution under the influence of noise is common to all imaging methods, and was initially mentioned with nautical visual observations using optical instruments. This led to the Rose criterion, which stipulates that in order to identify an object, we need to have a sufficient ratio between the signal and the noise (the "signal-to-noise ratio" or SNR), which is several units in magnitude and is found by examining the contrast.

Reducing the spectral width of the noise by filtering may also lead to a loss in power of spatial resolution. It is the level of noise which is responsible for the loss of power of spatial resolution rather than its spectral distribution, such as that which could result from prior treatment. It should then be noted that the technique of zero-filling, which is often used in the reciprocal space, does not alter the noise level and therefore does not improve the resolution. Indeed, it simply performs an interpolation of the signal by giving points whose values are intercalated between those of the initial points of the image, giving us a smoothed representation. This increased definition eases visual perception and merely offers the sensation of improved power of spatial resolution, even in the presence of noise.

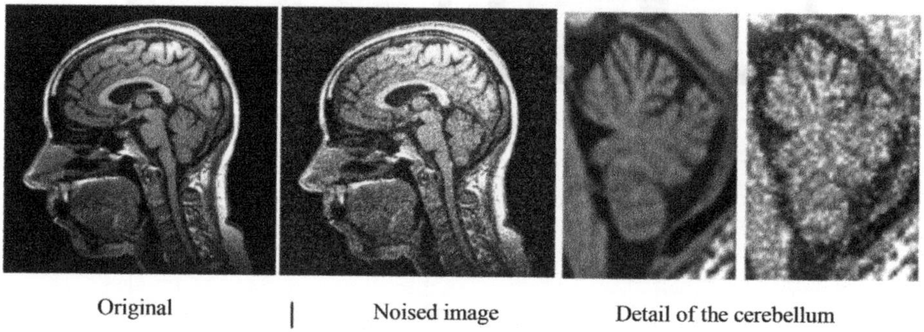

Figure 2.45. *Effect of noise on power of spatial resolution. The original image (represented in its absolute value – see Figure 2.20) has been modified by the superposition of "Rician" noise. The loss of power of resolution can be seen in particular on the zoomed images of the cerebellar region*

2.4.4. *Contrast and signal*

2.4.4.1. *Contrast and noise*

The contrast which exists between two neighboring regions within the image is obviously related to the difference in level of the image signal which exists between those elements. However, if the noise level is too high, we would not be able to easily tell the difference between those two regions. (Figure 2.46). It is for that reason that we use the notion of the SNR: for two neighboring regions of the image, represented by the subscripts *m* and *n*, the contrast-to-noise ratio between part *n* and part *m* is given by:

$$\left(\frac{C}{B}\right)_{nm} = \frac{I_n - I_m}{b} \qquad [2.34]$$

where the respective amplitudes of the image signals in n and m are I_n and I_m, and where b represents the mean square amplitude of the noise. This latter value usually has a Gaussian nature if the values of I_n and I_m are defined solely on the basis of the absorption signal (image formed from the real part of the values assigned to the pixels, for example). If they are defined on the basis of the modulus (or absolute value) of the signal, the distribution of the background noise usually assumes a "Rician" nature which, for SNR values greater than a few units, approximates a Gaussian signal.

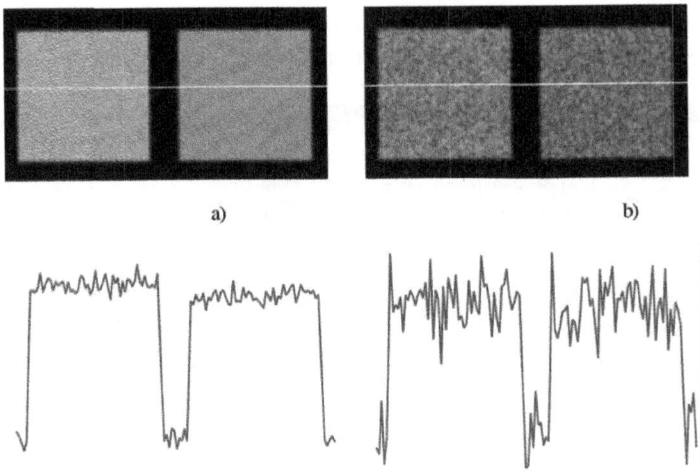

Figure 2.46. *Simulation to illustrate the principle concerning the necessity of using the notion of contrast-to-noise ratio: a) Common moderate noise level for two image elements presenting signals of similar amplitudes and a low contrast. b) The contrast tends to be reduced when the noise level becomes higher. The readings of the profiles correspond to the horizontal trace given in the images. Each rectangle (including the frame) is formed of 64 × 128 pixels*

2.4.4.2. *Rose criterion and limitations of the reciprocal space*

The previous section to some extent raises the question of the proportion between usable signal and (useless) noise which is acceptable in the display of an image. In fact, regardless of the dimension of the image, in each case we need to take into consideration the notion of volume, because we necessarily need a volume of material in order to be able to have a signal. This proportion of signal and noise, which we assess by taking the ratio between what represents the signal and what represents the noise (and which of course affects the contrast) needs to be taken into account in the exploration of the reciprocal space. The Rose criterion stipulates that, in order to be sure of achieving a limit of spatial resolution of around the width of a pixel, the ratio of amplitude between the signal part and the noise part therefore needs to be at least of the order of 5. If we refer to the expression of the SNR given in section 2.3, we can show that the ratio between the signal and the noise in amplitude of a pixel is of the form:

$$(S/N)_{PIXEL} = \sqrt{N_t} \cdot Vol_{PIXEL} \cdot P_{EX} \qquad [2.35]$$

This result stems from the fact that the power of the noise is distributed between the total number N_t of pixels which form the image, and which are each associated with a volume Vol_{PIXEL}. The term P_{EX} takes account of the experimental conditions which are: density of magnetization usable at the time of measurement, sensitivity of the recording coil, working frequency, losses caused by the equivalent resistance of the coil, noise factor linked to the transmission of the signal to the processing system and the bandwidth used. For a given imaging procedure, this term can be considered to be constant.

If we work with a parallelepipedic 1D sample, its volume is the product of its length (which is what interests us) by its thickness and then by its width. If it is a 2D sample, this volume is the product of its surface area (which interests us) by its thickness. Note that to an exactness of one coefficient, which depends precisely on the choice of the sample among the three proposed models, we have, for its volume $V_{SAM} \sim \frac{1}{\delta k^i}$, where i corresponds exactly to the type of iD image (i=1,2,3). By replacing Vol_{PIXEL} with the ratio between V_{SAM} and N_t, relation [2.35] gives us:

$$(S/N)_{PIXEL} \sim \frac{V_{SAM}}{\delta k^{i/2}} \cdot P_{EX} \tag{2.36}$$

Consequently, for a value of $(S/N)_{PIXEL}$ limited to 5 below, we would have:

$$(\delta k)_{\max} \sim 5^{-2/i} \cdot V_{SAM}^{2/i} \cdot P_{EX}^{2/i} \tag{2.37}$$

where $i = 1$, 2 or 3 if the image is 1, 2 or 3D respectively.

Consequently, although it is necessary to occupy a vast reciprocal space in order to ensure better power of resolution, the presence of noise necessarily limits the extent of that space. We may therefore be led to induce truncation, and the least penalizing way to do this is to use the technique of apodization for the weighting of the reciprocal space – e.g. preserving most of the central region and progressively limiting the high spatial frequencies. This operation therefore results in a weakening of the details of the image, whereas the opposite approach, which preserves mainly the high spatial frequencies, leads to an enhancement of these details.

2.4.4.3. *Physical parameters right to modify the contrast*

Leaving aside the issue of noise, although it is difficult to circumvent, we need to examine the basic acquisition sequences, unless we are unable to influence the noise level at all, to render the best possible contrast. As we cannot count on the variability of the tissues' proton content, apart from in a few exceptional cases for tissues or cavities, the contrast between two tissues must be accentuated by using the differences in relaxation times. Here, we shall limit ourselves mainly to the fundamental principles resulting from elementary sequences.

2.4.4.3.1. Contrast governed by the relaxation time T_1

In living tissues, the spin–lattice relaxation times T_1 of protons are relatively long (between tens of milliseconds and a second), and usually longer than the time needed for the execution of an elementary sequence which, in turn, is part of a complete MRI sequence. This characteristic stems from the fact that the protons observed belong to molecules which we may consider to be in a pseudo liquid state. We know that the translational and rotational mobilities of these molecules affect the value of T_1. Consequently, the difference in texture of two tissues leads to a difference in their respective proton relaxation times T_1. It is useful to exploit that difference to distinguish between those tissues, even if their proton concentrations are similar. However, we can see that if we apply a succession of identical RF excitations, the amplitude of the signal, notated as A, tends toward a permanent regime value given by the following relation:

$$A = A_\infty \cdot \sin(\theta) \frac{1 - e^{-T_R/T_1}}{1 - \cos(\theta) \cdot e^{-T_R/T_1}} \qquad [2.38]$$

where A_∞ is the amplitude observed if the spacing between RF pulses is sufficiently high for total recovery of magnetization to occur between each of the pulses of angle θ repeated every T_R seconds. Note that the A/A_∞ ratio is highest when θ has the value $\theta_E = \cos^{-1}(e^{-T_R/T_1})$, with θ_E known as the "Ernst angle".

Figure 2.47 gives a very general depiction of how the amplitude of a signal can be affected by the rhythm of the pulses, and reminds us that it is the components with short relaxation times T_1 which, when the density of nuclei is equal, yield the strongest amplitudes.

Of course, the variation law examined here, which could serve, for example, for an imaging sequence by projection/reconstruction or gradient echo (see section 2.5), necessitating only one RF excitation per elementary sequence, must be adapted to more complex sequences comprising many RF excitations. In any case, as indicated qualitatively by the curves in Figure 2.47, the decrease of the parameter T_R improves the difference in amplitude in favor of those media whose relaxation times T_1 are shortest. The contrast is then obtained if the difference observed is sufficiently high in comparison to the mean square amplitude of the background noise affecting the signals.

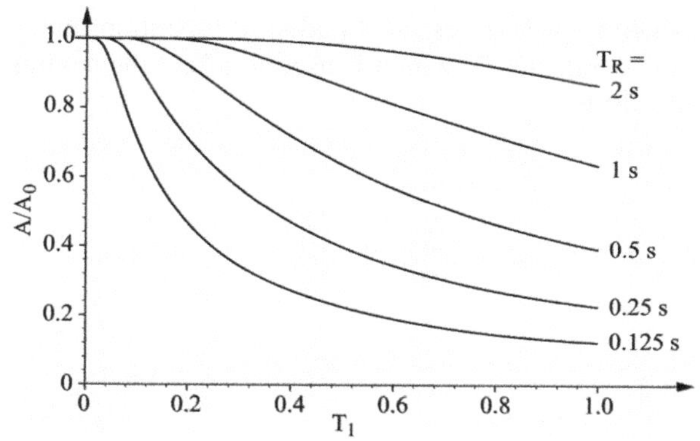

Figure 2.47. *Variations of the A/A_∞ ratio, with T1 ranging from 0.05 s to 1 s, and for several values of TR and 90° pulse. With lesser flip angles, the curves are all shifted downwards*

2.4.4.3.2. Contrast governed by the relaxation time T_2

With the exception of the protons of free water, where the two parameters T_1 and T_2 are generally equal and around 2 seconds, the spin–spin relaxation times encountered in biological tissues are generally much shorter than the spin–lattice relaxation times for the same tissue. Of course, molecular mobility also affects the value of T_2, which explains the fact – though this is not a hard-and-fast rule – that for two media, that which presents the higher value of T_1 may also have a higher value of T_2.

Without fully explaining the profound reason, which would require a more in-depth description, this difference between T_1 and T_2 can be accounted for by noting that spin–spin relaxation corresponds to the elimination of the transverse component of an isochromat – a component which cannot last beyond the return to thermodynamic equilibrium of its longitudinal magnetization. This behavior means that we will exploit this process of extinguishing to create "T_2 contrast" and the imaging sequence will then be based on a process inspired by the measurement of T_2, whereas for T_1 contrast, we use a process of recovery of the longitudinal magnetization, which is not directly observed during the experiment.

The procedure of heightening the contrast using T_2 therefore aims, still with the same density of nuclei, to accentuate the regions with the highest values of the spin–spin relaxation time. The basis of this approach is the spin-echo technique, recapped in Figure 2.48. A spin-echo sequence cannot impose a rapid rhythm on the succession of elementary sequences without the results being affected by T_1

relaxation. Consequently, it is necessary to adopt a repetition time T_R much longer than T_1, and it is the parameter T_E which is used to affect the contrasts, as indicated in Figure 2.48 and 2.49(b).

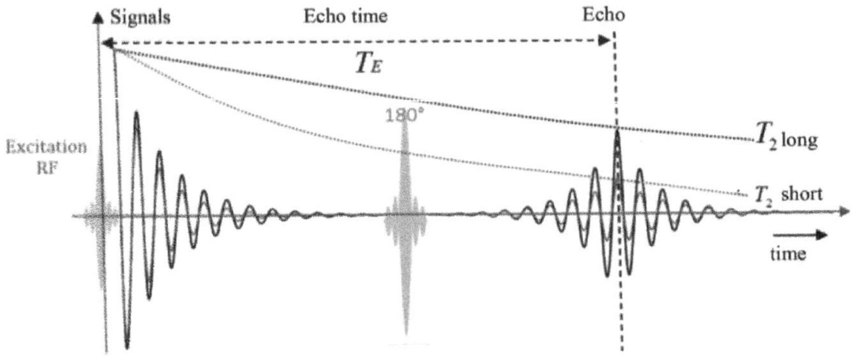

Figure 2.48. *The RF excitation pulse creates a transverse magnetization which will then be observed by examining echo signals – here a single 180° pulse. Dotted curves: expression of the independent evolutions of two isochromats with equal initial amplitudes, near-identical frequencies and different relaxation times T_2. The 180° pulse compensates for the heterogeneities of the static field. It must be admitted here that the excitation pulse includes rephasing of the magnetizations. The signals corresponding to the highest values of T_2 are strengthened at the echo time*

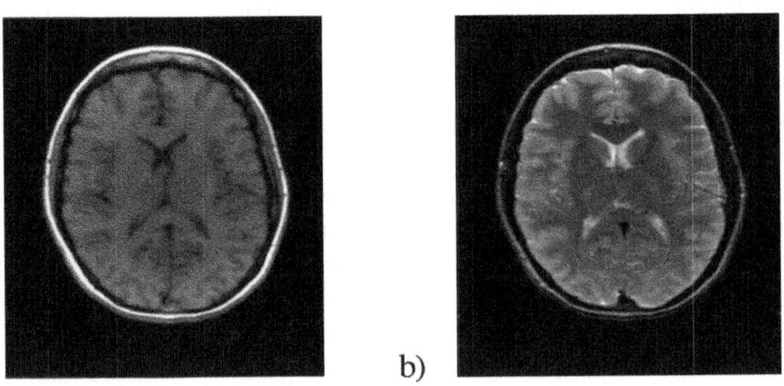

a) b)

Figure 2.49. *Inversion of the contrast between a T_1-weighted image in a) (T_R = 470 ms, T_E = 12 ms) and the image of the same slice using T_2 as a weighting criterion; b) (T_R = 2180 ms, T_E = 100 ms). Slice thickness 5mm. 256 × 256 image matrices. In a), the cerebrospinal fluid (CSF) (which has a long proton T_1) appears very dark, whilst the gray matter (medium-length T_1) and the white matter are relatively lighter. In b), the CSF appears very intensely (as it has a long protonic T_2); the gray matter is relatively light, whilst the white matter (which has a short T_2) is dark. Images provided by D. Sappey-Marinier, Cermep, Creatis*

In Table 2.3, we have indicated some orders of magnitude of proton relaxation times at 1.5 teslas. As the value of the field increases, the values of the relaxation time T_1 also tend to increase. If we increase the resonance frequency, it will gradually move beyond the maximum power spectral density of the molecular motions, causing relaxation which loses its efficacy. Variations of the values of T_1 are also observed with a rise in temperature, which can also lead to applications such as the use of MRI to monitor thermotherapy.

Relaxation time	Cerebrospinal fluid	Lipids	Gray matter	White matter	Muscle
T_1	2000	200	1000	700	600
T_2	200	90	100	80	50

Table 2.3. *Order of magnitude (in milliseconds) of some relaxation times of protons at 1.5 teslas in different tissues*

Whilst they have, hitherto, been considered as intrinsic properties of the medium being imaged, the parameters T_1 and T_2 can be altered if we dissolve paramagnetic complexes of metal ions or rare earth elements (Gd^{+++}, Mn^{++}, Fe^{+++}, etc.) in that medium. To express the effect of the presence of paramagnetic ions, we use the notion of relaxivity, which varies linearly with the inverse of the relaxation time, in relation to their concentration. The use of increasingly strong static magnetic fields is a cause for concern, for the reasons mentioned previously. The synthesis of ligands to manipulate the dynamic behavior of contrast agents on a microscopic scale is therefore becoming a very important field of research.

A very particular case relates to the local deterioration of the uniformity of the field in the presence of very small particles with very marked magnetic properties – they are said to be super-paramagnetic – based on iron oxides. The reduction of the proton signal in their vicinity results from a very local sharp decrease in the approximate time constant T_{tr}, introduced in section 2.2 (see section 2.2.4 and relation [2.9]). We tend to take T_{tr} as being equal to the time constant T_2^*, which can only be rigorously defined if the distribution of the static field throughout the volume of the sample, affected by the inevitable heterogeneity of that field, is Lorentzian in nature. This rule is very rarely understood, and even less respected, and we incorrectly speak of T_2^* contrast regardless of the mode of attenuation of the signal in the presence of inhomogeneity of the static field.

2.4.5. *Contrast elements in MRI practice*

From a practical perspective, the manipulation of the contrast greatly supplements the fundamental points discussed above, and it is useful to touch on this here. Indeed, conventional MRI is capable of providing large amounts of information about the morphology of organs and about the nature of tissues by detecting the signal from water molecules and lipids, which are present in high concentrations in the human body: nearly 50 mol/kg for water (this must not be confused with the concentration of water *protons*, which is twice as high) and approximately 1 mol/kg for lipids. Thus, the intensity of the image depends on the density of protons in the tissues, but this density does not differ greatly from one tissue to another. However, it also depends on, and is more sensitive to, the longitudinal and transverse relaxation times, which can be used to modify the appearance of the images – usually independently. In general, no use is made of the signal from the lipids, and in many situations it is better to remove their parasitic signal, using techniques which are fully understood and mastered today. In proton imaging, therefore, weighting methods are used to modify the signal from the water in order to draw physiopathological indications from it. Whilst appropriate adjustments of the parameters used in MRI sequences yield a certain level of contrast, this approach may still be insufficient on a practical level. Thus, it may be necessary to use external contrast agents, either to reduce the examination time or to obtain a specific observation. The methods developed are therefore based on two approaches: the use of an external agent (contrast agent) or simply the endogenous spin labeling of water.

– The first process relates to the use of agents that are introduced into the body, which, by their presence, modify the magnetic properties of the medium, leading to a visualization of certain compartments and their pathological alterations. It also relates to products which are able to reveal the presence of molecules by specific targeting, called “smart agents”.

– The second relates to acquisition methods that are able to visualize, and sometimes quantify, all the forms of transport or exchanges of molecules in a tissue or between tissues, without use of a contrast agent and by simple “magnetic” labelling of the spins: these methods are known as perfusion MRI, diffusion MRI and magnetization transfer MRI. We shall discuss this latter technique here, whilst diffusion MRI will be described in the final sections, on functional MRI (fMRI) techniques.

2.4.5.1. *External contrast agents*

The bio-distribution of a contrast agent depends essentially on its molecular size and its lipophilic nature, which makes it able to penetrate cell membranes. Thus, there is a distinction drawn between extracellular agents and intra-vascular products. Most commercial products are small-molecule extracellular agents. Injected intravenously, they diffuse quickly from the blood to the interstitial fluid and are mainly used for detecting ruptures of the blood–brain barrier (BBB) in the presence of cerebral lesions or tumors.

These changes in contrast can result in an increase in signal strength – here we speak of positive contrast, particularly when using paramagnetic agents – or a decrease in signal strength, when we speak of negative contrast, if super-paramagnetic agents are used.

2.4.5.1.1. Paramagnetic contrast agents

These contrast agents essentially include the paramagnetic elements which affect the relaxation time of water protons, and they have a twofold advantage. Their effect is mainly to decrease the relaxation time T_1, with the effect on T_2 being only moderate. Consequently, we obtain an enhanced signal from the tissues infiltrated by the slightly diluted product. Figure 2.50 gives a view of the mechanism by which the relaxation time is modified. A more detailed explanation than this would require a presentation of the processes of relaxation in the presence of chemical exchange, which is beyond our purposes here.

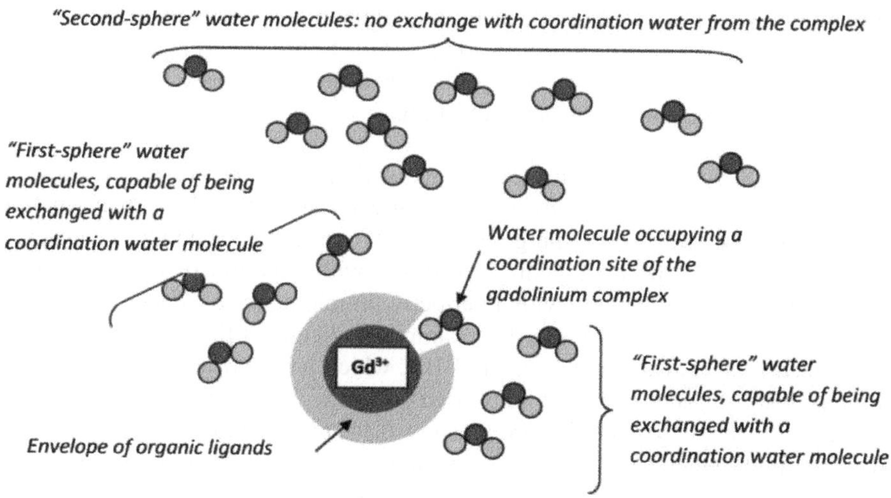

Figure 2.50. *Much simplified representation of the mechanism by which a contrast agent acts: the protons from the molecule represented here in one of the coordination sites are strongly coupled with the magnetic moment of electronic origin of the central ion (here Gd^{3+}) (see below)*

Paramagnetic agents belong either to the family of transition metals (chromium, manganese, iron) or the lanthanides (gadolinium, dysprosium). The gadolinium ion (Gd^{3+}) is particularly interesting, because it has 7 free electrons. This property lends it a strong magnetic moment and a high coordination number, meaning it is able to form a complex with 9 water molecules. These properties give rise to a strong dipolar coupling between the magnetic moment of the central ion and that of the protons in the nearby water molecules. Finally, it has only one degree of stable oxidation, preventing it from being metabolized *in vivo* by oxidation/reduction processes. However, the toxicity of the Gd^{3+} ion means that it must be complexed using organic ligands, which must possess very specific qualities: a high solubility in water at physiological pH, great thermodynamic stability and kinetics to prevent any "precipitation" in the organism, low osmotic concentration and finally a structure with at least one vacant coordination site, to facilitate exchanges with the water molecules. Thus, pharmaceutical laboratories have developed a variety of types of ligands based on $DOTA^{4-}$ (tetra-azacyclododecane-tetra-acetate), $DTPA^{5-}$ (diethylene-triamine-penta-acetate) or EDTA (ethylene-diamine-tetra-acetate) to form stable complexes, commercialized under the brand names Magnevist®, OmniScan®, Dotarem® and Gadovist®. All these agents, which are known as "first-generation agents", diffuse rapidly into the extracellular medium. In order to visualize the vascular system, new, slow-diffusing contrast agents are envisaged. Their retention time in the blood and the relaxivity depend on their molecular mass, which is increased either by the formation of hydrophobic interactions with the serum albumin such as Vasovist®, or by the adjunction of macromolecules (P760 and P792 – Guerbet) or dendrimers (Gadomer 17 – Schering). It is also possible to adjoin lipophilic groups to these molecules, in order to increase their membrane interaction and therefore their hepatic retention.

The last but by no means the least of the gadolinium complex's advantages is that it is well tolerated in the body and is not greatly allergenic. Less than 1 patient in every 1,000 experiences an allergic reaction. When it is injected intravenously, it diffuses rapidly into the extra-cellular medium, with a period of eleven minutes, before being eliminated renally with a period of 90 minutes in a subject whose renal function is normal. It can be used, for example, to heighten the contrast on highly vascularized tumors with rupture of the BBB, which will appear very intense. The optimum dose of gadolinium complex used today in clinical practice is 15 ml on average for an adult patient – approximately 0.1 mmol of Gd^{3+} per kg. With this dosage in mind (which can be adjusted depending on the patient or on the disease), it has been shown for example that the use of a dose equivalent to two times the normal dose would significantly improve the detection of lesions in multiple sclerosis or metastases.

The exchange between the coordination water molecules and the water molecules in the medium creates the mechanism known as the "inner-sphere" or first-sphere mechanism, which results in a significant decrease in the relaxation time of the protons overall. This contribution is joined by the lesser one of the so-called "outer sphere" or second sphere, which corresponds to the fact that the water molecules diffuse in the presence of disturbances of the magnetic field, caused by the presence of the edifices formed by the complexes of the paramagnetic ion. In this case, the modification of the proton relaxation time, as with any chemical exchange phenomenon, is strongly governed by the lifetimes at the different sites, which are represented here by these very specific notions of inner and outer spheres.

By way of example, the use of paramagnetic contrast agents in angiography (revealing cavities where the blood is circulating) and in tissue perfusion MRI (see section 2.2.6.1) is particularly interesting. In magnetic resonance angiography (MRA), it is usual to exploit the coherent motion of the protons from the blood in vessels so that only the signal generated by the moving spins is perceptible, whereas the signals from the stationary tissues are greatly attenuated. For this purpose, there are a variety of strategies which take account of the complex issues of filling and evacuation of the field of view by the blood flow. Thus, the imaging of blood vessels does not, *a priori*, necessitate the use of contrast agents. However, it is considerably improved by their presence in the blood circulating in those vessels – particularly if the repetition time of the sequence is short (see section 2.4.4.3.1). In the case of perfusion, we are interested in the motion of the blood in the capillaries, and the contrast stems from a decrease in the signal from the target organ. Following the injection of the paramagnetic contrast agent, there is a field gradient created between the intra-capillary and extra-capillary spaces, which temporarily attenuates the MRI signal. We therefore need to use a fast imaging method, such as echo-planar imaging (EPI – see section 2.5.2.6) to observe this variation, which is repeated a second time, after the bolus passes through the tissues. Applied to brain imaging, for example, we can use this to deduce physiological parameters such as the average transit time through the cerebral capillary network, or the volume of blood as well as the blood flow, mentioned above. Note, however, that like with angiography, there is an approach which does not use contrast agents, but involves "spin labeling" by inversion of the magnetization, using a 180° pulse: the so-called "Arterial Spin Labeling" (ASL) method, mentioned in section 2.2.6.1 (Figure 2.21).

2.4.5.1.2. Super-paramagnetic contrast agents

Primarily comprising nanoparticles of iron oxides (FeO, Fe_3O_4), the strong magnetic moments of these edifices cause a severe local disturbance in the static field, which results in a similarly localized loss of signal, giving rise to the improper – but employed – notion of T_2^* contrast (see section 2.4.4.3.2). As these nanoparticles have quite a tendency to clump together, we use two methods of stabilization. "Steric" stabilization involves covering the nanoparticles with long chains (dextran, carboxy-dextran, polyethylene-glycol, albumin, etc.). It is also possible to perform "electrostatic" stabilization, by covering the surface of the particle with carboxylate functional groups (citrate, aspartate, acetate, etc.). Thus, the coating used determines the size of the particle, its bio-distribution and its pharmacokinetics. This is also what happens with AMNPs (anionic maghemite nanoparticles). Obtained not from magnetite but from more complex spinels, maghemites and AMNPs are nanoparticles covered in an anionic coating of carboxyl ions of the type R–COO$^-$, which stabilize the edifices over a wide range of pH values, and help the substances to pass through cell membranes.

Super-paramagnetic nanoparticles are classified into four categories, depending on the size of their nucleus: the smallest, MIONs (Monocrystalline Iron Oxide Nano-preparations), whose nucleus is smaller than 10 nm; USPIOs (Ultra Small Particles of Iron Oxide), whose diameter is less than 50 nm; SPIOs (Small Particles of Iron Oxide), which range in size between 50 and 200 nm; and large particles, of diameter between 200 nm and several μm. These super-paramagnetic agents have a variety of domains of application, which can be distinguished on the basis of their distribution in the organism. Quick to be eliminated from the blood, SPIOs accumulate in the Kupffer cells of the reticuloendothelial system in the liver and spleen, helping to detect liver (hepatic) tumors. Thus, the distribution of SPIOs in normal tissues causes a decrease in signal with respect to the cancerous tissues, whose post-injection signal remains relatively strong. The smaller USPIOs present a longer intravascular retention time, and can be used for imaging of the lymphatic system, angiography or monitoring of cerebral ischemia. However, it is difficult to control the toxicity of these particles. Finally, administered orally, the largest molecules are used for the exploration of the gastro-intestinal tract (the intestines, the pancreas, etc.). Thus, SPIOs cause a reduction in the signal from the intestines, favoring the signals from other abdominal and pelvic tissues. We must be vigilant of the effects caused by these particles, which are much more substantial than those of gadolinium and cause inhomogeneities over a region extending far beyond their true location. Finally, we should highlight the interest of MIONs for animal experimentation by functional MRI, which amplifies the BOLD contrast (see section 2.6.4) three to five times.

2.4.5.1.3. New contrast agents

In the same way as the labeling of the serum albumin, which helped to increase the specificity of vascular agents, other molecules which are overexpressed by certain organs or certain cells (e.g. cancer cells) can be used to target specific receptors or specific cells. Numerous cellular receptor made up of sugars, such as sialic acid, are often overexpressed on the surface of cancerous cells. Consequently, contrast products have been developed which bind to the sialic acid molecules. Other approaches are based on the ability of a peptide to penetrate the cell and bind to the DNA. Thus, if it is attached to a gadolinium III complex, such an agent can be located by MRI.

These so-called "smart" contrast agents have the property of being bio-activatable by an environmental stimulus such as the pH or the presence of an enzyme. The first system was created to activate a substance similar to Gd-DOTA using a substitute for β-galactosyl. In the presence of its enzyme, β-galactosidase, this substitute can liberate a coordination site on the Gd^{3+} complex for a water molecule, and thereby modify the local relaxivity. The activation of contrast agents by modification of the pH is also interesting, in that, under the influence of pathological mechanisms, the extra- and/or intracellular pH is often modified, as happens with ischemic cerebro-vascular accidents (strokes) or in tumors.

2.4.5.2. *Magnetization transfer*

The magnetization transfer technique provides an additional source of contrast for certain tissues which are the site of rapid exchanges between the bound protons of the water molecules, in the tissue network, and the free protons. The term "bound protons" denotes those protons whose degree of mobility is more or less restricted, because of an association with macromolecules or membrane insertion. This compartment of bound protons is characterized by a very short relaxation time T_2, which completely prevents detection by MRI because the corresponding very wide line means that the amplitude of the signal is at the same level as the background noise. However, exploitation of the magnetization exchanges between these two compartments enables us to modify the signal observed from the free protons. As the bound protons exhibit a very broad spectrum, they can be excited by an RF pulse placed at 1 or 2 kilohertz from the resonance frequency of the free water protons. Saturation of the magnetization of the bound protons, which exchange with the free protons, generating the free water signal detected, results in a reduction of the signal observed. This imaging technique may be shown to have two advantages: firstly, it improves image contrast by reducing the parasitic signal from the surrounding tissues, and secondly, it provides quantitative information about the tissue composition in terms of exchanges with macromolecules. The first advantage is primarily used to eliminate background noise from the stationary tissues in angiography. The second advantage relates essentially to the physiopathological characterization of membrane tissues such as white matter, made up of myelinated axons. As the degree of magnetization transfer of a tissue is governed by the concentration of macromolecules in that tissue, any tissue destruction such as demyelination leads to a decrease in the amount of magnetization transfer which can be evaluated. Thus, this technique is widely used for characterizing alterations to white matter, which are diverse and are invisible with conventional MRI. In the case of multiple sclerosis, magnetization transfer is able to provide a qualitative follow-up of the demyelination/remyelination processes of white-matter lesions. However, magnetization transfer MRI is limited to clinical research for reasons of its technical complexity and difficulty to interpret. In epilepsy, it has been suggested that a decrease in the effect of magnetization transfer could reflect gliosis or indeed the presence of microdysgenesis.

2.5. Imaging sequences and modes of reconstruction

2.5.1. *Introduction*

The elements peculiar to the image obtained by MRI (resolution/contrast) would undoubtedly be different if the imaging method stemmed from an approach such as the "sensitive point" method – which was envisaged at the very beginnings of MRI. This method involved profiling the static field so that a small volume satisfied the condition of resonance, and then moving that "point" through the sample – an operation which was necessarily time-consuming. However, the performance expected of an imaging technique is to provide the best possible sensitivity and contrast per unit observation time. With regard to standard MRI, the existence and evolution of a high number of MRI sequences tends to serve this essential criterion whilst retaining an element of specificity; a general trait is the fact of giving direct access to a signal in the reciprocal space of the image. In this section, we shall therefore review the main basic sequences, tracking their progression toward ever-better performances in terms of the duration of the exam. This progression is accompanied by ever-increasing complexity, so we shall begin by mentioning the features of the fundamental or "classic" methods, before going on to discuss a number of improvements leading to fast imaging, which is necessarily more elaborate. Without going into detail about the calculations, we shall also give an overview of the reconstruction methods. In this view of the "jungle" of sequences, we shall, as far as possible, avoid using their esoteric names, which are usually the property of the various manufacturers.

2.5.2. *Overall view of acquisition sequences*

2.5.2.1. *Sequences and their associated trajectories in the reciprocal space*

In our discussion here, without fear of loss of generality, we shall use a 2D representation. In this section, there is no need to reproduce diagrams, as we shall refer to the definitions relating to the reciprocal space which are summarized by expressions [2.16] and [2.17]. However, for a brief reminder, we can consider the phase, which plays a part in expression [2.16], counted in relation to the direction y in the rotating frame (R'). In addition, if we refer to Figure 2.6, we note that immediately after an RF excitation directed along x, the so-called initial phases of the different isochromats are distributed and measured by angles around the direction y because of the spreading of the resonance frequencies. This spreading results from the non-uniformity of the static field. If we allow the isochromats to evolve freely, each of them experiences an additional phase shift, which we modify with a new pulse – this time of 180° and still directed along x. This operation leads all of the magnetizations to occupy directions with respect to the plane xOz, in relation to the axis Ox in (R') and at the end of the echo time T_E, symmetrical to those which they occupied before the free evolution (Figure 2.10). It is easy to verify that a 180° pulse along y creates a similar situation, but this time with respect to the plane yOz. Consequently if, between the first excitation pulse (whose angle is *a priori* random, but different to 180°) and the second 180° pulse, we maintain a static field gradient, we can, at the end of the echo time T_E, restore the initial phase of the isochromats with the same sign if the second pulse is along y; to this phase is added the value π if the second 180° pulse is along x.

2.5.2.1.1. Origin in the reciprocal space of the trajectory associated with an elementary sequence

The spreading of the frequencies of precession of the isochromats (generally linked to the inhomogeneity of the static field) is usually very slight compared to that caused by the application of a gradient. Consequently, the values of the different phases observed after an excitation (or a slice excitation followed by refocusing – Figure 2.11) are very similar to one another with respect to the phase excursions caused by the gradients which serve to encode the phase of the signal. Expressed in the space (k_x, k_y) – the 2D phase space – we consider that these differences in phase define a spot centered on the origin. This spot will be smaller when the uniformity of the field is good during the excitation pulse, and we accept that the refocusing of the slice selection is entirely effective. In summary, we can consider that the representation in the reciprocal space of an elementary MRI sequence initialized by an RF excitation which creates transverse magnetizations begins at a representative point placed at the center of that space. This point can move in the plane (k_x, k_y) under the influence of the gradients and additional RF pulses used during the elementary sequence.

2.5.2.1.2. Modification of a trajectory under the influence of a 180° pulse

In view of the above remarks, and also based on equation [2.16], we can establish relations between the values of the phase angles on the one hand, and on the other, relations between the expressions of the signal $s(\vec{k})$ in the reciprocal space, placing ourselves on both sides of the 180° pulse:

180° pulse along y in (R') $\qquad \varphi(\frac{T_E}{2}+\varepsilon) = -\varphi(\frac{T_E}{2}-\varepsilon) \quad s(\vec{k})_{\frac{T_E}{2}+\varepsilon} = s(\vec{k})^*_{\frac{T_E}{2}-\varepsilon}$

180° pulse along x in (R') $\qquad \varphi(\frac{T_E}{2}+\varepsilon) = \pi - \varphi(\frac{T_E}{2}-\varepsilon) \quad s(\vec{k})_{\frac{T_E}{2}+\varepsilon} = -s(\vec{k})^*_{\frac{T_E}{2}-\varepsilon}$

where ε can be on either side of the 180° pulse centered at time $\frac{T_E}{2}$.

These results indicate that a 180° pulse along y can be expressed in the k-space by replacing the representative point with its symmetrical equivalent through the origin. The 180° pulse along x in (R') involves adding an extra phase term equal to π, but to have the same signal as previously, because $e^{j\pi} = -1$. Consequently, when a 180° pulse is applied, the continuity of the trajectory of the representative point is broken, and the representative point becomes its symmetrical mirror through the origin in the reciprocal space. This property will be illustrated in section 5.2.2.1, on the spin-echo sequence, and in Figure 2.52 in particular.

2.5.2.1.3. Shape of the trajectories and filling of the reciprocal space

The application of the gradients (here G_X and G_Y) shifts the representative point whose coordinates are $k_X = \frac{\gamma}{2\pi}\int_{ti}^{t} G_X(t')dt'$ and $k_Y = \frac{\gamma}{2\pi}\int_{ti}^{t} G_Y(t')dt'$, where t_i is the time when the evolution in question begins. The components $\frac{dk_X}{dt} = \frac{\gamma}{2\pi}G_X(t)$ and $\frac{dk_Y}{dt} = \frac{\gamma}{2\pi}G_Y(t)$ determine an instantaneous slope $\frac{dk_Y}{dk_X} = \frac{G_Y}{G_X}$. We can therefore modify the shape of the trajectory: for instance, we could reverse the path by simultaneously changing the sense of the gradients, or eliminate the influence of a gradient so as to achieve a rectilinear trajectory parallel to the k_X or k_Y axis. With gradients of a uniform value in time and space, the trajectories in the reciprocal space are straight-line segments whose lengths are conditioned by the type of sequence used. Let us take two examples, which readers can, if they wish, easily plot:

In "radial" imaging, we adopt the condition $\sqrt{G_X^2 + G_Y^2} = G_0$ (constant value). The trajectories associated with each of the elementary sequences (e.g. RF excitation followed by a measurement over a time period Δt) are radii, with length $\gamma \frac{G_0}{2\pi} \Delta t$ in the plane (k_X, k_Y). This leads to a non-uniform exploration along radii of the reciprocal space around a central point $\left|\vec{k}\right| = 0$. The relative under-representation of high spatial frequencies affects the clarity of the contours (Figure 2.51).

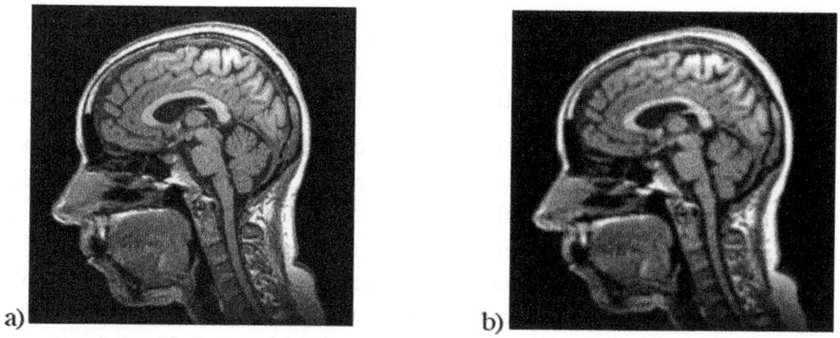

Figure 2.51. *a) Reference image (see Figure 2.20) previously used, obtained at 1.5 teslas by a 2DFT in 256 × 256 points with Cartesian scanning of the reciprocal space in echo mode. b) Image of the same slice, reconstructed by projection/reconstruction of 403 centered profiles (diameters) angularly equidistant over 180° and each containing 362 sample points in echo mode; dimensions of the image matrix: 442 × 442. The details are less clear than in the left-hand image because of less dense sampling on the periphery of the reciprocal space of the image*

Cartesian description prevents this problem by preserving the possibility of performing uniform sampling of the signal. The trajectories are parallel and equidistant straight-line segments whose spacing can be evaluated if we know the number of reading points, N_m; the duration of this reading – e.g. Δt – corresponds to the constant value of the gradient applied. If G_X is the reading gradient and ΔG_Y the step of the orthogonal phase-encoding gradient applied for identical durations T_ϕ, the condition of identical sampling in the two directions k_X and k_Y is written

$$\frac{G_X \Delta t}{N_m - 1} = \Delta G_Y T_\phi$$

Finally, it is entirely possible to envisage other trajectories to read the signal. Thus, we can use sinusoidal gradients in order to be able to describe the Fourier plane, either using concentric circles or oscillating gradients, damped so as to obtain spiral trajectories. In all these cases, an interpolation of the measuring results into the reciprocal space can be envisaged, in order to reconstruct the image by way of the Fourier transform. We shall come back to these issues related to sampling later on (see section 2.5.3). To begin with, we shall focus primarily on techniques for Cartesian description of the reciprocal space.

2.5.2.2. *Classic sequences based on the use of spin echoes*

2.5.2.2.1. Basic sequence and its associated representation in the reciprocal space

In Figure 2.52, particular note should be taken of the parts of the reading gradient represented by the heavy line. We shall use the 2D model with slice selection. In the interests of simplicity, we shall express the sequence as follows: $(\theta - T_E/2 - 180° - T_E/2 - Echo - T_R)_n$, where the terms in parentheses correspond to the so-called elementary sequence, which need not be redefined here. The presence of the subscript n reflects the fact that this sequence needs to be repeated, each time with a recurrence interval of duration T_R (or repetition time), in order to facilitate a new trajectory in the reciprocal space or possibly the repeating of the same measurement several times. The repetition time is therefore necessarily longer than the time T_{elem} shown in Figure 2.52. Of course, this sequence could be "incorporated" into a broader approach: for instance, it could be preceded by different preparations of the magnetizations for the permanent regime imposed by the pace necessitated by the value of T_R. Note that the surfaces of the lobes of the reading gradient before and after the refocusing pulse and the echo peak must be equal. Similarly, the additional surfaces described during incrementation of the phase-encoding gradient must be equal to one another. Thus, the rectilinear trajectories corresponding to the acquisition of the signal are equidistant (shown in Figure 2.52 – lower part). By adjusting the sampling period to the distance between them, we perform uniform "scanning" of the reciprocal space, which prevents image distortion. Repetition of the elementary sequence with modification of the phase-encoding gradient enables us to characterize the k-space. The directions of these gradients may be any, so long as they remain mutually orthogonal and the slice gradient is also orthogonal to the plane defined by the gradients (Gphase,Gread). This is why the gradients are not assigned any particular direction in the laboratory space: the direction can be obtained by linear combinations of the currents passing through the three coils X, Y and Z. These combinations must yield three orthogonal gradients.

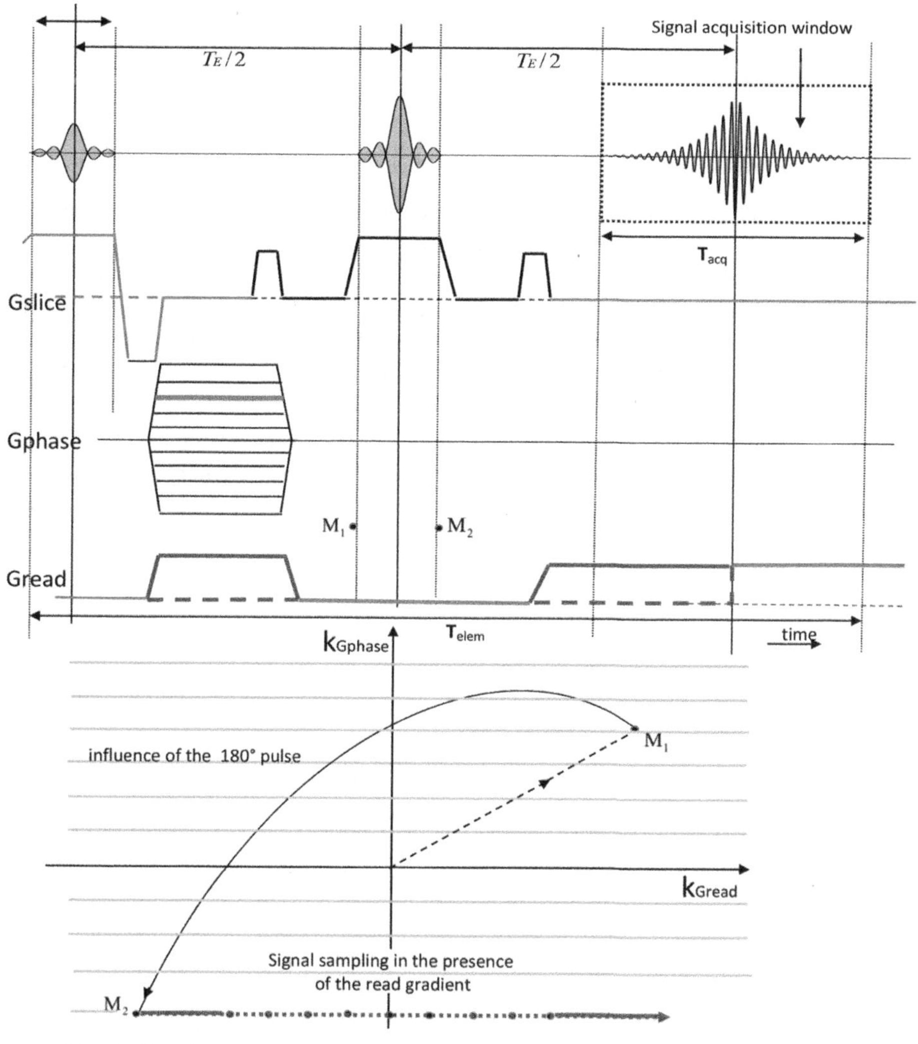

Figure 2.52. *Top to bottom: Spin-echo sequence with slice selection to create a 2D image. Only one elementary sequence is represented. Gradient pulses; shown in black on the slice selection gradient line, the gradient pulses intended to disperse the unusable terms created by the RF pulses (see Figure 2.13 in section 2.2 and its associated text). Bottom: only the trajectories corresponding to the variations of the phase-encoding gradient and reading gradient are shown; the effects of the spreading gradients are automatically compensated*

Finally, note that the compensation gradients, here assigned to the Gslice, do not affect the trajectory during the acquisition, which only takes place when the reading gradient is applied.

2.5.2.2.2. Contrasts linked to the relaxation times and constraints

The analysis of the contrast and of the contrast-to-noise ratio (CNR) involves modulation of the expression of the signal after the action of the relaxation times and also the value of the angle of the initial RF pulse. As, firstly, the spin–lattice relaxation has the time to act during the recurrence time T_R, and secondly as the observation takes place in echo mode T_E seconds after the excitation, the amplitude of the signal observed in the permanent regime, imposed by the speed of the overall sequence, will depend on the ratios T_R / T_1 and T_E / T_2. We observe, first of all, that during its execution, the elementary sequence has no effect on the recovery of the longitudinal magnetization. Also, if we accept that when an elementary sequence is repeated, it is repeated without any transverse magnetization, we obtain an echo signal whose amplitude is of the following form, which is similar to expression [2.38], given in the previous section:

$$A = A_0 \sin\theta \cdot \frac{1 - e^{-T_R/T_1}}{1 - \cos\theta \cdot e^{-T_R/T_1}} \cdot e^{-T_E/T_2} \qquad [2.39]$$

where A_0 is the amplitude which would be obtained with extremely widely spaced 90° pulses ($T_R \gg T_1$). Note that in this expression, the fact of writing T_2 is merely a formulation of principle because the diffusion phenomena, which have the time to take effect in the T_E seconds during which several gradients are switched, mean that the time constant to be considered is shorter than the spin–spin relaxation time T_2.

In a scenario where θ is 90°, the expression becomes simpler and shows how, with a constant noise level, for regions with similar or identical proton population densities, we can use the relaxation times to alter the perception of the signals. Without the need to perform calculations, note that if we wish to obtain a contrast which strengthens the signal from those regions with a short T_1, we need to adopt a similarly short repetition time T_R. In addition, as we need to avoid additional weighting linked to T_2 as much as possible, we then need to choose as short an echo time T_E as possible. T_2 weighting tends to heighten the contrast in favor of the regions where the values of that parameter are highest. In order to avoid the influence of T_1, we have to take relatively high values of T_R (the rule adopted is

$T_R \geq 5T_1$), which means the acquisition times are longer. With the maximum levels of these values, on the condition that we choose as short an echo time as possible, we finally obtain a weighting function based on the density of nuclei. Hence, this sequence presents options which, unfortunately, prove too costly in terms of machine time, which explains the quest for economic (and therefore fast) methods. The multiple-slice technique exploits the resting time $T_R - T_{elem}$ to apply the elementary sequence to neighboring slices in order to deliver a whole series of slices "for the price of one" in terms of duration. However, a number of slices $Nc = \text{integer part of} \dfrac{T_R - T_{elem}}{T_{elem}}$, is possible for the duration $T_R - T_{elem}$, but remains limited. The total duration of recording of these Nc slices is therefore approximately $Np(T_R - T_{elem})$ where Np is the number of steps of phase encoding. To this duration must be added (independently of any adjustments relating to the setting of the measurement) a preliminary stage when the magnetizations are placed in the permanent regime by a succession of a few "blank" sequences.

2.5.2.2.3. Some options

Into the elementary sequence, we can incorporate additional actions which increase the duration of its execution. For example, in order to ensure that no transverse component remaining from previous elementary sequences skews the measurements by re-forming a transverse component at the time of the first pulse (see Table 2.2, section 2.2, situation corresponding to the third row), we can apply gradients to eliminate the transverse components by creating a temporary inhomogeneity, just before the initial RF excitation. These gradients must be distinguished from the spreading gradients (represented in Figure 2.13), which are made necessary by the influence of the excitation itself. Of the various manipulations which may precede the elementary sequence, we can mention the creation of saturation bands (tagging) which, by eliminating the signal in a grid pattern, enable us to observe deformations which have occurred in organs by comparing two images of the same field of view. We can also mention the use of a preparatory 180° excitation to flip the magnetizations, which will then recover their value with spin–lattice relaxation rates which are different depending on the nature of the sample, used in the technique of ASL (arterial spin labeling) mentioned in section 2.2.6.1. Over the course of the "inversion time" T_I, between the preparatory 180° excitation pulse and the first observation pulse, the signals from tissular components "recover", with T_1 relaxation times of varying length. By placing the measuring RF excitation at the moment when a given tissue contribution touches zero, that contribution can be eliminated from the image. The list of these fine-tuning strategies and improvements is long, but they are not necessarily specific to the spin-echo sequence.

2.5.2.3. *Fast sequences derived from the spin-echo sequence*

Time-saving represents a significant economic constraint. There are at least three ways to approach it: (i) obtain images of distinct (and if possible, contiguous) slices in order to explore a volume in only the time necessary to record a single slice. We can process multiple slices almost simultaneously, given that the repetition of the sequence takes place on independent volumes during the repetition time; (ii) use a series of successive echoes produced by 180° pulses of different T_2 weighting; (iii) reconstruct an image on the basis of successive stimulated echoes produced also during a single sequence (Figure 2.55) a single sequence. Finally, the property of hermiticity of the reciprocal space, which can be expressed by the relation $s(\vec{k}) = s^*(-\vec{k})$, offers a possibility for time-gain if we are capable of contenting ourselves with a little over half the data in that space.

2.5.2.3.1. Dual-echo sequence

We shall place ourselves in situation (ii). There is no need to give another figure to indicate the principle, which consists of adding a second 180° pulse once the recording of the first echo is completed. The interval between the second 180° pulse and the previous one might be T' seconds, for example. We shall leave it up to the reader to verify that if the first echo takes place at time T_E, the second is formed at time $2T'$ (times counted from the middle of the initial excitation pulse). Hence, the gradient used during the reading of the first echo must efficiently play the role of a dephasing and then re-phasing gradient, to acquire the next echo. The amplitude available is governed by the ratio between the repetition time and the relaxation time T_1. Note that if the two 180° pulses are identically calibrated, we will of course observe the same process of transverse decrease with each echo. By preference, we shall use a scenario where θ is 90°, and where the repetition time is long in comparison to T_1. In this way – on condition that we choose a short T_E and a long T' – it is possible to respectively obtain a somewhat poorly weighted image (using the nucleus density as the weighting criterion), which is sometimes called an anatomical image, and a T_2-weighted image called a physiological image, or indeed a pathological image, because differences in the values of T_2 would show a pathological situation. Thus, we gain two results in a time only very slightly longer than that taken to gain one.

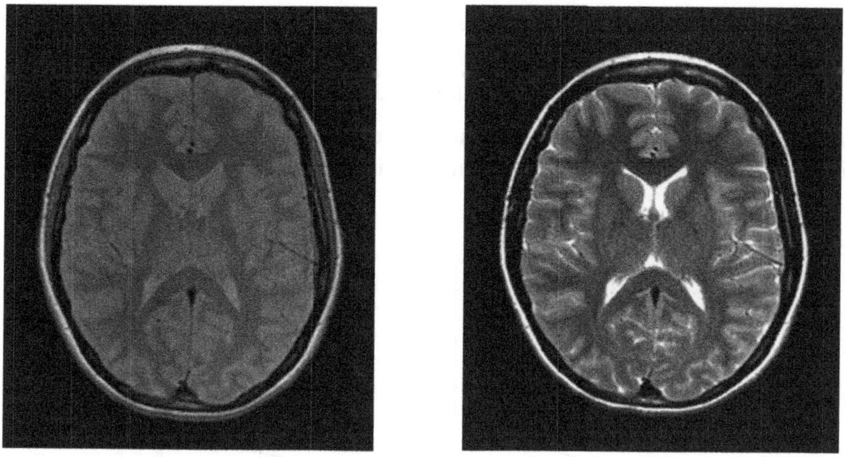

a) b)

Figure 2.53. *Images of the brain of a volunteer, obtained with a dual-echo sequence. a) "Proton-density"-weighted image; echo time TE = 12 ms. b)* T_2*-weighted image; echo time TE = 85 ms. (256 × 256; slice thickness 1.5 mm at 1.5 teslas).Image provided by D. Sappey-Marinier, Cermep, Creatis*

2.5.2.3.2. Elements relating to multiple spin-echo sequences

In an attempt to gain more time, it is possible to perform a series of regularly spaced 180° pulses and use each echo formation to characterize as large an extent as possible of the reciprocal space. As with the previous case, we need a single initial RF excitation, preferably giving a 90° flip angle. After the application of the reading gradient, we can bring the representative point to the origin of a new row. The most economical method is therefore to reapply the phase-encoding gradient in the opposite sense before the 180° pulse as indicated in Figure 2.54 and its legend. An important precaution relates to the choice of the phase for the 180° pulses, i.e. their orientation in the plane Oxy in the frame (R') with respect to the orientation of the initial excitation. The echo-train technique was indeed designed with the aim of reducing the effects of diffusion on the apparent relaxation time T_2 (see Figure 2.82 in section 2.6.3.1).

Another difficulty, entirely relative: we can show by analysis (and know from experience) that the imperfections of the excitation field lead to a regime where the amplitudes of the even-ranked echoes are stronger than those of the odd-ranked echoes. We can remedy this difficulty by modifying by 90° the phase of the 180° pulses with respect to the excitation pulse. This phase condition was proposed by the designers of the sequence which bears their names (Carr, Purcell, Meiboom and Gill), referred to as CPMG. The fast sequence will therefore be written $90°x - T_E/2 - (180y - T_E/2 - Echo)_{nE}$, where *nE*, the repetition indicator, is called the "turbo-factor". As the physically possible and acceptable values in terms of SNR are limited, we are forced to cover a sufficient reciprocal space in several steps.

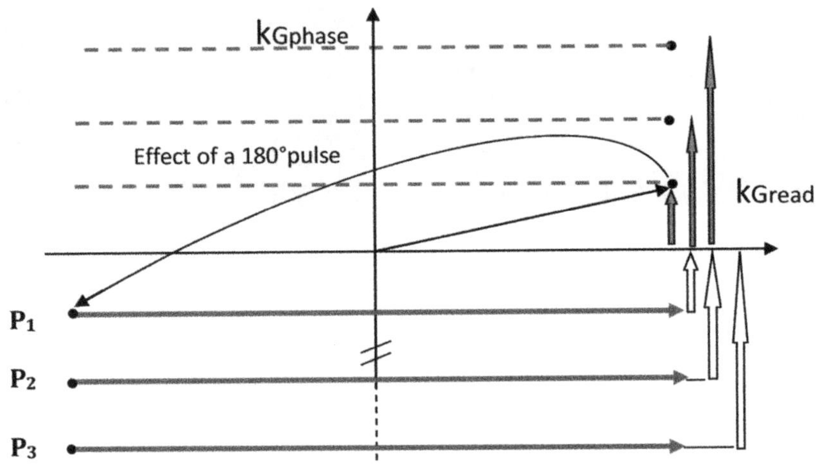

Figure 2.54. *In order to describe the straight-line segments beginning at P1, P2,..., Pn, we need numerous excursions in the direction kGphase. These excursions are indicated by gray arrows which correspond to values successively taken by the phase-encoding gradient, taking account of the action of the 180° pulses. After each segment described along kGread, we shift the figurative point in the sequence so as to obtain kGphase = 0. This operation is performed by applying a phase-encoding gradient pulse (white arrows) of same amplitude that employed previously (gray arrow). The process is represented here in a single half-plane*

The linking of the different "segments" of the reciprocal space may prove to be the source of problems, causing artifacts in the image because of discontinuities in signal strength. The measurements in the reciprocal space are affected by the relaxation times. This situation needs to be managed, by reserving the first echoes in each fast series – the most intense – for the most central regions in the reciprocal space. Indeed, these signals greatly influence the contrast (low spatial frequencies corresponding to the largest objects). Consequently, this organization of the acquisition segments can have a positive impact on truncation effects, but necessarily causes a certain limitation of power of spatial resolution.

Given the significant risk of energy deposition caused by the repetitions of 180° pulses during the fast spin-echo sequence, we can replace these pulses with pulses of a lesser amplitude leading to smaller flip angles. We can verify that each of these RF pulses gives series of numerous echoes, called "stimulated echoes". However, the application of spreading gradients enables us to take only those which are located at the center of the intervals between RF pulses. We shall not go into a detailed description of this variant, which can be written as $90°x - T_E/2 - (\alpha y - T_E/2 - Echo)_{nE}$, illustrated in Figure 2.55.

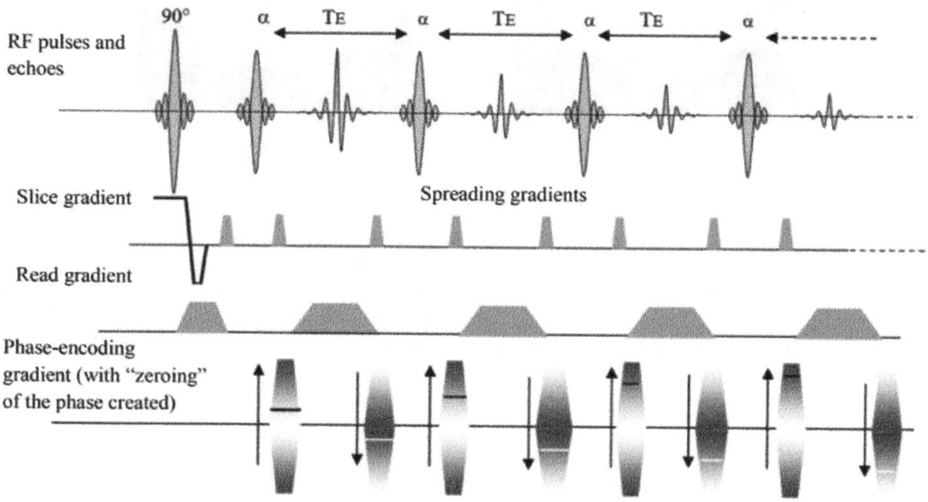

Figure 2.55. *Representation of a rapid spin-echo sequence using an a priori given "refocusing" angle α and of unspecified direction in the rotating frame* (R'). *Note that alterning sign of Gphase pulse is due to its position after the alpha refocusing pulse*

When α ≡ 180°y, it is useless to place spreading gradients because the original pulse comes into play in accordance with row 2 in Table 2.2 from section 2.2. When α is smaller than this, the "refocusing" pulses no longer respect the previous condition, which corresponds to the two cases in row 4 of Table 2.2 and the spreading gradients are imposed. In addition, the formation of stimulated echoes (a subject which will not be discussed here) whose amplitude is also governed by the spin–lattice relaxation time, has the consequence of creating a contrast depending simultaneously on T_2 and T_1. As Figure 2.56 shows, this family of fast sequences can yield differentiated contrasts.

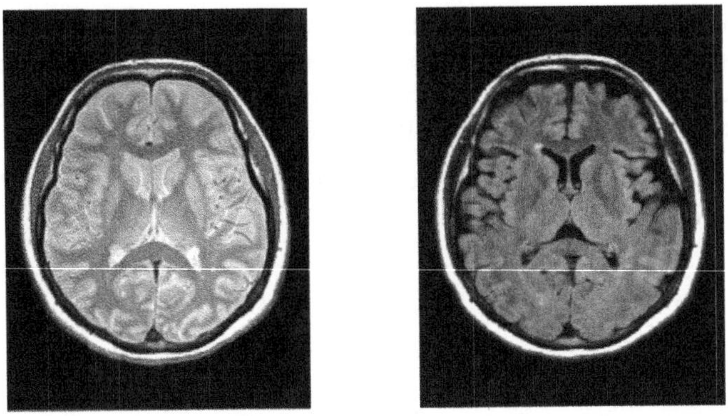

a) b)

Figure 2.56. *"Turbo spin-echo" images; turbo factor = 8 a). The regions with the highest proton density and T_2 are strengthened, which heightens the appearance of the cerebrospinal fluid, the gray matter and lipids in comparison to that of the white matter. b) "FLAIR" ("FLuid Attenuated Inversion Recovery") sequence using prepared magnetization inversion, which reduces signals from the regions with long spin–lattice relaxation times T_1. These regions appear black and yield clear delineation of the cerebral circonvolutions. 256 × 256 images. D. Sappey-Marinier, Cermep, Creatis*

2.5.2.4. *Classical sequences based on the use of gradient echoes*

We shall denote the sequence as follows: $(\theta$ - $T_E/2$ – *Gread* - $T_E/2$ – *Gread* – *Echo* - $T_R)_n$, because we are going to replace the spin echo with a gradient echo. The sequence needs to be repeated numerous times in order to cover the reciprocal space. As in the case of spin echoes, this sequence can be "incorporated" into a broader experimental setup, leading to more or less fast rhythms depending on the value of T_R. We are also led to permanent regimes which are just as diverse as those already encountered. Finally, the gradient-echo sequence, which presents the advantage of limited RF exposure, opens the door to ultra-fast techniques such as the "echo-planar" method, which is almost a real-time imaging method. The 2D model with slice selection is still the initial model and can be fairly easily generalized.

2.5.2.4.1. Basic sequence and its representation in the reciprocal space

The arrangements in Figure 2.52 must be modified to take account of the fact that there is only one RF pulse in each elementary sequence. The "value" of that pulse, which may, *a priori*, be any value except for 180°, can be determined by the rhythm imposed (T_R) and the spin–lattice relaxation time, so it affects the contrast.

Note that the amplitude of the gradient echo signal depends very strongly on the uniformity of the static field, whereas for spin-echo sequences, this dependence is very slight. Its formation takes place at a time, again notated *TE*, when the reading gradient, once it is reversed, fulfils the condition $\int G_{read} dt = 0$ along the interval here notated T_0 and where G_{read} is time-dependent. This condition corresponds to the equality of two surfaces represented in gray in the diagram associated with that gradient. As in the case of spin-echo, the difference between T_{elem} and T_R can be exploited for the observation of other slices. Note that in order to decrease T_{elem}, it is useful to reduce the time interval denoted T_0 in Figure 2.57, and at the start of which is positioned the point M_1. In fact, here, the operator is a hostage to the existence of the time necessary for the gradient switching. One way to deal with this limitation is to begin the acquisition as soon as the reading gradient starts to rise again. Of course, we then need to reposition the values thus obtained on the grid of the reciprocal space in order to render them usable.

2.5.2.4.2. Advantages and disadvantages of the gradient-echo sequence

The main disadvantage – all other things being relative – stems from the fact that the attenuation of the echo results from a complex process essentially involving the non-uniformity of the static magnetic field. In comparison to spin-echo, where the refocusing of the echo is to a certain extent forced by the 180° pulse, the amplitude of the echoes is relatively less marked in this case. The non-uniformity of the field may be attributable to numerous factors: the quality of the static field, the stability of the reading gradient, the deformation of the field lines by various materials – particularly those with pronounced magnetic properties; all these faults usually cause deformations in the image. These defects cause an attenuation of the signal which, in the simplest model, is represented by a coefficient $e^{-\frac{T_E}{T_2^*}}$, if we accept this model – which is fairly improbable but is generally accepted – for the transverse relaxation in a field whose non-uniformity covers the whole of the subject. Remember that the existence of highly localized variations of magnetic susceptibility accelerates the decrease of transverse magnetization, resulting in a degradation of the signal amplitude, also shown with the gradient-echo technique.

An advantage to the sequence is that the limitation of the RF energy deposition can be improved by using pulses with a small flip angle. It has shown itself to cope well with fast repeat rates to improve T_1 contrast, which is what is needed when using contrast agents. Finally, it is at the heart of a fairly wide variety of fast sequences.

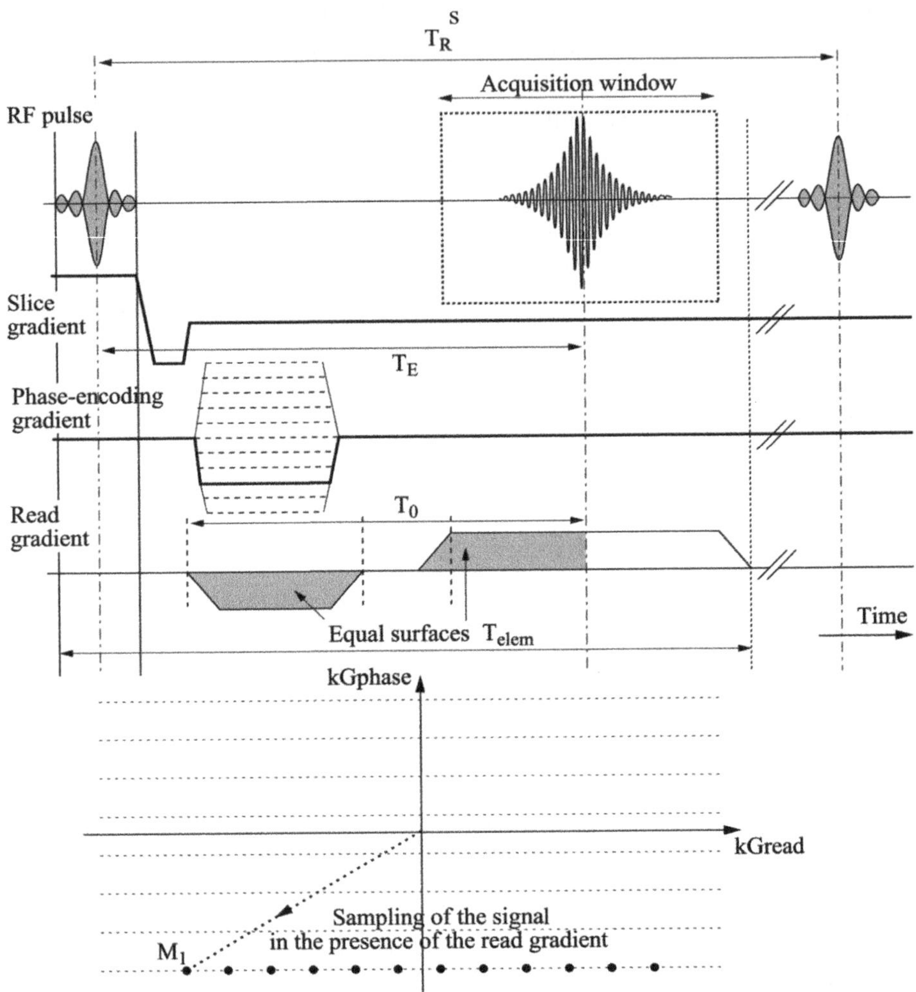

Figure 2.57. *Gradient-echo sequence with slice selection to create 2D images. Only one elementary sequence is shown here. In the description performed with the reciprocal space (bottom diagram), only the trajectories corresponding to the variations of the phase-encoding and reading gradients are indicated. The elementary time corresponding to the duration of execution of the basic sequence can be rendered shorter than in the case of a spin-echo sequence*

2.5.2.5. *Fast sequences based on the use of gradient echoes*

Although a complete study goes beyond the few cases discussed here (interested readers can find in-depth analyses and numerous examples in the existing literature), we can, for simplicity's sake, mention two main approaches. After measuring the echo, we can dynamically recover a certain amount of transverse magnetization which can be re-used in the next echo, at the risk of having a rather complex contrast. The other option is to do the opposite – i.e. completely eliminate all transverse components and thereby return to a contrast governed by the relaxation time T_1. We shall comment on these examples after making a brief detour to look at the notion of the permanent regime – created by the repetition of sequences.

2.5.2.5.1. Notion of permanent regime

The general idea is to exploit the properties of permanent regimes combined with a sustained repetition of the elementary sequences. We have already used this technique previously, employing relation [2.39]. The successive application of identical pulses to a system of nuclei (repetition time T_R), for a given isochromat, leads to a permanent regime including the flipping of its magnetization, recovery of its longitudinal component and decrease of its transverse component. A simple model involves examining the behavior of an isochromat whose pulsation of precession and angular frequency in the rotating frame (R') are equal, when it is subjected to two actions: (1) a series of regularly-spaced excitations, of flip angle θ, with the direction Ox being that of the effective field in (R'); (2) a fluctuation of static field denoted ΔB, acting identically on the phase of the isochromat in question during each interval between excitations.

These constraints are illustrated in Figure 2.58.

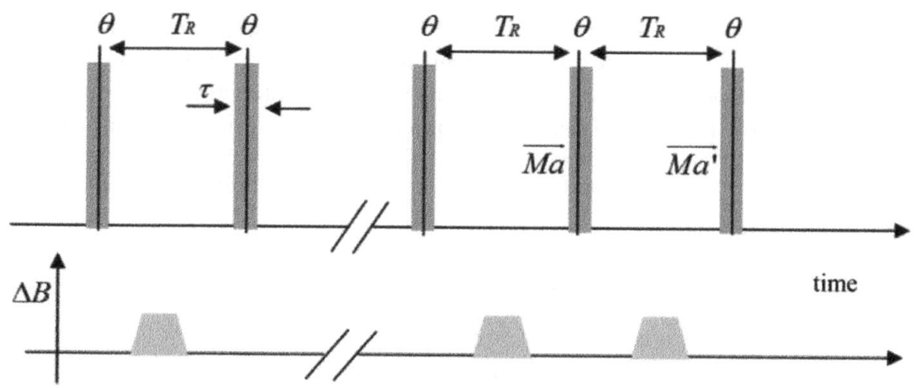

Figure 2.58. *Sequence of RF pulses of durations τ and flip angle θ, equidistant in time, applied along the direction Ox in the rotating frame (R'). These pulses are associated with static field pulses ΔB placed between the RF excitations. The magnetization of the isochromat under consideration is notated $\overrightarrow{Ma}$ before each pulse, so in the permanent regime, $\overrightarrow{Ma} = \overrightarrow{Ma'}$*

From the analytical point of view, reaching the permanent regime involves saying that the three components of the isochromat's magnetization return to the same value every T_R seconds. The duration of the pulses is supposed to be sufficiently short so that relaxation can be taken into account only during the time intervals between the pulses – intervals when the static field pulses are activated, causing a de-phasing Φ. Thus, for the isochromat considered in (R'), the permanent regime condition is written:

$$\begin{aligned} Ma_x &= Ma_x \cdot \cos\Phi - [Ma_y \cdot \cos\theta \cdot \sin\Phi - Ma_z \cdot \sin\theta \cdot \sin\Phi] \cdot e^{-\frac{T_R-\tau}{T_2}} \\ Ma_y &= Ma_x \cdot \sin\Phi + [Ma_y \cdot \cos\theta \cdot \cos\Phi - Ma_z \cdot \sin\theta \cdot \cos\Phi] \cdot e^{-\frac{T_R-\tau}{T_2}} \\ Ma_z &= M_0 + [Ma_y \cdot \sin\theta - Ma_z \cdot \cos\theta - M_0] \cdot e^{-\frac{T_R-\tau}{T_1}} \end{aligned} \quad [2.40]$$

The left-hand members of these equations are the components of $\overrightarrow{Ma}$, and the right-hand members are the components of $\overrightarrow{Ma'}$, obtained by flipping $\overrightarrow{Ma}$ by an angle θ, followed by rotation by an angle Φ around the Oz axis. M_0 is the value of magnetization of the isochromat in the absence of any excitation. It should be noted that this expression is established with respect of the conventional sense of rotations around the axes of the trihedron (R'): for nuclei with a positive gyromagnetic ratio, the angle θ is negative if we apply an RF pulse in the sense of Ox; the same is true for Φ if ΔB is positive.

Solving the system of equations [2.40] gives us the expressions of the components before each RF excitation: expressions which we can use to find the expressions corresponding to any point in time during the interval of duration $T_{R-\tau}$, provided we know the characteristics of the interactions between the isochromat and ΔB. If we use $E_1 = e^{-\frac{T_R-\tau}{T_1}}$ and $E_2 = e^{-\frac{T_R-\tau}{T_2}}$, immediately after the RF pulse the expressions of the components are then such that:

$$\begin{aligned} M_x(T/2) &= K_{MDE_1} \cdot E_2 \cdot \sin\theta \cdot \sin\Phi \\ M_y(T/2) &= -K_{MDE_1} \cdot \sin\theta \cdot (1 - E_2 \cos\Phi) \\ M_z(T/2) &= K_{MDE_1} \cdot [(1 - E_2 \cos\Phi)\cos\theta - E_2 \cos\Phi + E_2^2] \end{aligned} \quad [2.41]$$

where $K_{MDE_1} = \dfrac{M_0(1-E_1)}{D}$ and $D = (1 - E_1 \cos\theta) \cdot (1 - E_2 \cos\Phi) - E_2 \cdot (E_1 - \cos\theta) \cdot (E_2 - \cos\Phi)$, D being the determinant of the linear system formed by equations [2.40].

From equations [2.41], indicated after the beginning of these calculations, we find more particularly the notion of the "Ernst angle" for the permanent regime of the isochromats, whose precession frequency is shifted with respect to that of the frequency field. The problem becomes more complicated, because in the rotating frame (R'), the action of the pulse causes a rotation around an effective field resulting from the field applied along *Ox* and the fictitious field along *Oz* (see Figure 2.5). One way of simplifying the problem is to place ourselves in the case where the term E_2 is very small, whilst *TR* is significantly larger than *T2*, on the one hand, and on the other, to observe that the overall magnetization (the sum of the magnetizations of all the isochromats), is flipped by an angle on average equal to θ. However, we do not ignore the fact that the spreading of the values of the angle Φ attenuates the statistical weight of the $\sin\Phi$ and $\cos\Phi$ terms. Thus, the component $M_x(T/2)$ has a small value, whilst the component $M_y(T/2)$ can be assimilated to $-K_{MDE_1}\cdot\sin\theta$. The determinant D can then be approximated by $(1-E_1\cos\theta)$. Thus, we have a signal whose amplitude is proportional to the term $\dfrac{M_0(1-E_1)}{(1-E_1\cos\theta)}\cdot\sin\theta$, where M_0 is the thermodynamic equilibrium magnetization of the collection of isochromats (sample). This amplitude is maximum when θ has the value θ_E of the Ernst angle, such that $\cos\theta_E = E_1$. This expression of the amplitude can be exploited in imaging where the sequences are repeated and where the measurements are necessarily taken in the permanent regime.

2.5.2.5.2. Gradient echo spoiler sequence to eliminate residual transverse magnetization

As we increase the rate at which the elementary sequence in Figure 2.57 is repeated, artifacts appear in the images and the contrast is modified. These phenomena are due to the fact that any residual transverse magnetization could later cause longitudinal components and new transverse components after an RF pulse (see Table 2.2, section 2.2). This phenomenon is all the more pronounced when the repetition time is short, whilst for a gradient-echo sequence with a sufficiently high repetition time, the spreading of the transverse magnetizations is generally significant enough for this problem to resolve itself.

Consequently, if we want fast execution of the image with relaxation time T_1 weighting – i.e. if we want to use only signals obtained from longitudinal magnetizations which have recovered their value by spin–lattice relaxation, and then been transformed into transverse components (to exploit the effect of T_1) –, it is advantageous to, as fully as possible, use spreading gradients to eliminate the transverse components which remain after the acquisition before triggering the next

RF excitation. In summary, we need to avoid having each RF pulse restore a superfluous component to the direction of the static field, which would skew the dependence of the amplitude of the echo on T_1. Remember, also, that the RF pulses generally used have an angle less than or equal to 90°.

The role of the so-called "spoiler" gradients is to destroy any residual traces of transverse magnetization by prolonging the reading gradient and introducing, in the direction of slice selection, a gradient pulse with a random value applied just before each new RF excitation.

Put simply, we exploit the presence of the reading gradient so that the superposition of these two gradients creates a relatively high resulting gradient, acting effectively on the de-phasings. This gradient is, in turn, random in order to prevent the creation of a permanent regime which would host coherent phenomena causing alterations in the T_1-weighted signal. In order to obtain the expression of the amplitude of the echo, we can use relation [2.39], in which we replace the relaxation time T_2 with T_2^*, knowing that in both cases these are approximate formulations: for spin-echo, we need to take account of the diffusion; for gradient-echo we need to be certain that the T_2^* formulation satisfies the conditions of validity imposed by the distribution of the static field throughout the volume of the sample. This does not prevent the experiment from working from a qualitative point of view – particularly when the uniformity of the field is affected by the presence of products which degrade it locally by the effect of magnetic susceptibility, and thus indicate the position of those substances, making it a potential labeling technique.

Note that there is another way – which can be superposed on that described in Figure 2.59 – to decorrelate the residual transverse magnetizations, which is to use successive RF excitation pulses of an equal angle but whose phase varies randomly (Figure 2.60). In this figure, the dark gray line shows the evolution of the phase of a given isochromat under the influence of the reading gradient; the passage through zero corresponds to the formation of the gradient echo. The light gray line indicates the variation of the dispersed phase of the magnetization of a transverse component created by pulse no. 2 (see Table 2.2, section 2.2) on the basis of the residual transverse magnetization left by the isochromat. Also in Figure 2.60, the phase ϕ indicated results from the application of pulse no. 2 to the residual magnetization left at the end of the previous reading – a pulse which creates a transverse component which, *a priori*, is different for each isochromat. All of the spreading therefore leads to the elimination of the resultant of the undesirable transverse components once the permanent regime has been set up, provided we maintain directions of pulses oriented at random in the rotating frame (R').

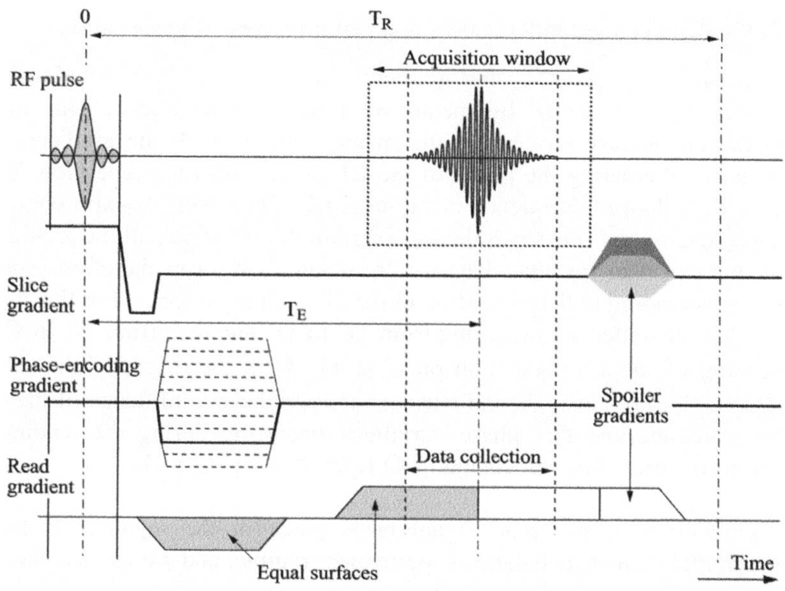

Figure 2.59. *2D elementary gradient-echo sequence with the introduction of a spoiler gradient to eliminate the residual transverse magnetizations. This gradient may comprise a prolongation of the reading gradient, combines with a gradient of a random value, here applied in the direction of the slice gradient. This choice of direction of the slice is related to the fact that in order to ensure proper spoiling in the volume of each voxel, we need to act in the direction of the largest dimension of the voxels – i.e. that which corresponds to the slice thickness in 2D imaging*

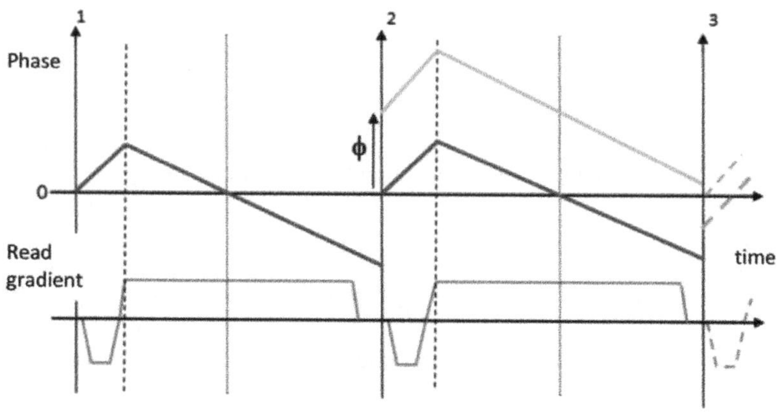

Figure 2.60. *Very simplified representation using the plot of the phase (upper trend) and the reading gradient (lower trend) in the gradient-echo technique with spoiling of the residual magnetizations with RF pulses numbered 1 to 3, exhibiting random phases (not represented but simply numbered at the time of their application)*

2.5.2.5.3. Gradient echoes with reuse of residual transverse magnetization

In the permanent regime of a fast gradient-echo sequence, the residual transverse magnetization can be reused by means of a recovery technique. This involves finishing the elementary sequence with an application of all the gradients in the opposite sense, alternating the phase of the RF pulse with each repetition. We can represent this "balanced" sequence in Figure 2.61, along with its expression in the reciprocal space. As we can see in the top diagram in this figure, all the gradients are balanced; we can also say that they are "re-wound". We can therefore produce a diagram corresponding to that operation in the 3D reciprocal space. For this purpose, we accept that the dotted arrows going from D_1 to D_2 and then from D_2 to O show the refocusing of the slice selection pulse at O (O is the origin of the trihedron (kslice, kphase, kread). Next, the defocusing in P_1 under the influence of the phase-encoding gradients, and that phase variations occurring during the reading, will usually be compensated by refocusing in O from P_2 (Figure 2.61).

The acquisition of the echo signal takes place in the segment $P_1 P_2$. The sustained rhythm creates a "balanced" permanent regime, and we can use the results indicated in section 2.5.2.5.2, but modified to take account of the phase alternation. This alternation entails replacing Φ with $\Phi+\pi$ in the numerator and denominator of these expressions. We obtain an approximate formula for the echo signal by replacing E_1 and E_2 with $1-{T_R}/{T_1}$ and $1-{T_R}/{T_2}$, because $T_1 \geq T_2 \gg T_R$; this gives us an echo signal in the center of the interval T_R, whose amplitude can be expressed in the form:

$$A = A_0 \frac{\sin\theta}{1+\frac{T_1}{T_2}+(1-\frac{T_1}{T_2})\cdot\cos\theta} \cdot e^{-\frac{T_R}{2T_2}} \qquad [2.42]$$

It is interesting to note that the coefficient of amplitude $\frac{\sin\theta}{1+\frac{T_1}{T_2}+(1-\frac{T_1}{T_2})\cdot\cos\theta}$ presents a maximum for $\cos\theta = \frac{T_1-T_2}{T_1+T_2}$, which gives us $\frac{\sqrt{T_2}}{2\sqrt{T_1+T_2}}$. If T_2 is negligible in comparison to T_1, the amplitude of the signal is then modulated by the ratio $\frac{\sqrt{T_2}}{2\sqrt{T_1}}$, whose value is fairly low because $T_1 > T_2$.

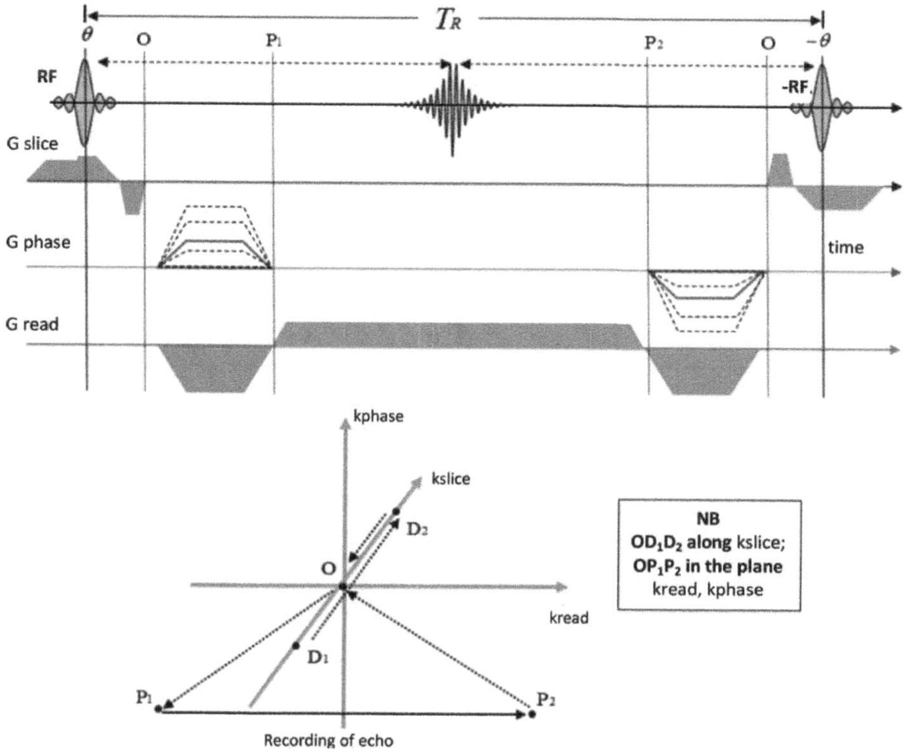

Figure 2.61. *Principle of the balanced gradient-echo sequence in the permanent regime. Top: diagram of the elementary sequence. The important times are represented by points* D_1, D_2, P_1 *and* P_2, *and* O *is the origin of the frame (in gray) in the reciprocal space which occupies the bottom diagram. The trajectories in the reciprocal space are shown in black. That which corresponds to the slice selection and which takes place along the direction of the kslice axis results in the shift* $D_1 \rightarrow D_2 \rightarrow O$, *which represents the action of the slice selection pulse. Note the alternation of the sign of the phase of the RF pulse with angle* θ

A fuller examination of this sequence shows that this response obtained in the absence of a fictitious field in the rotating frame (the isochromat said to be at resonance) is acceptable for the neighboring isochromats on condition that the uniformity of the static field is very high. This is the main drawback to this approach. There are two variants, which essentially use the instauration of a permanent regime (or stationary state), described here but involving only the re-winding of the phase-encoding gradient. We therefore need to return to the use of spreading gradients in all three directions (slice, phase and reading). The corresponding trajectories in the reciprocal space are indicated in Figure 2.62.

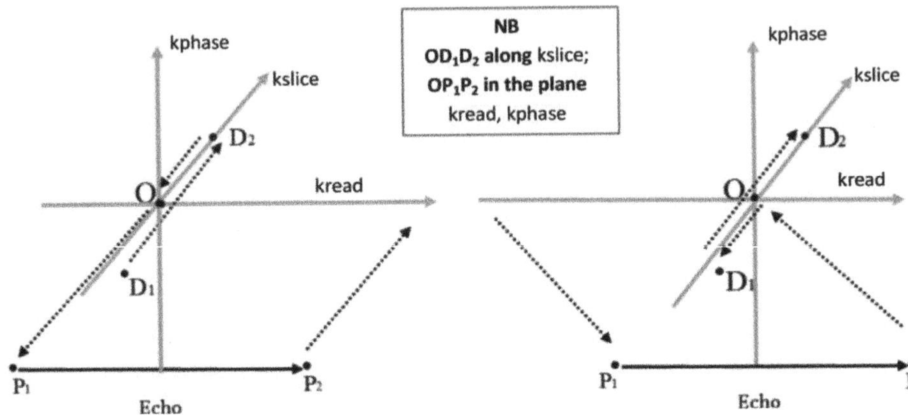

Figure 2.62. *Representation in the 3D reciprocal space of the stationary-state regime trajectories of two non-balanced sequences where only the phase-encoding gradient is "re-wound". Left: the sequence facilitates the elimination of the residual magnetizations; right: the sequence facilitates exclusive use of those residual magnetizations. Note that unlike the analogous diagram in Figure 2.61, the trajectories (dotted black arrows) of the vertical plane (O kphase, kread) are not closed on themselves: here the missing horizontal part, along the axis determined by the reading gradient, corresponds in a manner of speaking to the application of the spreading gradients in the three directions (slice, phase and reading)*

In terms of the description of these sequences, which give rise to applications in biomedical MRI, we shall not go into detail and refer the reader to recent works. Let us simply highlight that the computation of the signal is still based on the use of the expressions [2.41], as is the case with the result [2.42]. Thus, we are led here to attenuation in T_2^* instead of attenuation in T_2, as for the closed-trajectory sequence, which corresponds to a less pronounced constraint on the uniformity of the static field.

What is important in these families of sequences based on the gradient-echo technique relates to the fact that the relaxation time T_1 always has an important role in the expression of the signal *via* the amplitude coefficient. Thus, the rapidity of their execution can be exploited in order to observe dynamic variations of that parameter – hence their medical usefulness. This also explains why they often appear in the form of very numerous acronyms, associated with the companies which use them. It is difficult to orientate oneself in this jungle of approaches and patent-holders.

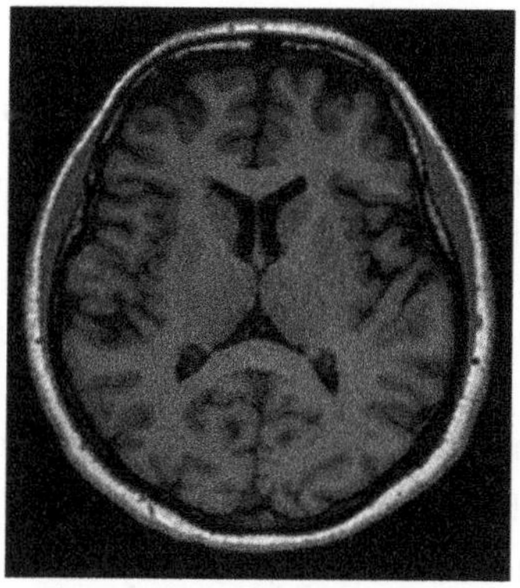

Figure 2.63. *Example of an image obtained with a rapid gradient-echo sequence: the sequence MP RAGE (Magnetization Prepared Rapid Gradient Echo) can be used to obtain 3D images with "preparation" of the magnetizations – essentially a 180° pulse intended to strengthen the appearance of tissues with short proton relaxation times T_2, and therefore short T_1 times as well. Here, the repetition time is 1970 ms, the inversion time 1100 ms and the echo time 3.9 ms. The gradient echo signals are observed in residual transverse magnetization destruction mode (see section 2.5.2.5.2). Slice thickness: 1 mm. Number of slices: 176. 256 × 256 image matrix. Image provided by D. Sappey-Marinier, Cermep, Creatis*

2.5.2.6. *Ultra-fast "echo planar" sequences*

The term "echo planar" stems from an original method developed in 1974 by P. Mansfield, whose initial idea was to gather all the information for an image in a unique profile and then re-attribute the pieces of information on the basis of their spatial origin, drawing inspiration from crystallographic techniques. The implementation of this method in practice led to the use of successive gradient echoes after preparation of the magnetization, such as the inversion of the signal or use of a spin echo; the simplest process is still to apply a single excitation (Figure 2.64). Enabling us to use the image of the same region of interest multiple times a second, the "echo-planar" sequence represented the first real-time MRI approach. In order to determine the arrangement of the sequence, we can imagine it on the basis of its trajectory in the reciprocal *k*-space and we shall consider the case of a 2D image. The reading, performed from bottom to top, can be replaced by a top-to-bottom reading by exchanging the sign of the initial phase-encoding gradient and that of the "blip" gradients (Figure 2.64). These line-changing gradients are interleaved between the successively alternated "lobes" of the reading gradient, which are supposed to be very short. We can then envisage scanning methods which are easier to handle on the plane of the switching of the reading gradient as the use of a sinusoidal variation. This leads to non-Cartesian sampling, which needs to be corrected in the interests of reconstruction. Finally, in order to improve the definition in the reciprocal plane (number of measuring points per unit surface area), we can capture the image by local recordings which can be juxtaposed, or indeed by repeated scans of the whole surface, but shifted so as to increase the density of the samples. We can also restrict the reciprocal domain to be explored to a half-plane, because of the properties of hermiticity of the signal in the Fourier space. The main advantage, which is the fast execution, comes at the cost of the existence of various types of constraints. For instance, we can cite the dependence of the amplitude of the signal on the order of the echo: if $T_E(n)$ is echo time number n (counted from the middle of the RF pulse) the amplitude of the echo is then proportional to $e^{-\frac{T_E(n)}{T_2^*}}$, where T_2^* is, *a priori*, totally empirical because it must take into account the non-uniformity of the static field, essentially joined by additional phase-changes caused by the presence of a particular reading gradient. As this gradient is high in order to create the echo signals as fast as possible, drift effects appear during the switching. In a manner of speaking, this drift corresponds to a shift in time, and therefore in the reciprocal k-space, of the echo time in a direction parallel to kread, leading to a potential splitting of the image during reconstruction. Another problem caused by these high-value gradients is the notable role played by the Maxwell terms in the modification of the value of the static field for which the undesirable components, orthogonal to the geometric direction OZ, influence the slice selection.

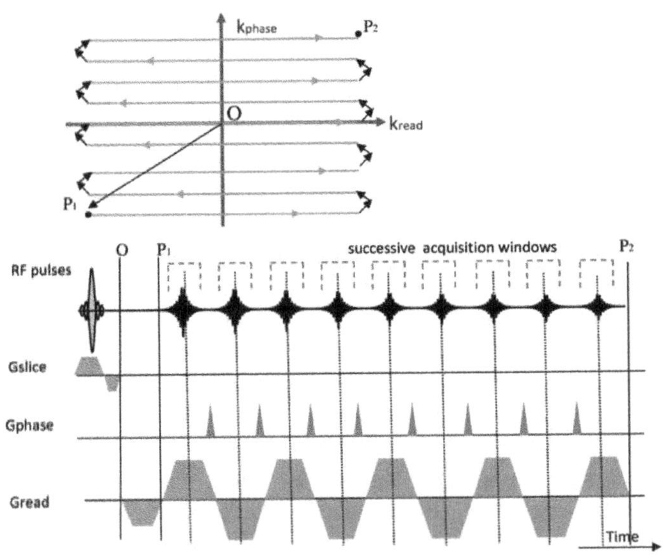

Figure 2.64. *Illustration of the principle of the "echo-planar" method for 2D acquisition. Top: choice of a trajectory from P1 to P2 covering the reciprocal space (kphase, kread). Bottom: order of sequence imposed by the trajectory chosen. The "jump" from one line to another is then caused by a brief application of the phase-encoding gradient – an incrementation called a "blip"*

To this still-inexhaustive list of technical problems are added the physiological problems caused by the switching of intense gradients. It must be pointed out that for non-invasive observation of organisms smaller than humans, such as small animals, the performances must be even better, leading to higher spatial resolution. The use of an echo-planar sequence will be illustrated in section 2.6.4.3, with the case of functional neuroimaging.

2.5.2.7. *Other possibilities for description of the reciprocal space*

With a Cartesian coverage of the reciprocal space, the aforementioned sequences are essentially founded on the principle of reconstruction by Fourier transform, and correspond to rectilinear trajectories, parallel to the kread axis. *A priori*, there is no restriction of the choice of trajectories, provided we are able to follow them when choosing the gradients. Thus, two other descriptions can be mentioned: radial rectilinear trajectories and spiral trajectories.

2.5.2.7.1. Radial trajectories

The 1D model which we used to introduce the notion of MRI in section 2.2 can be considered as an element of a radial sampling of the k-space. Changing the orientation of the reading gradient with a fixed modulus, in a plane or in the k-space, and thus defining a variety of radial trajectories, can in turn lead us to perform 2D or 3D sampling of the k-space. The primary advantage is that the corresponding sequences are relatively less sensitive to the effects of motion blurring than Cartesian sampling sequences. The main disadvantage stems from the non-uniform sampling of the reciprocal space, which can only partially be compensated by interpolation of the data on a Cartesian grid. Note that these approaches, which can be used in "Free Induction Decay" (FID) mode or in "echo" mode, are not specifically suited for fast imaging. However, if the FID signal is observed "immediately" after the RF excitation, we can image samples whose nuclei exhibit very short relaxation times. On the other hand, the main difficulty lies in the application of the reading gradient (note the absence of phase-encoding gradient): if we apply the RF pulse in the presence of the reading gradient, this skews the definition of the slice. Otherwise, the time necessary for its instauration is wasted. In order to prevent this rise time of the gradient becoming a dead time, which is always tricky to deal with when correcting the 1D profile associated with the acquisition, it is possible to sample the signal in a non-uniform manner during that rise. In the case of the echo signal, the main advantage is that, to a certain extent, we describe a diameter of the profile space, so simultaneously two angular orientations spaced 180° apart. It is therefore possible to reduce the duration of the acquisition session. The risk common to these two modalities stems from the fact that local variations between the RF excitation frequency and the frequencies of precession (also called "offset difference") can lead to an alteration of the PSF and thus cause image distortions. The presence of non-uniformities of the field, of varying origins (inhomogeneities of the static field, local variations in susceptibility, chemical shift, etc.) has repercussions for the clarity of the image. Finally, note that this radial approach is not limited to exploration in a plane, but rather that it also offers the possibility for 3D, spherical description, with use of the single reading gradient reoriented in the three dimensions.

2.5.2.7.2. Spiral trajectories

This approach exhibits significant interest by the rapidity of its execution in a single scan or in a number of successive scans (segmentation) of the reciprocal space. It therefore proves well suited to observations in dynamic mode: cardiac observation and kinetics, in particular. A complete study of this technique would occupy a large chapter in its own right. We shall examine it very briefly, simply viewing the principle on the basis of a 2D model created with slice selection. The trajectory in the reciprocal space, which in this case needs to be covered by a reading gradient with two dimensions U and V, orthogonal and in the frame (R), can be determined by its components which are, for instance, $k_U = K \cdot \alpha(t) \cdot \cos(\alpha(t))$ and $k_V = K \cdot \alpha(t) \cdot \sin(\alpha(t) - \pi / 2)$, where K is a numerical coefficient. The simplest case is that of the Archimedean spiral where $\alpha(t)$ varies linearly over time.

Using the relation of definition $\vec{k} = \dfrac{\gamma}{2\pi} \int_0^t \overrightarrow{G(t')} \cdot dt'$, which gives $\overrightarrow{G(t)} = \dfrac{2\pi}{\gamma} \dfrac{d\vec{k}}{dt}$, we can determine the expression of the gradients to be applied. These expressions define the control of the gradients, knowing the time-dependency of $\alpha(t)$. At a constant sampling frequency (Figure 2.65 (a)), the density of measuring points decreases rapidly along the trajectories, suggesting a parabolic variation, as shown by the complete study. This involves imposing a variation of the angle $\alpha(t)$ proportional to $\sqrt{t}$, which gives an equidistant distribution of measuring points along the trajectory (Figure 2.65 (b)). The gradients are ordered by way of a transitory phase. That which is shown here (Figure 2.65 (c)) corresponds to an exponential rise of the amplitude of the gradients. The sampling obtained is shown in Figure 2.65 (d). The number of revolutions must be approximately $Npix / 2$ if $Npix$ is the number of pixels required along the side of an image of dimensions $Npix \times Npix$. The acquisition is triggered simultaneously with the application of the two gradients. As the duration of a complete acquisition can be up to 50 ms, with 160 measuring points, we have only presented here what happens in the first few milliseconds. In order to enhance the definition of the exploration of the reciprocal space, we can use multiple, intercalated loose spiral scans to relieve the constraints to which the instrumentation is subject, but this prolongs the duration of the exam. Without going into detail, we can say that radial techniques benefit from their low motion sensitivity, but depend on the offset differences, and for these reasons, can potentially be superior to the echo-planar technique. Yet as we shall see, there are also other very fast solutions which are based on phased-array coils and new modes of reconstruction (so-called "parallel imaging") rather than on the loading of the material.

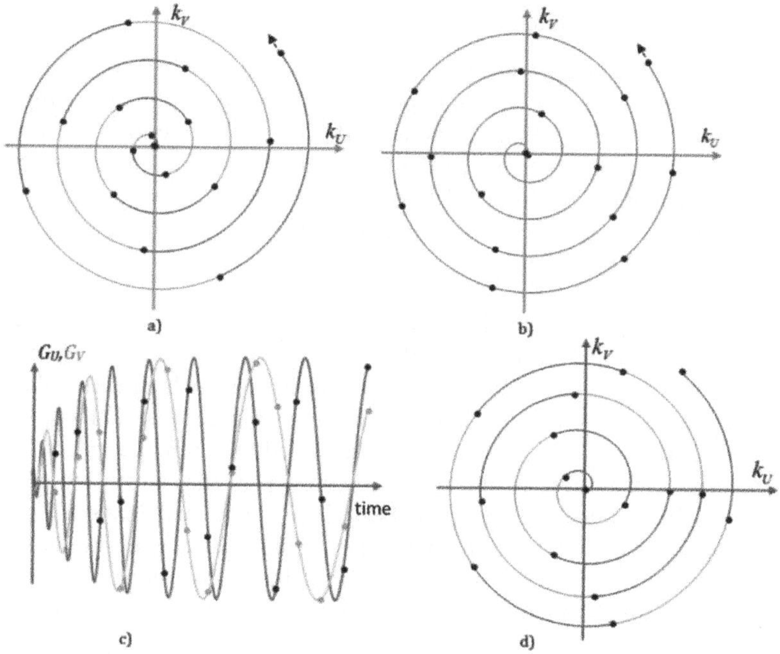

Figure 2.65. *Spiral scanning at a constant rate; example with 15 intervals: a) at constant sampling frequency and constant angular scanning frequency, the sampling density is higher at the center of the reciprocal space than at the edges of the domain explored. b) The obtainment of samples regularly spaced along the trajectory offers a better distribution of sampling density. c) Compromise solution facilitating an increase of the gradient acceptable on the instrumental level and progressively leading to equidistant sampling along the corresponding trajectory in the reciprocal space. The sampling times on the plots of the gradients in c) correspond to the measuring points shown in d)*

2.5.3. *Modes of reconstruction*

In terms of the organization of this section, it is possible to follow the historic progression of the development of MRI. Modes of image reconstruction appear which are linked to the method of acquisition, and the use of the Fourier transform has a center-stage role. We shall not linger over the different types of calculations, as interested readers can find these elements in numerous specialized works. Our discussion here looks at the modes of reconstruction as elements associated with the imaging strategy, which leads us to place parallel imaging at the end of this brief summary.

2.5.3.1. *Projection-reconstruction*

2.5.3.1.1. Brief reminder of the principle

The earliest MRI images were obtained only shortly after the earliest X-ray tomography images. The fact of having projections whose amplitude is proportional to the nuclear magnetization accumulated in the direction of projection (perpendicular to the direction of a reading gradient) then soon led P. Lauterbur to a reconstruction technique based on "filtered profiles", also known as Filtered Back-Projection (FBP). This technique was developed in Lauterbur's laboratory from the mid-1970s, using radial sampling.

We shall not recap the principles of these reconstructions, which are based on the "Central Slice Theorem" and are carefully discussed in works on tomographic approaches. Remember that in MRI, we normally have signals directly taken in the reciprocal space. We can therefore prepare the signals directly in that *k*-space using weighting functions. Once they have been transcribed to the image space by inverse Fourier transform, they give us the so-called filtered profiles which are needed for reconstruction. Indeed, the presence of noise (which is not considered in Figure 2.66 (a,b,c)) means the weighting function must reduce the contributions of the components with high spatial frequencies, where the relative noise level is highest and therefore imposes an additional limitation on the power of spatial resolution.

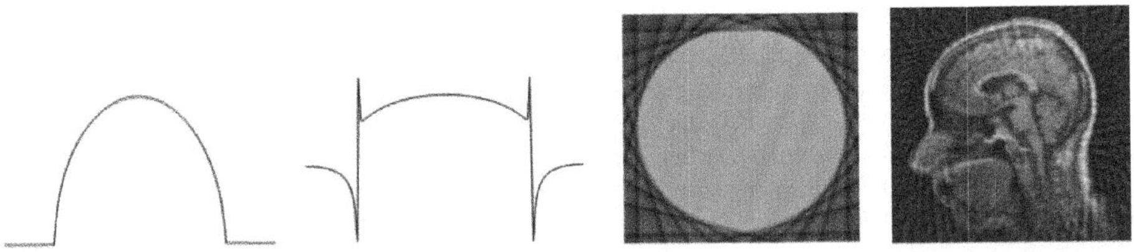

a) b) c) d)

Figure 2.66. *a,b,c): representation by an elementary simulation of the principle of backprojection of the "filtered profiles", 2D model. a) 1D projection of a circular object, obtained at 128 points; b) filtered profile; c) reconstitution of the circle simply with 12 profiles equally distributed across 180°. The lines shown correspond to azimuthal under-sampling (an insufficient number of diameters), which can also be seen in d) where the image is reconstructed with only 40 diameters. (Reference image, see Figure 2.20)*

2.5.3.1.2. Limitation of the reciprocal space and influence of sampling

From the standpoint of power of spatial resolution (capacity to separate details in the image), an essential element to take into consideration is the limitation of the reciprocal space, which affects the PSF. This capacity may be lessened by the initial choice of a small width of the acquisition window or – which is essentially the same thing – the use of too restrictive a weighting function, designed to attenuate the effects of noise. In the examples given in Figure 2.67, the windows applied to the profiles can be considered as approximately rectangular. Each window, depending on its width, accentuates to a greater or lesser extent the convolution effect created by the variation in amplitude of the temporal signals under the influence of the transverse relaxation, whether with free precession or with an echo. Summarily, the effects shown in Figure 2.67 stem from modifications of the PSF, which result in stretching of the central lobe of the function PSF (0); this stretching lessens as we progress from situation (a) to situation (d), altering the value of $k_{r\ \max}$.

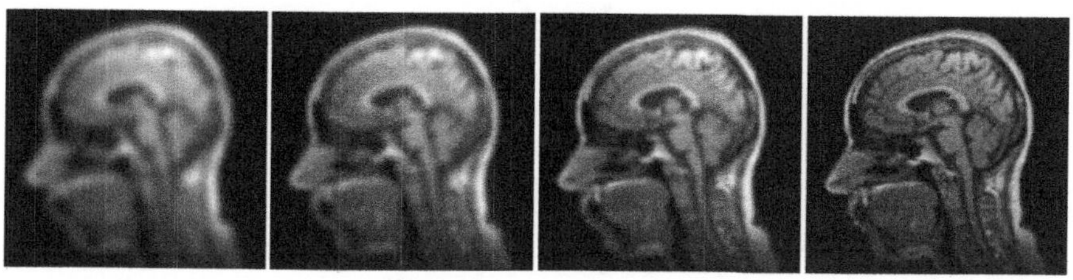

a) b) c) d)

Figure 2.67. *Consequence of limitation of the k-space sampled with a Hanning window which transmits a) ~20%, b) ~30%, c) ~50%, d) ~100% of the width of the acquisition window. The details become increasingly clear as we move from a) to d). Projection/Reconstruction of 403 diameters, sampled in 362 points before application of the window. (For reference image, see Figure 2.20)*

As the coordinate k_r corresponds to each azimuthal direction of the profiles, we need to envisage a reciprocal space delimited circularly in 2D imaging. The sampling of this circular space is not uniform (example given in Figure 2.68). The criterion which needs to be respected along the profiles, or if readers prefer, along the diameters, is the usual criterion of sampling qualified as "radial" (or "diametral"), satisfying the Nyquist–Shannon condition. It can be expressed by stating that the sampling step in the k-space must be such that $\delta k_r \leq \frac{1}{Dc}$, where Dc is the diameter of the field of view – also circular. Figures 2.66 (c and d) indicate that we need to increase the number of profiles whilst reducing the angular step $\Delta\theta$ of their distribution. This step will be chosen as constant. A criterion can be established by comparing the distance between the two sampling points situated at the neighboring extremities of two radial trajectories, consecutive with the inter-sample distance along the diameters (Figure 2.68). In the space of the field of view, we would have the condition $R_C\Delta\theta \leq \frac{2R_C}{N_s - 1}$, where N_s is the number of points measured along a diameter, which gives us the final condition $\pi(N_s - 1) \leq 2N_r$, where N_r is the number of diameters (or, if readers prefer, the number of profiles). Note that if N_s is much greater than the unit value, it can be used in place of $N_s - 1$. This result can also be established by basing our reasoning on the reciprocal space. The condition is not satisfied in Figure 2.66 (c and d) or in Figure 2.68. Nor is it satisfied in Figure 2.67 (d), which is actually satisfactory, meaning that the strict rule about azimuthal (or angular) sampling may be tempered.

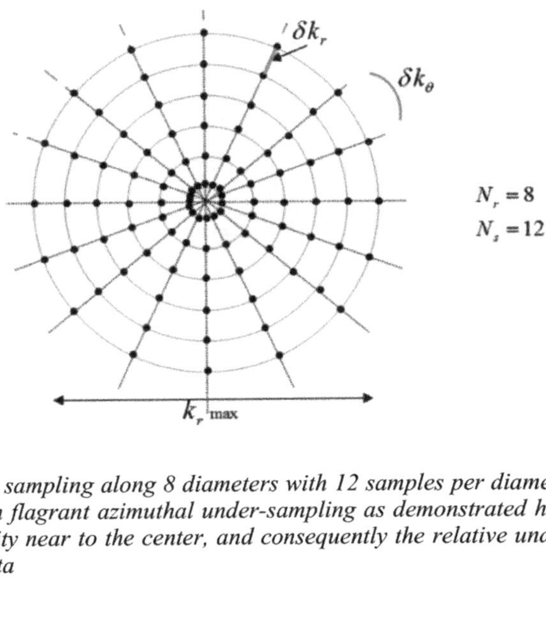

Figure 2.68. *Radial sampling along 8 diameters with 12 samples per diameter, viewed in the k -space. Even with flagrant azimuthal under-sampling as demonstrated here, note the very high sampling density near to the center, and consequently the relative under-representation of the peripheral data*

The azimuthal under-sampling results in the apparition of stria in the reconstructed image (Figure 2.66 (c and d)). The more usual "radial" (or "diametral") under-sampling causes aliasing of the profiles, as shown by the simulation in Figure 2.69 (a). The two aforementioned types of artifacts (radial and azimuthal) can be interpreted by considering the two forms of sampling individually, and then combining them. This leads us to modify the expression and consequently the representation of the function PSF(0) determined above, with a continuous model. Interested readers will find models, calculations and results in the various works published on this subject. Note simply that when sampling is discrete along the radial dimensions but the representation is continuous along the angular dimension, we obtain a PSF which depends only on the diameter $k_{r\ \max} = N_s \cdot \delta k_r$, where N_s is equal to twice the number of sampling circles such as those shown in Figure 2.68. Radial under-sampling is then illustrated in Figure 2.69 (b), where it simply appears as the result of the deformation of the profile presented in Figure 2.69 (a).

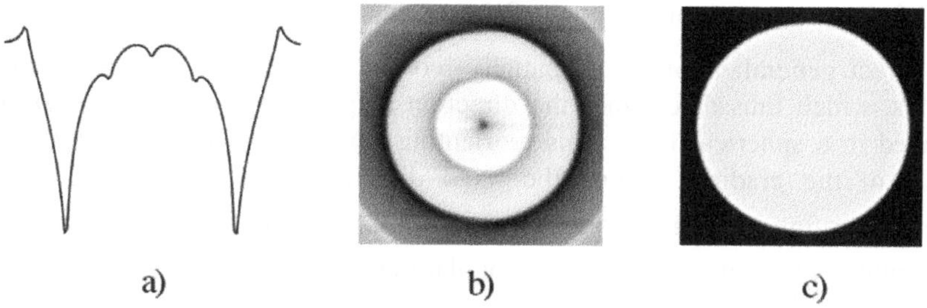

Figure 2.69. *a), b) radial under-sampling in the case of the circle reconstructed in Figure 2.66 c): left to right, example of an aliased profile represented at 128 points in a); then the reconstructed image in b) using azimuthal sampling such that* $2N_r > \pi N_s$, *with* $N_s = 128$, $N_r = 402$. *This grid can also be used in the presence of radial sampling without aliasing, as in c), to supplement the simulation in Figure 2.66(c)*

A "radial" spread function PSF(r) is represented in amplitude image mode in Figure 2.70 (a). Corona-type artifacts appear when the diameter of the circular field being explored attains the value $1/\delta k_r$. Figure 2.70 (b) shows an amplitude image of the function PSF(θ) obtained with under-sampling of the angle θ for a limited value of the variable r, which cannot reasonably be allowed to increase to infinity.

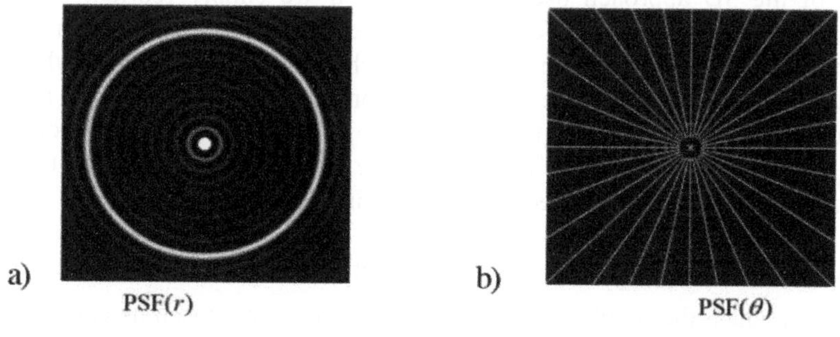

Figure 2.70. *a)* PSF(r) *simulated with 24 points per diameter. b)* PSF(θ) *simulated with 16 angular orientations for 180°. Representations by 128 × 128 simulations, with calculations performed directly on a square grid*

The simple examples in Figure 2.70 (a) and (b) result from purportedly continuous distributions of samples in the angular and radial directions respectively. More generally, the PSF appears to be the inverse Fourier transform of the sampling function. These elements specific to the method of projection/reconstruction, which we have simply touched on above, are also to be found in 3D techniques, for which we shall now briefly discuss the two approaches employed in MRI.

2.5.3.1.3. Projection/reconstruction in 3D

A direct generalization is to spatially reorient a gradient vector with constant modulus, which thus determines the direction of projection of all the information contained in a spherical field. This reorienting can be envisaged, in particular, by rotation of the gradient around the fixed direction OZ of the frame (R) by discretely varying its longitude φ in a circular manner, and then the colatitude θ of its extremity. There are therefore two angular parameters (θ and φ) to be sampled, so that:

$G_Z = G_0 \cos(l\delta\theta)$, $G_X = G_0 \sin(l\delta\theta)\cos(m\delta\varphi)$, $G_Y = G_0 \sin(l\delta\theta)\sin(m\delta\varphi)$ where l and m are integers ranging from 0 to L_θ and M_φ respectively, which gives us $(L_\theta + 1)(M_\varphi + 1)$ reading directions associated with the gathering of K values of k_r along a radial distance whose Cartesian coordinates are given by:

$$k_z = k_r \cos(l\delta\theta),\ k_x = k_r \sin(l\delta\theta)\cos(m\delta\varphi),\ k_y = k_r \sin(l\delta\theta)\sin(m\delta\varphi).$$

Another way of creating an image of volume, proposed by P. Lauterbur, is to use only 2D projections/reconstructions, as illustrated by Figure 2.71. This interesting variant of the 3D approach uses the fact that if we simply reorient the reading gradient whilst keeping it parallel to a plane, we can reconstruct a 2D image which is the projection of the object on that plane. Such planes are distributed cylindrically around the object, and therefore have a common direction. For each side along that direction U in Figure 2.71, and for each 2D projection, there is a corresponding 1D profile which can be used to reconstruct the corresponding slice at that level. Consequently, it is possible to reconstruct a 3D image, one slice at a time.

The recording therefore requires the acquisition of N_{pro} profiles for each of the projection planes, the number of which may, in turn, be chosen as equal to N_{pro}. This leads to a duration of measurement proportional to $(N_{pro})^2$, as in 3D MRI with two phase encodings. Note, finally, that the "2D profiles" are obtained on the basis of 1D profiles covering the whole of the object being imaged. These 1D profiles can be calculated by Fourier transform of the received signal, and then "filtered" – i.e. weighted with a view to backprojection.

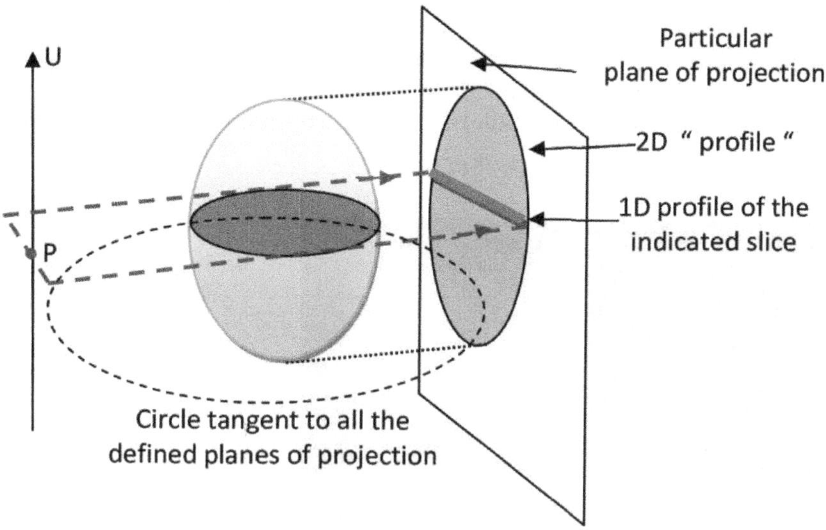

Figure 2.71. *2D projection/reconstruction used for 3D imaging. By rotating a plane of projection around the direction PU, and therefore around the object, we collect 2D "profiles" of the object. In order for this to be done, a radial reading gradient must be reoriented in a plane parallel to each new plane of projection. Only one is represented here. At the elevation U of point P, we can therefore read the corresponding 1D profile. The 1D profiles collected on a given elevation are combined to give the 2D image of the slice at that level. The set of images thus obtained creates a 3D image*

2.5.3.2. *Uses of the Fourier transform*

By introducing a strategy for Cartesian scanning of the k -space with orthogonal gradients, R. Ernst in some way brought MRI into line with NMR spectroscopy using Fourier transform – a technique of which Ernst himself had previously been the main pioneer. In terms of the essential nature, we shall take from this mode of reconstruction the fact that the extent of the spatial domain explored is the inverse of the digital resolution in the k -space and that, "reciprocally", the digital resolution in the image space is the inverse of the extent of the reciprocal domain used. The relation applies along each spatial direction and its reciprocal direction. The use of the Fourier transform therefore depends on the presentation of the grid of numerical data (odd- or even-number dimensions of the sides) and on the way in which that operation is defined depending on the flexibility of the processing tool being used. We shall not go into detail on these issues; however, we shall look at three interesting and independent elements.

The first of these is the technique known as zero-filling: understandably, the number of sampling points in the reciprocal space is limited; zero-filling (also known as Fourier interpolation) before reconstruction helps improve the definition and the appearance of the image (Figure 2.72). In this way, the dimension of the image shown can be rendered greater than the dimension of the image acquired. A simple and effective way to look at the contribution of zero-filling is to examine its influence on a 1D image (or profile) as shown, later on, by the example in Figure 2.73.

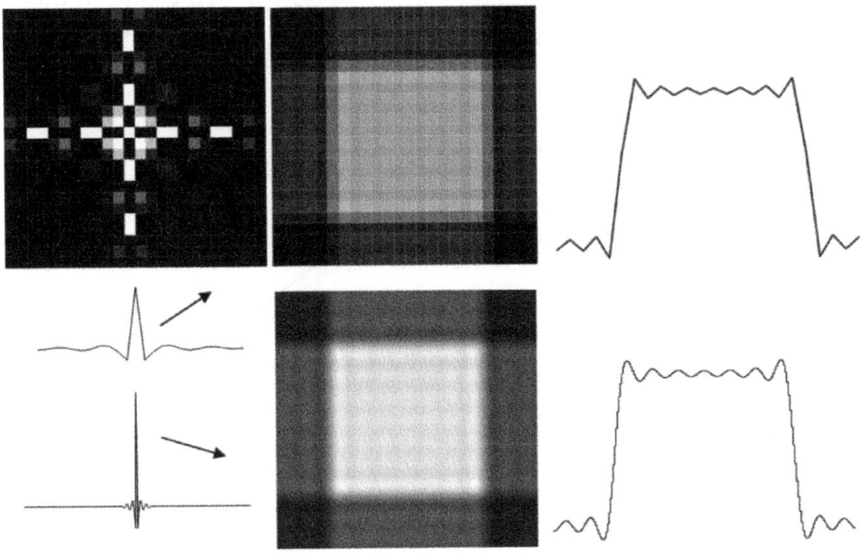

Figure 2.72. *Example of zero-filling in the case of an image heavily artifacted by the truncation effect from an acquisition at 24 × 24 points in the reciprocal space. Top, left to right: image of the samples in the k-space evidentiated by linear grayscale; corresponding image with modulus reconstructed by Fourier transform, and grayscale through a horizontal transverse segment (24 points) at halfway height of the image. Bottom, left to right: echo signal in the absence of phase encoding with sampling at 24 points for the upper plot and after zero-filling at 88 points on both sides of the echo for the lower plot; modulus image interpolated at 200 × 200 points; grayscale through a horizontal transverse segment (200 points) at halfway height of the image. Simulation without weighting or damping by transverse relaxation. As there is an even number of points and the sampling enables us to see the peak of the echo, note the resulting dissymmetry of the representations. This dissymmetry can be corrected by shifting of the sampling and correction of the phase then introduced*

The second element we shall look at relates to the SNR: while it is able to modify the definition of the image, zero-filling has no effect on the SNR. However, the dimension of the matrix of points actually sampled during the acquisition of the signal plays an important role. With a 3D matrix, resulting from $N_1 \times N_2 \times N_3$ measurements, the SNR per pixel becomes proportional to $\sqrt{N_1 N_2 N_3}$, which justifies the interest of 3D techniques in terms of sensitivity.

Thus, another advantage stems from the fact that because of its formulation, a Fourier transform redistributes the values of the original samples in the final samples. Consequently, the noise belonging to the samples of the temporal signal is redistributed in the samples of the image signal. When this process is used, the noise in the image results from the superposition of different elements of noise. As a result of the Central-Limit Theorem, the final noise observed in the real part or in the imaginary part of the image is practically Gaussian once the values of each of the numbers (N_1, N_2, N_3) reach a few tens. Of course, the noise of the modulus image $(\text{Re}^2 + \text{Im}^2)^{1/2}$ will therefore be Rician in nature, in this case.

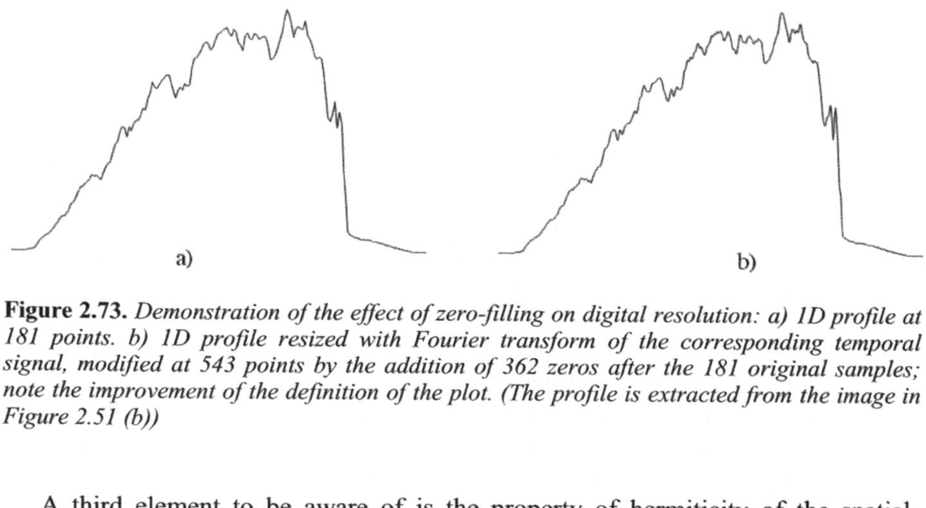

Figure 2.73. *Demonstration of the effect of zero-filling on digital resolution: a) 1D profile at 181 points. b) 1D profile resized with Fourier transform of the corresponding temporal signal, modified at 543 points by the addition of 362 zeros after the 181 original samples; note the improvement of the definition of the plot. (The profile is extracted from the image in Figure 2.51 (b))*

A third element to be aware of is the property of hermiticity of the spatial-frequency *k*-space, which has long been exploited in Fourier optics, resulting in the fact that $s(-\vec{k}) = s^*(\vec{k})$ by virtue of relation [2.16]. This makes it possible to use a half-reciprocal space to reconstruct the image.

Note that separating the samples into two "half" spaces depends on the evenness of the number of samples in the direction perpendicular to the line dividing the two sub-*k*-spaces. In Figure 2.74 (a), this number is odd, and the vertically median samples are used in the reconstruction of image (b).

This remark leads us on to look at approaches which are able to reduce acquisition times and limit the duration of certain examinations.

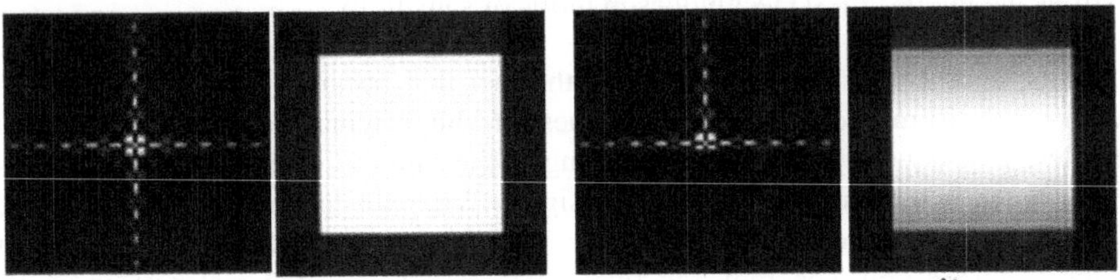

a) b)

Figure 2.74. *a) Image reconstructed using the whole of the reciprocal space sampled at 65 × 65 points; b) image reconstructed only with the upper part of the k-space comprising the median samples. In order to prevent the undulations relating to truncation which appear in the image b), we can also reuse the upper part of the data in b) to replace the missing measurements. Simulation with a 65 × 65 initial matrix and filling with 64 zeros (129 × 129 image)*

2.5.3.3. *Particular or incomplete scans of the reciprocal space*

The major advantage to the Fourier transform stems from the rapidity of its digital execution with modern computation techniques. Consequently, its use as a reconstruction tool can be envisaged for *a priori* random sampling, if it is possible to reconstitute a Cartesian grid of data in the *k*-space. However, certain descriptions of that space, which may contain rather non-uniform (and therefore incomplete) sampling, as happens with radial or spiral approaches, lead to an appreciable time gain in terms of image acquisition. It is for this reason that techniques for reconstitution of Cartesian data grids ("*gridding*" techniques) have developed in all domains of tomography; these techniques are, of course, applicable to MRI. These techniques using "incomplete" scanning of the reciprocal space can include the irregular samplings imposed by various conditions encountered in imaging procedures. Consequently, compensations are frequently necessary in fast imaging, e.g. when we need to sample the signal during the instauration of the reading gradient. In light of the nature of the subject, the corresponding developments have a predominant place in signal processing methods.

2.5.3.4. *Parallel imaging*

Capturing images in the shortest time possible is a point of competitiveness in the use of MRI as a medical investigative tool. We can exploit the linking of resonators in a network, operating simultaneously on limited and complementary portions of the field of view. The effectiveness of this parallelism can be enhanced by an additional saving in terms of recording time, achieved by reducing the number of lines of the reciprocal space that are scanned. This necessarily imposes a mode of reconstruction which is able to take account of these two conditions: using multiple sensors whilst limiting the recordings. Hence, the majority of the publications on this subject have focused on calculations appropriate for this situation, and therefore based on different algorithms to those envisaged in our discussion up until now. Calculations in the image space and calculations in the reciprocal space then define two distinct orientations, which we can view intuitively based on what we know about MRI using the 2D Fourier transform.

The basic principle of parallel imaging can be viewed simply as follows: (i) without modifying the total width explored in that space, it is possible to reduce the time taken to scan the reciprocal space by eliminating the steps of phase encoding. If we "skip" one in every two lines, then the "acceleration factor", also known as the reduction factor, is $R_a = 2$. Reconstruction by Fourier transform reduces the field of view by a factor of two. For a receiver coil which sees the whole of the object being imaged, the reconstructed image is aliased in the direction of the phase-encoding gradient. If (in the ideal scenario) the coil sees only strictly half of the object – the half in the direction of the phase encoding – that half appears in the image. (ii) If, for this experiment, we use a second resonator which sees strictly (ideally) only the other half of the object, the corresponding image shows that half. Using this array of two very specific receiver coils, we can reconstruct the whole of the object from an acquisition which is twice as quick as normal. Before expanding this discussion to multiple coils and a reading of increasingly widely spaced lines in the reciprocal space, it must immediately be mentioned that the scenario indicated above is merely an example which is not remotely possible in practice, because the field of view of each sensor greatly overlaps those of its neighbors (and vice versa) on the one hand, and on the other, that the line of phase encoding which are absent from the field of view for one sensor may exist for some of the neighboring sensors. Consequently, the correct reconstruction of an image will necessarily require specific processing.

2.5.3.4.1. Parallel imaging using the SENSE (*SENSitivity Encoding*) technique

This too uses the "Fourier" model, where each sensor yields an image which is aliased in the direction of phase encoding, with a small number of lines recorded and a large number of lined ignored. The processing is performed by beginning with the images taken by the sensors distributed around the field of view along the direction of the phase-encoding gradient, here taken to be parallel to OY, whilst the reading gradient will be taken as parallel to OX in the laboratory frame.

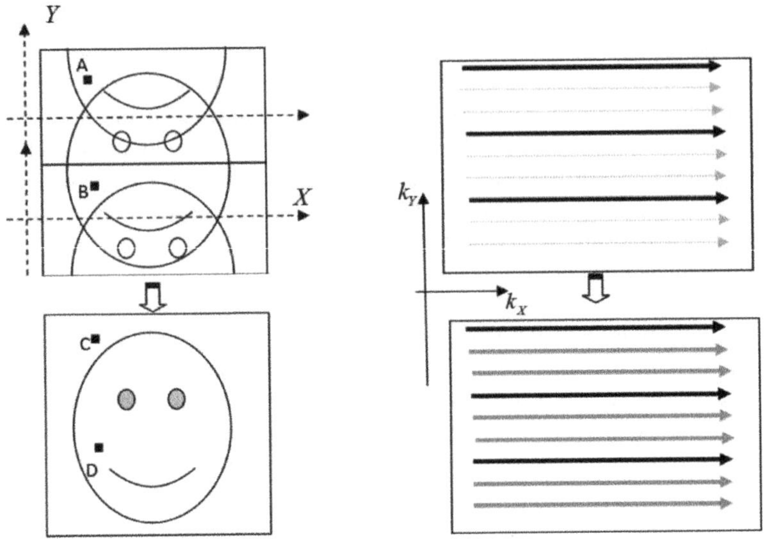

Figure 2.75. *Representation of the principles used in the SENSE method (left) when $R_a = 2$, where all the data gathered and superposed at each of the pixels A and B are distributed correctly at C and D. SMASH method (right) when $R_a = 3$, where we suppose a complete reconstitution of the Fourier lines missing originally (wide, solid gray lines) in the direction of phase encoding given by k_X. The computations are simply represented by the symbol* ⇩

We can therefore consider the image signal taken at the same pixel characterized by its coordinates *X* and *Y* in each of the images obtained by the sensors, of which there are a number NC. Thus, for that point, we have as many different signals, *a priori*, as there are sensors. These signals result from aliasing which differs when we switch from the image from one sensor to the image from another. The composition of that aliasing depends on the spatial sensitivity distribution of that sensor. For X and *Y*, this enables us to write NC relations similar to relation [2.43] – relations which are acceptable in view of the problems of noise, provided the coils of the different sensors are appropriately decoupled:

$$s_n(X,Y) = \sum_{r=1}^{R_a} C_n(X,Y_r) \cdot S(X,Y_r) \text{ where } n = 1,\ldots,NC \qquad [2.43]$$

where $C_n(X,Y_r)$ corresponds to the spatial sensitivity distribution of the resonator n, $S(X,Y_r)$ is the image signal sought at the point (X,Y_r), knowing that r can take R_a values as a consequence of the aliasing. The aim of the processing is therefore to use our knowledge of the sensitivity distribution to work back to the values of $s_n(X,Y)$, i.e. to the complete image, by inversion of relations such as relation [2.43], which can be obtained with the set of points provided by the set of NC aliased images available to us. The conditions of realization of that operation will not be analyzed here. It should be mentioned that we can show that the SNR obtained is inversely proportional to the value fixed for R_a.

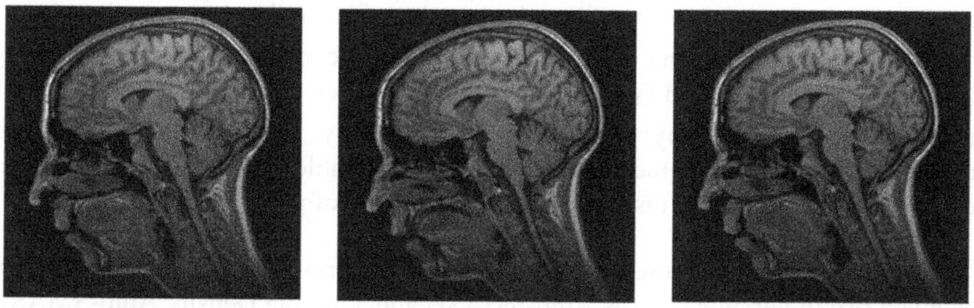

a) b) c)

Figure 2.76. *Three slices extracted at the same level of 3D images on the same subject, obtained with different reconstruction techniques: a) Parallel imaging using the GRAPPA technique (a variant of the SMASH approach); b) Parallel imaging using the SENSE technique. In both these situations, R = 2 and the time taken to acquire the images is 4.37 minutes. c) Case where R = 1, the time taken to acquire the images is 8.74 minutes. The basic sequence employed in all three cases is an MP RAGE sequence identical to that used in Figure 2.63 (TR = 1970, TI = 1100 ms, TE = 3.93 ms, flip angle on excitation 15°, 176 slices of 1 mm thickness). System 1.5-tesla field and birdcage headcoil with 8 channels. There is no appreciable difference between these three images. Images provided by D. Sappey-Marinier, Cermep, Creatis*

2.5.3.4.2. Parallel imaging with SMASH (*SiMultaneous Acquisition of Spatial Harmonics*) technique

The notion of the spatial sensitivity of each resonator still plays an important part, but this time we will consider its value $C_n(X,Y_r)$ for each point (X,Y) of the definitive field of view which therefore encapsulates the geometry of the whole object (e.g. slice selected if we use 2D MRI). Consequently, with the usual notation employed in Fourier imaging, the signal collected by all the sensors, viewed in the reciprocal space, must – using the general formulation of terms in the k-space seen previously in section 2.2 – be written as:

$$s(k_X,k_Y) = \int_X\int_Y \rho(X,Y)\cdot\sum_n^{NC} c_n C_n(X,Y)\cdot e^{j2\pi k_X X}\cdot e^{j2\pi k_Y Y}\,dXdY \qquad [2.44]$$

Suppose that we can approximate the weighting term $\sum_{n}^{NC} c_n C_n(X,Y)$, where c_n represents the contribution of each resonator, by a harmonic formula $e^{j2\pi\cdot\delta k_Y\cdot Y}$. With this condition accepted, we then have $e^{j2\pi\cdot\delta k_Y\cdot Y}s(k_X,k_Y)=s_0(k_X,k_Y+\delta k_Y)$, which is tantamount to expressing a term from the line $k_Y+\delta k_Y$ in the k-space. As soon as we have recordings coming from each resonator, we can modify the weighting coefficients c_n so as to modify the value of δk_Y, and thus find new trajectories in the reciprocal space simply by calculation. Consequently, this method is effective, provided we have arrangements of coils which are able to create the harmonics necessary in the sensitivity distribution in the direction of phase encoding.

In spite of this restriction, the SMASH principle proves relatively fruitful, as it has given rise to variants such as Auto-SMASH and Variable Density Auto-SMASH (VD-Auto-SMASH), which lead to the densification of the k-space on both sides of the line $k_Y=0$ facilitating the reconstruction of the images. A generalization of these variants is therefore possible, using the method known as GRAPPA (GeneRalized Autocalibrating Partially Parallel Acquisition) which, in summary, repeats the process of sensitivity calibration on a coil-by-coil basis, which of course complicates the reconstruction process but provides greater flexibility in the configuration of the sensor array.

Parallel methods such as SENSE or GRAPPA, which thus prove useful for observations in the dynamic regime, are also envisageable with radial or spiral sampling. Readers interested in the methods briefly mentioned here can consult a very abundant body of literature on the subject, published since the late 1990s.

In addition to the significant progress made in the field of MRI over the past three decades, techniques in spectroscopy have also progressed, having also developed with a view to clinical applications. In the targeted applications discussed below, in the final part of this chapter, an important part is given to magnetic resonance spectroscopy (MRS). This constitutes a major axis which exploits the progressive rise of the values of magnetic fields and borrows aspects from imaging procedures, as shown by the example given in section 2.1.

2.6. Application of MRI: uses and evolution in the biomedical field

2.6.1. *Introduction*

Beyond the imaging of the protons of water (and lipids) which generally provide information about the anatomical representation of the human body and its tissular and compartmental makeup, NMR can be used to obtain other information, relating to a biological or cognitive function, a metabolic pathway or a physiopathological mechanism. Today, it is possible to acquire this information thanks to the development, over the past two decades, of a set of methods which can be categorized as "functional imaging". While some of these techniques are now being used in the radiology departments of hospitals and universities, their main use relates to fundamental questions which require complex methodology and constant evolution. This leads us to classify them as advanced research methods and applications.

Among these functional MRI methods, we can distinguish a number of main axes of research depending on the type of information obtained and the physical parameter(s) used to generate contrast. In keeping with the techniques used in conventional MRI, one group of activities relates to the acquisition methods employed to visualize, and sometimes quantify, all the forms of transport or exchanges between water molecules, with or without the use of a contrast agent. These methods relate primarily to the techniques of perfusion MRI, diffusion MRI or magnetization transfer MRI. A number of points have already been touched on, briefly, such as perfusion MRI with a contrast agent, or Arterial Spin Labeling discussed in section 2.2.6.1 and in section 2.4.5.2 for magnetization. In the final part of this chapter, we shall linger a little further over diffusion MRI, which is an extremely promising technique because of the essential data it provides to help elucidate the micro-architecture and connectivity of the brain, and to aid the characterization of neurological diseases.

Another important group of research activities relates to the methods used to examine the mechanisms of brain oxygenation. Associated with the mechanisms of capillary perfusion, hemodynamic variations linked to neuron activity can cause variations in the NMR signal which are measurable using the techniques of so-called "functional" MRI (fMRI). Based on the principle of BOLD (blood oxygen level dependent) contrast, variations of deoxyhemoglobin in the blood can be demonstrated during neuron activation. This technique which, today, is well known and mastered in terms of acquisition by the EPI (echo planar imaging) sequence, enables us to examine the cognitive functions, from the very simplest (the motor or sensory functions), through the exploration of language and working memory, to the most complex of functions (planning, emotions, social relations, etc.).

Finally, a third group relates to the detection of molecules participating in cellular metabolic phenomena, or in the mechanisms of neurotransmission. This is the domain of magnetic resonance spectroscopy (MRS), or indeed its offshoot in the form of "spectroscopic" imaging or "chemical shift" imaging. This metabolic imaging is capable of creating images revealing the distribution of metabolites in a particular organ. Estimation of the concentration of these intracellular metabolites in turn gives us access to fundamentally important information such as intracellular pH, membrane or lipid metabolism, cell or neuron viability, degree of ischemia or tissue inflammation, or to quantify the concentration of certain neurotransmitters.

Thus, we propose to focus, in this final part, on three major methods of functional MRI, first giving an overview of magnetic resonance spectroscopy (MRS), then an analytical description of diffusion MRI, followed by a brief presentation of cerebral activation MRI, known as fMRI. To conclude, we shall touch on multimodal MRI, taking the example of the combination of MRI and positron emission tomography (PET), which is the most advanced hybrid system available today for clinical applications.

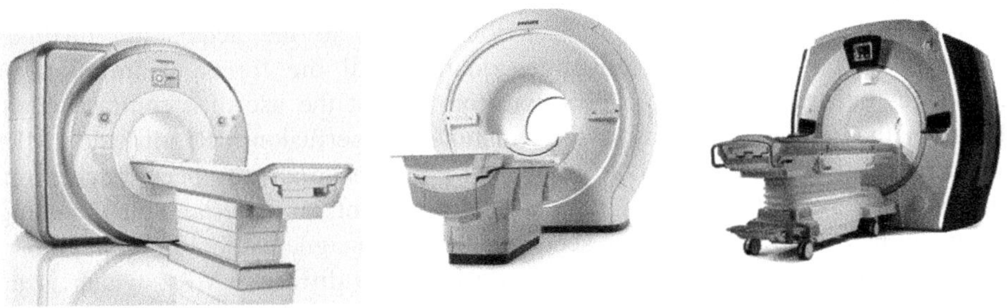

a) b) c)

Figure 2.77. *Photos of 3-tesla MRI systems marketed by: Siemens a); Philips b); and General Electric c). Documents provided by the manufacturers*

2.6.2. *Spectroscopy and imaging: technical and clinical complementarity*

Of the medical applications of NMR, imaging is, beyond compare, far better known and much more widely used than spectroscopy. Why is there this difference? Is it a problem of medical advantage, technical complexity or scientific culture? In order to answer this question, let us briefly examine the technical peculiarities and additional constraints encountered with the application of spectroscopy.

2.6.2.1. *Technical aspects*

The earliest clinical systems for magnetic resonance spectroscopy (MRS) using a full-body magnet operating at 2 teslas were installed around 1985 – about five years after the earliest imaging systems. Finally, it was possible to take metabolic observations of the human brain. Some years earlier, preliminary studies, carried out with magnets whose air gap was limited to between 20 and 30 cm, had already facilitated significant experimentation on the muscle of the arm or the leg by phosphorus-31 spectroscopy. Its spectrum, obtained by Fourier transform of the NMR signal observed in a uniform magnetic field, is composed of the resonance peaks of the different chemical groups of phosphate molecules: essentially Inorganic Phosphate (Pi) Adenosine triphosphate (ATP), Phosphocreatine (PCr), which are present in sufficient concentration (respectively around 4, 8 and 35 milli-moles per kilogram at a pH of 7 in muscle) and other metabolites such as phosphate mono-esters (PMEs) and di-esters (PDEs), which are respectively the precursors to and derivatives of membranous phospholipids. The relative mobility of these molecules, mostly present in the cell cytoplasm, can be seen by values of the relaxation time T_2 of the phosphorus nucleus which are sufficiently high to lead to resonance line widths that are not excessive. Also, the spreading of the ^{31}P spectrum of the aforementioned metabolites facilitates the detection of the phosphorus nucleus *in vivo*, although this nucleus is significantly less sensitive than protons are. The efficacy of the spectroscopic method improved with the increased value of the magnet's static field which, proportionally, increases the spectral spreading and heightens sensitivity. In order to increase the power of spatial resolution, we need to maintain a sufficiently small line width, at the cost of a correction of the homogeneity of field; this correction must perform better in proportion to the value of the field. Thus, proton spectroscopy in humans can be performed at 1.5 teslas (resonance frequency 64 MHz) and at that value, with brain imaging, we are able to obtain a line width at halfway height of around 15 Hz, which is equivalent to 0.24 ppm (parts per million), while the extent of the proton spectrum is less than 10 ppm. The switch to 3 teslas and then to 7 teslas provides a significant improvement in terms of the absolute spreading of the spectrum, and of sensitivity, which increases a little more quickly than does the value of the field. The development of NMR with very strong fields therefore proves very useful for delivering cerebral spectra comprising ten or more metabolites, whose identification and quantification are essential for understanding and monitoring numerous diseases.

Proton spectroscopy (^{1}H) requires a technical setup very similar to that used in MRI, as we are observing the same type of nucleus. However, unlike with MRI, where the signal is essentially generated by the very abundant water molecules, and also lipids in high concentration in the tissue, spectroscopy seeks to evaluate the tissue's content in terms of molecules whose metabolic role is very important but whose concentration is far less. For this reason, we have to use magnetic fields stronger than (or equal to) 1.5 teslas in order to achieve limits of detection below a milli-mole/kg (Figure 2.78). However, as the effects of susceptibility caused by the patient are also accentuated with a higher value of the field – particularly at 7 teslas – the expected gains in terms of sensitivity and spectral resolution can sometimes be attenuated by these phenomena, which affect the uniformity of the static field.

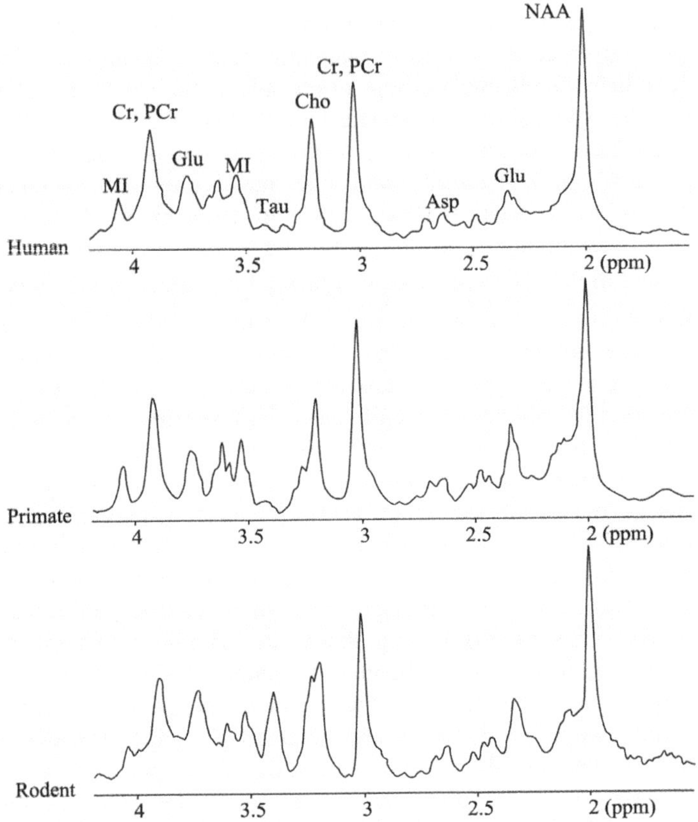

Main metabolites: NAA = N-Acetyl Aspartate ; Glu = Glutamate/Glutamine ; Asp = Aspartate ; PCr = Prosphocreatine ; Cr = Creatine ; Cho = Choline ; Tau = Taurine ; MI = Myo-inositol

Figure 2.78. *Proton spectra for the brain in a human being, a primate and a rat, obtained with a 7-tesla field, a frequency of 300 MHz in the region of 1.5 to 4.2 ppm. Note the similarities and differences between these spectra. Documents kindly provided by F. Boumezbeur, CEA/Neurospin and J. Valette, CEA/MIRCen*

There are two possible options. With the first, unlike what happens in MRI, the NMR signal comes from a small volume of interest, selected by a set of selective pulses: very briefly, the selection of three orthogonal slices allows us the possibility of examining the voxel situated at the intersection of these three volumes (this is the so-called CHESS (*CHEmical Shift Selective*) sequence). As with imaging, we have different sets of sequences based on gradient switching in order to create this localization of observation where the field is as spatially homogeneous as possible. With the second option, which was mentioned in section 2.2 with the introduction of additional dimensions to the image, we use the discriminating capability of imaging to record the spectrum of each pixel in 2D imaging, or of each voxel in 3D imaging (which is a considerably longer process). This gives rise to spectroscopic imaging, where each element in the image (pixel × slice thickness or voxel thickness) exhibits a particular spectrum. The two approaches used in MRS are illustrated for 2D modeling in Figure 2.79. In metabolic imaging, the distribution of each spectral component can appear in the form of an image (see Figure 2.23, section 2.2.6.3) and in principle there are the same number of images as there are resonances, each characterizing a particular metabolite.

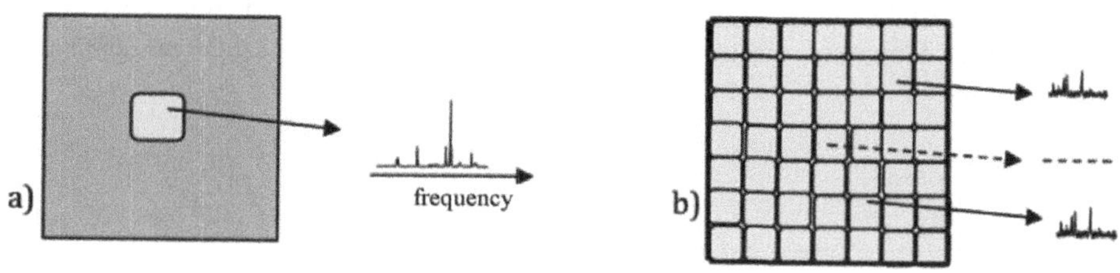

Figure 2.79. *Difference between Localized Spectroscopy and Spectroscopic Imaging: a) Localized Spectroscopy using a volume of interest selected by gradients as in the Point RESolved Spectroscopy (PRESS) sequence, where detection takes place in double spin-echo mode. b) Spectroscopic Imaging leads to the obtention of a single spectrum for each voxel (pixel multiplied by the slice thickness) in a 2D slice. The sequence is based on two phase-encoding gradients, with the second replacing the reading gradient. Its overall duration is proportional to the product of the number of lines by the number of columns in the image grid. Only two spectra are illustrated here*

The common point between these two approaches is that it is necessary to eliminate the signal from the water and lipids, whose molecular concentrations (respectively 50 and around 1 mole/kg) are too high in comparison to those of the metabolites, and therefore cannot be represented on the same scale. With this in mind, certain strategies stem from NMR spectroscopy methods such as the combination of pulses to attenuate the signal from the water and lipid protons, and/or appropriate digital data processing methods, the difficulty of which can be seen in Figure 2.80. By measuring the intensity of the resonances of each spectrum, we are able to estimate the average tissue content in each voxel, and represent that result in the form of metabolic images of each significant spectral component (Figure 2.79 b). For reasons of sensitivity and also of measuring time, the digital resolution of these images is reduced (16 × 16, 24 × 24 or 32 × 32).

The information contained in a voxel of the metabolic image may therefore come from distinct tissue elements (partial volume effect), caused by the spreading of the information between neighboring voxels under the influence of the PSF (see section 2.4). This aspect makes it difficult to quantify the concentration of metabolites with absolute certainty. Therefore, we perform relative quantification by calculating the concentration ratios between the metabolite in question and a reference metabolite. If we wish to obtain absolute quantification, it is necessary, firstly, to have an absolute reference, be it internal – such as the creatine or water peak, acquired previously – or external, set by using a test object (a "phantom"). The first approach, using an internal reference such as creatine, has the advantage of being very simple, but it is founded on the debatable hypothesis that the concentration and relaxation times do not change depending on the anatomical regions and diseases under examination. On the other hand, the use of water as a reference enables us to circumvent variations of this type, because of its high and relatively stable concentration. However, tissue- and disease-related variations of 10 to 20% may occur.

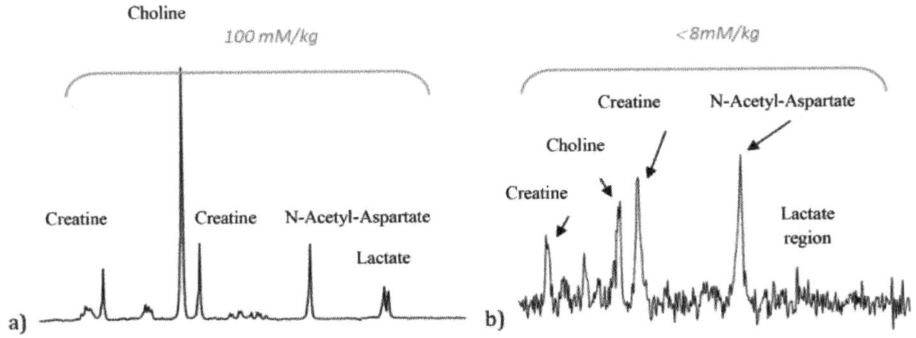

Figure 2.80. *Main cerebral metabolites of proton spectrm (shown between 1 and 4.3 ppm), observed at 200 MHz: a) spectrum of a water solution of Creatine, Choline, N-Acetylaspartate and Lactate at equal concentrations of 100 mM/kg. b) Spectrum obtained in vivo in a healthy rat brain, over the same sensitive volume (4 × 4 × 4 mm³), with the same acquisition parameters. The lactate signal is swamped by noise in b), whilst the two peaks are clearly separated in the solution spectrum. PRESS sequence is used here*

Finally, absolute quantification based on the measurement of a reference of known concentration based on an external phantom necessitates the taking into consideration of the variations in relaxation time and acquisition parameters between the two measurements. Thus, quantification is a step where great caution must be exercised. The processing of spectra and metabolic images is still not easily accessible for clinical use. However, analysis methods capable of the identification and quantification of the signals have developed over the course of the past two decades. These methods have gained widespread recognition – as happened, for instance, with the software tool jMRUI – through international programs involving research hubs and manufacturers.

2.6.2.2. *Observation of different nuclei and complementarity of MRI and MRS*

Thus, proton MRS does not require any highly specific technical equipment in addition to that used in MRI, with the exception of the acquisition sequences and a sufficient guarantee as regards the homogeneity of the field. Its use in clinical practice is increasing for a number of reasons. Cerebral observation techniques are capable of detecting numerous metabolites – here we shall restrict the number of examples. Thus, they represent one way of evaluating neuron integrity using the N-acetyl-aspartate (NAA) signal; monitoring energy metabolism by measuring the creatine (Cr) signal, or membrane metabolism by looking at the signal from molecules containing choline (Cho); and detecting the presence of anaerobic glycolysis which causes an increase in lactate levels, indicating cell hypoxia or ischemia. By altering the acquisition parameters, e.g. by adjusting the echo time, it is possible to monitor the variations in concentration of other molecules, such as certain neurotransmitters, glutamate, gamma-amino-butyric acid (GABA) and amino acids such as glutamine, alanine or taurine, or finally myo-inositol, which is specific to inflammatory activity of glial cells. Such acquisitions with a short echo time – $T_E < 30$ milliseconds – are tricky, and require optimum experimental conditions: in particular, excellent homogeneity, good visibility of the spectral region of interest and effective elimination of the signals from water and lipids, which are liable to mask the lactate signal.

Although its use for imaging is very infrequent, phosphorus-31 (spin 1/2, resonance frequency ~ 0.4 times that of the proton, abundance 100%) is used predominantly in spectroscopy. Facilitated by the installation of horizontal magnets whose aperture diameter (30 cm) was only sufficient to examine an arm or a leg, the earliest clinical applications of phosphorus MRS studied muscular metabolism. In addition, ^{31}P MRS is the only technique used for *in vivo* analysis of the intracellular pH, obtained by measuring the chemical shift of the inorganic phosphate (Pi) peak in comparison to that of phosphocreatine (PCr). The presence of phosphorus in phospholipids can also be exploited to study membrane metabolism. This information is particularly useful for early therapy follow-up of tumors, whose very fast membrane metabolism leads to a high demand for metabolic precursors such as choline, and causes an increase in the peak of phosphomonoesters (PMEs). In the brain, phosphodiesters (PDEs), derived from the metabolization of phospholipids, also present an advantage for monitoring alterations in the level of myelin. Although it is relatively easy to detect the phosphorus nucleus, the low value of the relaxation times T_2 of the metabolites necessitated fast acquisition of the FID signal, which occurs after the excitation, and the development of specific localization sequences such as ISIS (Image Selective In Vivo Spectroscopy).

Spectroscopy using Carbon 13 (spin 1/2, resonance frequency ~ 0.25 times that of the proton, abundance 1.1%) is also highly advantageous for monitoring intermediary mitochondrial metabolism (Krebs Cycle) and understanding cell oxidative metabolism – particularly in brain diseases such as stroke or epilepsy. However, in view of its rarity (1.1%) in this isotope form, it is necessary to use molecules synthetically enriched with Carbon 13, or employ techniques of dynamic polarization to increase its sensitivity, and gradually build up to the imaging of that isotope.

NMR observation of sodium 23 (spin 3/2, resonance frequency ~ 0.26 times that of the proton, abundance 100%) is a relatively little-used spectroscopy method. Its use for imaging, which has been envisaged since the 1980s, is today experiencing renewed interest with the clinical installation of 3-tesla magnets (Figure 2.81). Present in significant concentration in the extracellular medium (140 mM/kg) by contrast to the intracellular medium (5 mM/kg), sodium constitutes a marker of the mechanisms of diffusion of water, which seeps back into the intracellular medium following the slowing or stopping of the NA/K–ATPase pumps by a pathological process.

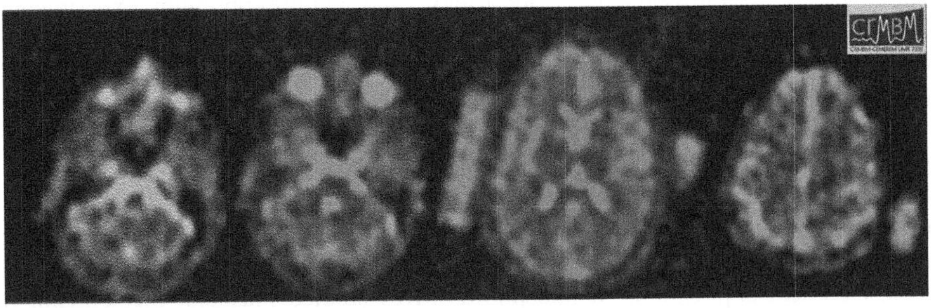

Figure 2.81. *Brain images obtained by sodium MRI at 3 teslas (resonance frequency 33.8 MHz). Radial acquisition sequence (T_E = 200 µs). The two right-hand images show the trace of a phantom image containing a sodium chloride solution. Images kindly provided by W. Zaaraoui, CRMBM/CEMEREM, Marseille*

The occurrence of these mechanisms in numerous diseases such as stroke, multiple sclerosis (MS), epilepsy or Alzheimer's disease, make them a potential marker for characterizing and monitoring these brain diseases. In spite of the fact that its sensitivity is less than proton imaging, sodium imaging offers the potential for heightened contrast because of its intra/extracellular mobility, and the relaxivity due to its quadrupole moment represents an excellent probe for the functional imaging of tumors.

Fluorine-19 (Spin 1/2, resonance frequency 94% that of the proton, abundance 100%) offers a high degree of sensitivity. It is not found in the human body except in the teeth and bones – two structures whose solid state makes it difficult to detect because of an extremely short relaxation time T_2 . Playing a role in the composition of perfluorocarbon molecules and oxygen vectors, its imaging *in vivo* has been performed on animal models with a view to studying pulmonary ventilation. Fluorine spectroscopy is also useful for monitoring fluorinated drugs such as 5-flourouracil or gemcitabine (2'-deoxy-2',2'-difluorocytidine) which are used in chemotherapy.

To these nuclei which are candidates both for MRI and MRS, and consequently are the subject of particular attention with a view to biomedical applications, we can add helium 3 (spin 1/2, resonance frequency ~ 76% that of the proton, and abundance only 0.000137%). It is used in gaseous form, and has the peculiarity of having a particularly long relaxation time T_1 (between a few minutes and a few days depending on the environment). In its so-called "hyper-polarized" form, it produces a significant and long-lasting signal. When introduced *in vivo* into the lungs and observed by dynamic imaging, it enables us to analyze the performances of ventilation and characterize functional shortcomings. It can also be contained in microcapsules and used as a marker for blood vessel imaging.

The examples briefly discussed here show that the complementarity between MRI and MRS manifests itself by way of the functional aspect of the observations. The first, morphological approach to MRI is well known because it is practiced on a daily basis in radiological routine, so we shall devote greater depth of discussion to its newer, functional applications.

2.6.3. *Diffusion MRI: a morphological and functional approach*

Diffusion is a phenomenon of molecular transport resulting from the thermal agitation that exists in a solid, liquid or gaseous medium. It is therefore random in nature. When molecular diffusion takes place in a very extensive medium, it is possible to approximate this phenomenon by the simple "random walk" model in a limitless domain, i.e. very large with respect to the dimensions of the molecules and to the elementary displacement in which they take part. With a very simple one-dimensional model, if X is the direction of motion, belonging *a priori* to the fixed laboratory frame, analysis of the process yields the following differential equation:

$$\frac{\partial}{\partial t}P(n,k/m,0) = \frac{a^2}{2\tau}\frac{\partial^2}{\partial X^2}P(n,k/m,0) \qquad [2.45]$$

In this formulation, $P(n,k/m,0)$ represents the conditional probability of a particle which occupied the position ma at time zero occupying the position na at the time $k\tau$, (m,n,k) being positive, negative or null integers. Here, a is the amplitude of the diffusion steps, and τ is the time interval between two successive displacements supposed to be infinitely short and equiprobable in all senses. This entity, which models diffusion on a microscopic scale, is called a "propagator".

Equation [2.45] is very interesting, because as we can immediately see, it is very similar to Fick's first law – a phenomenological equation representative of the phenomenon of diffusion at the macroscopic scale:

$$\frac{\partial}{\partial t}C = D\frac{\partial^2}{\partial X^2}C \quad [2.46]$$

where C represents the concentration of a substance diffusing in the direction X through another medium, and D the corresponding diffusion coefficient. The diffusion current J_X, used in the formulation of Fick's law, is the opposite of the term $\frac{\partial}{\partial t}C$. Thus, the diffusion coefficient is interpreted as the ratio $\frac{a^2}{2\tau}$ (which is expressed in m^2/s). In the case of diffusion of water molecules in water (here we speak of self-diffusion), the diffusion coefficient (or diffusivity) is equal to $2\times10^{-9}\ m^2/s$ at 15°C. In addition, it must be noted that the normal distribution law which can be assigned to the "propagator" $P(X,t/X_0,0) = \frac{1}{\sqrt{4\pi Dt}}\cdot e^{-\frac{(X-X_0)^2}{4Dt}}$ is indeed a solution to the differential equation [2.45] if we adapt the formulation. Hence, when diffusion takes place with no particular constraint in a given direction, we can operate in the context of this Gaussian model. If we look at 3D diffusion considered as a Gaussian process, the propagator or conditional probability density is written in such a way as to reflect the distribution of the random motions which occupy all three spatial directions and whose statistical properties are described by a 3 × 3 covariance matrix. This gives us the following formulation:

$$P(\vec{r},t/\vec{r_0},0) = \frac{1}{\sqrt{(4\pi t)^3 \det(\mathrm{D})}}\cdot e^{-\frac{(\vec{r}-\vec{r_0})^{\mathrm{T}}\mathrm{D}^{-1}(\vec{r}-\vec{r_0})}{4t}} \quad [2.47]$$

where D^{-1}, the covariance matrix of the distribution of the random motions performed during the time period t, enables us to define a diffusion ellipsoid in this Gaussian model.

This Gaussian model has to be abandoned when the molecules encounter obstacles – particularly when they are confined in cavities whose dimensions are commensurate with the amplitude of the diffusion motion. Furthermore if, in a given medium, the diffusion constant of the water molecules is less than the accepted value in a very extensive and isotropic aqueous medium, this indicates the presence of restricted diffusion. This is what happens in most living tissues.

2.6.3.1. *Role played by diffusion in NMR experiments*

With isotropic liquids, we can use a 1D description. The qualitative explanation of the role played by diffusion in a spin-echo experiment is provided by the simple phase diagram in Figure 2.82. Indeed, diffusion adds extra attenuation of the amplitude of the echo to that which results from spin–spin relaxation. The same is true for the FID signal or for any signal observed in these conditions. The variations of the magnetization of an element of volume of the sample under the influence of diffusion can be taken into account if we modify the Bloch equations.

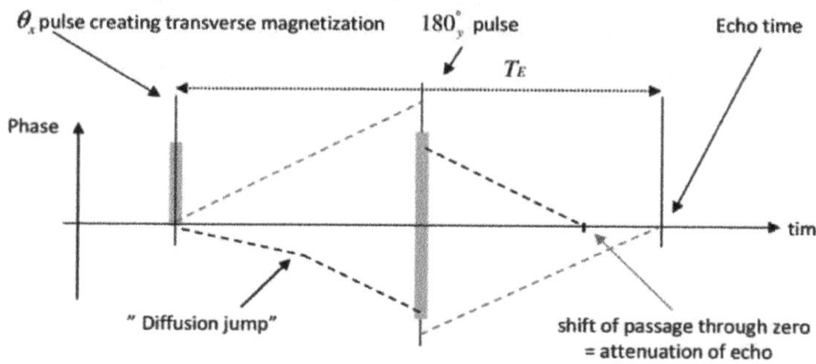

Figure 2.82. *Phase diagram qualitatively illustrating the degradation of the amplitude of a spin echo under the influence of diffusion in a non-uniform static field. Remember: the 180°y pulse changes the sign of the phase. The spins of a particle associated with the light gray plot having undergone a "displacement" in terms of diffusion values thus avoid the predicted echo time in the absence of any motion*

Torrey thus showed that the Bloch's equation satisfied by each component had to be modified identically by the addition of the term $\frac{\partial M_Q}{\partial t} = \vec{\nabla} \cdot D \vec{\nabla}(M_Q - M_{Q0})$; $Q = X, Y, Z$. In his example, the heterogeneity of the static field is considered to affect the three components of the field $\vec{B}$ counted in the laboratory frame as follows: the main field, directed along OZ exhibits a gradient with the expression $G \cdot Z$, the two components along OX and OY are due solely to the heterogeneity and have the expressions: $-\frac{1}{2} G \cdot X$ and $-\frac{1}{2} G \cdot Y$ in accordance with the law $\vec{\nabla} \bullet \vec{B} = O$. In these conditions, the amplitude of the free precession signal takes the form $A = A_0 \cdot e^{-t/T_2} \cdot e^{-(\gamma^2 G^2 t^3/3) \cdot D}$. The amplitude A_0 is proportional to the number of resonating nuclei present in the volume being observed, all affected by the RF pulse which triggers free precession in the presence of the static field and of the gradient G. We generally use the notation $A = A_0 \cdot e^{-t/T_2} \cdot e^{-b \cdot D}$, with $b = \gamma^2 G^2 t^3 / 3$, the dimension of b being a second per square meter. In practice we use one second per square millimeter. For a spin-echo sequence, Torrey's model gives us $b = \frac{1}{12} \gamma^2 G^2 T_E{}^2 t$ after the 180° pulse and consequently $b = \frac{1}{12} \gamma^2 G^2 T_E{}^3$ at the echo time.

2.6.3.1.1. PGSE (Pulsed Gradient Spin Echo) sequence, and the elementary notions of q-space

In the PGSE sequence, also known as the "Stejskal and Tanner" sequence (Figure 2.83), two identical gradient pulses are placed on both sides of the 180° pulse in a spin-echo sequence and thus form the "diffusion gradient".

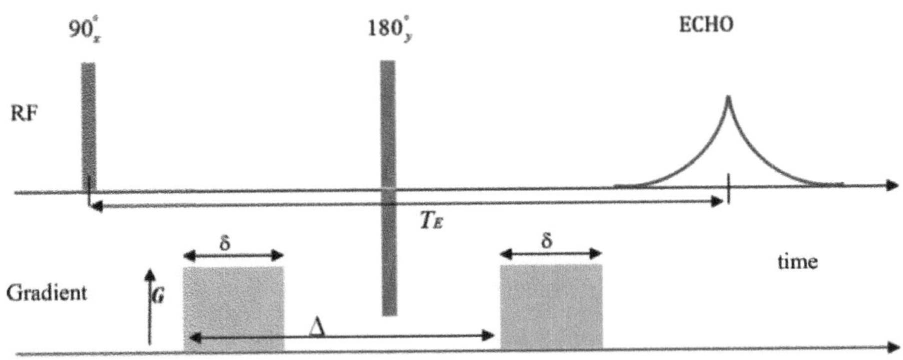

Figure 2.83. *PGSE sequence with RF pulses and idealized pulses of the diffusion gradient: the direction of the diffusion gradient **G** will not necessarily be a reference direction for the lab. Note that we can replace the 90° pulse with a pulse whose flip angle is lower*

The gradient ***G*** is intended solely to reveal and measure the diffusivity. This revelation is made possible by MRI if we superimpose on ***G*** the gradients necessary for a slice selection, a phase encoding and a reading one, thus yielding a diffusion-weighted image. By comparing the amplitudes of the echo with and without the gradient ***G***, we are able to circumvent the effect of the other gradients and envisage a measurement of the diffusion coefficient in the direction of ***G***. The 180° pulse leads, at the time of the echo, to compensation of the phase evolution relating the lack of uniformity of the field. However, it is perfectly possible to use a sequence where we simply change the sense of the gradient (Figure 2.84). In both cases, if the observation is taken once the second diffusion gradient pulse has been applied, the expression of b becomes $b = \gamma^2 G^2 \delta^2 (\Delta - \frac{\delta}{3})$.

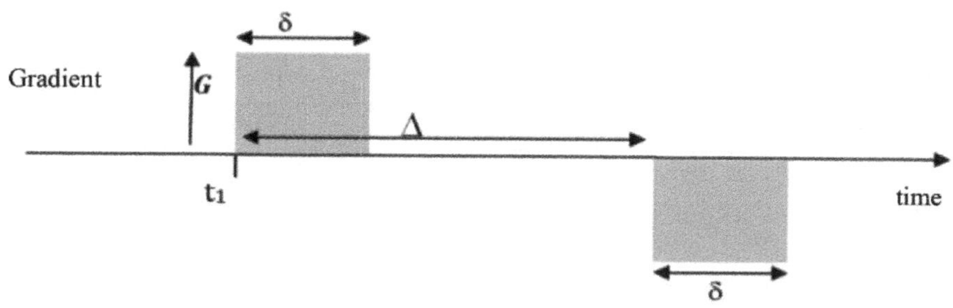

Figure 2.84. *Ordering of the idealized diffusion gradient pulses if we do not use the spin-echo technique to examine the effect of diffusion*

The interesting element defined on the basis of the PGSE sequence is the notion of the q-space. We can indeed determine the phase evolution of an elementary transverse magnetization caused by the application of supposedly ideal diffusion gradient pulses. If, at the moment of the first application of duration δ of the diffusion gradient, taken to be oriented in the sense OX, the magnetization stays at X_0, and the phase evolution acquired is $\varphi_1 = \gamma \cdot G_X \cdot X_0 \cdot \delta$. If the magnetization has shifted to $X_0 + \lambda X$ at the time of the second pulse, the phase acquired during the application of the gradient becomes $\varphi_2 = \pm\gamma \cdot G_X \cdot (X_0 + \lambda X) \cdot \delta$. The sign + corresponds to the case of the sequence shown in Figure 2.83; the sign – to that shown in Figure 2.84. By finding the overall balance $\varphi_1 + \varphi_2$ of the phase evolution caused by the diffusion gradient, for the first sequence (Figure 2.84) we obtain $\lambda\varphi_1 = \varphi_2 + (-\varphi_1) = \gamma \cdot G_X \cdot \lambda X \cdot \delta$. The sign given to φ_1 results from the presence of the 180° pulse.

With the second sequence (Figure 2.84), we would have $\lambda\varphi_2 = \varphi_2 + \varphi_1 = -\gamma \cdot G_X \cdot \lambda X \cdot \delta$. Apart from the sign, we obtain the same value of the phase acquired. Note that this phase evolution is proportional to the displacement λX which took place during the period Δ and which is therefore encoded by the product $\gamma \cdot G_X \cdot \delta$, in the same way as a position X is encoded in imaging by the product $\gamma \cdot G_X \cdot t$ when we seek to obtain a profile in the direction OX. Consequently, the term $q_X = (2\pi)^{-1} \gamma \cdot G_X \cdot \delta$ (along with the sign corresponding to the chosen sequence) plays the part of a reciprocal value of λX, similarly as $k_X = (2\pi)^{-1} \gamma \cdot G_X \cdot t$ plays the role of a reciprocal value for X in conventional imaging. We can therefore define a "q-space", reciprocal to the space of the diffusion displacements during a given period of time and put in place a technique for imaging these motions by repeating the sequence for different values of gradient G_X applied (see Figure 2.85) during the two gradient pulses of duration δ.

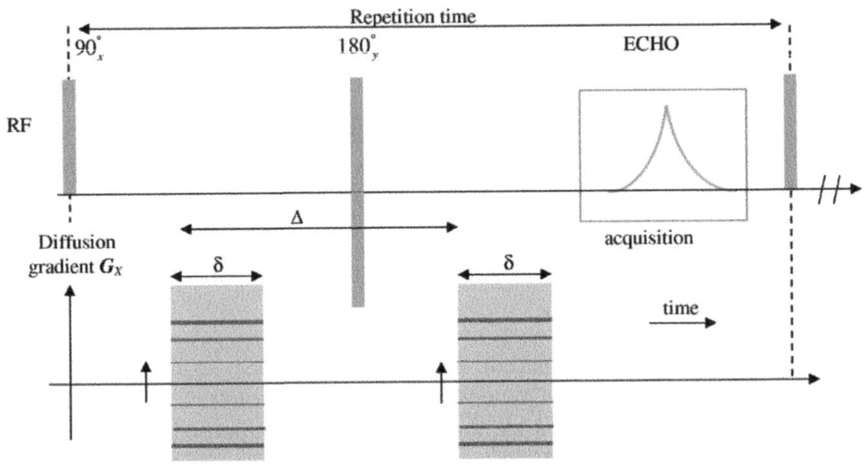

Figure 2.85. *PGSE sequences in spin-echo mode with encoding of the diffusion displacements in the direction OX: in an elementary sequence, the diffusion gradient assumes the same values on both sides of the 180° pulse*

If, for simplicity's sake, we do not take account of the line width due to transverse relaxation, we can then express the signal in two dimensions (q and time), such as:

$$s(q,t) = [\int_{\lambda X} \rho(\lambda X) \cdot e^{j2\pi \cdot q_X \cdot \lambda X} \cdot d(\lambda X)] \cdot e^{j \cdot 2\pi \cdot f \cdot t}$$

whose Fourier transform with respect to time gives only the frequential spectrum of the whole sample (a single band) but is advantageous if we combine position imaging with local evaluation of diffusivity. The distribution of spectral amplitudes in the direction of the motion is given by the following Fourier transform: $\rho(\lambda X) = \int_{q_X} f(q_X) \cdot e^{-j2\pi \cdot q_X \cdot \lambda X} \cdot dq_X$ where $f(q_X) = \int_{\lambda X} \rho(\lambda X) \cdot e^{j2\pi \cdot q_X \cdot \lambda X} \cdot d(\lambda X)$. It indicates that for each abscissa λX representing a value of displacement during the interval Δ, the signal amplitude will be weighted by the value $\rho(\lambda X)$, which represents the population of nuclei having made a movement λX during the time interval Δ.

The concept of the parameter q is associated with any gradient, whatever its orientation. We can therefore use a 3D q-space in the same way as it was possible to define a 3D k-space. These two spaces are measured with the same unit: the meter^{-1}. As the aim is to monitor diffusion displacements, the field of applications becomes vast (Physics, Chemistry, Biology, Medicine and Industry). Finally, it is not necessary to restrict ourselves to a single dimension and a gradient $\vec{G} = G_X \cdot \vec{I} + G_Y \cdot \vec{J} + G_Z \cdot \vec{K}$, where $(\vec{I}, \vec{J}, \vec{K})$ are the unit vectors of the three axes of the frame (R), defines a vector $\vec{q} = (2\pi)^{-1} \gamma \cdot \vec{G} \cdot \delta$. With any given spatial direction, we can associate a k and a q value, in the same direction for the combination of "k imaging" and "q imaging" as suggested above.

2.6.3.1.2. Diffusion-weighted imaging and apparent diffusion coefficient in a given direction

We consider only one direction of $\vec{q}$ and take account of the fact that the additional attenuation, caused by the diffusion gradient, is governed by a law of the type $e^{-b.D_{app.}}$, where $D_{app.}$ is called "Apparent Diffusion Coefficient" (ADC). We can observe a modification of the image signal in the regions of the object where diffusivity is high. The principle of its quantitative determination is therefore given by the following formula:

$$Log_e \left(\frac{\text{Amplitude of echo in presence of gradient}}{\text{Amplitude of echo in absence of gradient}}\right) = -b.D_{app.} \qquad [2.48]$$

The method will be briefly discussed in section 2.6.4.2.3, when anisotropy of diffusion is examined through the lens of the diffusion tensor. Changing the direction of the diffusion gradient (or, if readers prefer, the direction of the vector $\vec{q}$) modifies the ADC. In practice, as the acquisition time of MR imaging exams is counted, we limit ourselves to three measurements, each in one of the three orthogonal directions defined by the laboratory frame (R), and these results differ when diffusion is anisotropic.

2.6.3.2. *Anisotropic diffusion in the three dimensions, diffusion tensor and its determination by MRI*

The anisotropy of diffusion is conditioned by the properties of the medium, such as biological tissue, wherein the molecules observed are diffusing. It is easy to see that when these molecules diffuse along a long, narrow tube, the highest diffusivity value will be observed along the axis of the tube. We can represent the anisotropy using a tensor constructed from the diffusion coefficients.

2.6.3.2.1. Diffusion tensor

As the diffusion of the molecules in all possible directions in space may not obey the same probability law, there is necessarily a corresponding ellipsoidal spatial distribution of the values of the diffusion coefficients when we use the Gaussian model expressed by relation [2.47]. The values characterizing diffusivity can then be represented in the laboratory frame (R) by a 3×3 matrix, called the "diffusion tensor", formed by the components of the inverse of the term D^{-1} introduced in equation [2.47]:

$$\mathbf{D} = \begin{pmatrix} D_{XX} & D_{XY} & D_{XZ} \\ D_{YX} & D_{YY} & D_{YZ} \\ D_{ZX} & D_{XY} & D_{ZZ} \end{pmatrix} \quad [2.49]$$

The interpretation of these terms relates to the fact that the diffusion "current", in the sense of Fick's first law, associated with the variation in concentration is not necessarily collinear with the concentration gradient. Consequently, the interpretation of the terms indicated here tells us that a "diffusion current" in direction OX can result from a concentration gradient along OY, so in general, $J_Q = -D_{QP} \cdot \dfrac{\partial C}{\partial P}$ where $(Q, P = X, Y, Z)$. In view of the fact that the probabilities of evolution are the same in both senses of a direction, this symmetry is found in the table representing **D**, and we have the relation $D_{PQ} = D_{QP}$ where $(Q, P = X, Y, Z)$, which reduces the number of independent coefficients in this table to six – all positive. The diagonal elements D_{PP}, $(P = X, Y, Z)$ correspond to diffusion along the three axes of the laboratory taken for a frame. The frame (R) in which we defined the elements of the matrices D_{PP} and D_{PQ} is not usually the frame of reference for the study of diffusion. In that new frame of reference, the diffusion tensor $\mathbf{D}'$ is diagonal:

$$\mathbf{D}' = \begin{pmatrix} D'_{X'X'} & 0 & 0 \\ 0 & D'_{Y'Y'} & 0 \\ 0 & 0 & D'_{Z'Z'} \end{pmatrix} \quad [2.50]$$

We switch from the specific frame of the diffusion tensor to the laboratory frame by an operation of rotation, using the operator $\mathbb{R}$. On the other hand, the diffusion tensor $\mathbf{D}$ is made diagonal $\mathbf{D}'$ by the opposite operation of rotation: $\mathbf{D}' = \mathbb{R}^{-1}\mathbf{D}\mathbb{R}$.

Finally, the eigenvalues D'_{PP}, where $(P = X', Y', Z')$, are usually represented as $\lambda_1, \lambda_2, \lambda_3$ in decreasing order. They are evaluated by diagonalization of the matrix **D** if we know its elements. The three eigenvalues $\lambda_1, \lambda_2, \lambda_3$ associated respectively with the three eigendirections of the diffusion tensor define three axes of an ellipsoid of diffusion whose surface area is given by $\mathbf{u}^{\mathrm{T}}\mathbf{D}\mathbf{u} = 1$, where $\mathbf{u}^{\mathrm{T}}$ is the transpose of the random unit vector **u**. This surface is a sphere in the case of isotropic diffusion.

Thus, the lengths of the ellipsoid's axes are given by the values λ_1, λ_2 and λ_3, which correspond to the relative amplitude of diffusion in their respective directions. In particular, an ellipsoid which is significantly elongated in a given direction indicates the most effective direction of diffusion, as suggested by Figure 2.86.

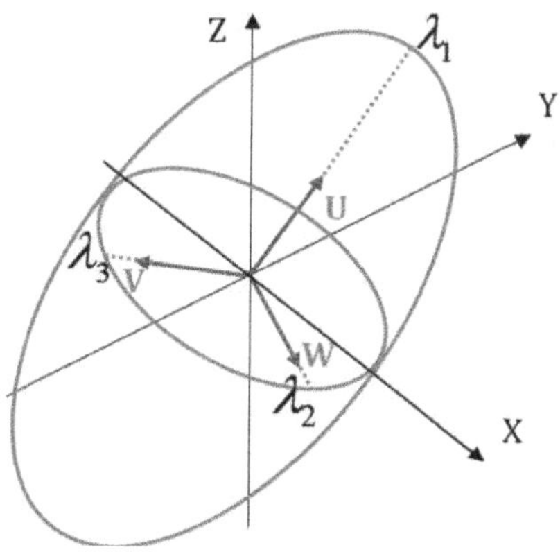

Figure 2.86. *Diffusion coefficients are represented in the laboratory frame (X, Y, Z) by an ellipsoid model, based on the unit vectors $\vec{U}, \vec{V}, \vec{W}$ associated with the three eigenvalues λ_1, λ_2, λ_3. The prevailing direction of diffusion is that given by the direction of $\vec{U}$*

2.6.3.2.2. Diffusion Tensor Imaging – DTI

If we take account of anisotropy, we get the following expression of the amplitude of the signal at the echo time PGSE $A = A_0 \cdot e^{-t/T_2} \cdot e^{-\gamma^2\delta^2(\Delta - \frac{\delta}{3})\mathbf{G}^{\mathrm{T}}\mathbf{D}\mathbf{G}}$, where $\mathbf{G}^{\mathrm{T}}\mathbf{D}\mathbf{G}$ is the matrix product of the gradient vector by the diffusion tensor and then by the transposed gradient vector. Consequently, if we preserve the formulation in terms of $e^{-b \cdot D}$ in order to take account of the damping of the echo caused by diffusion, the expansion of the exponent gives us:

$$\gamma^2\delta^2(\Delta - \frac{\delta}{3}) \cdot [G_X^2.D_{XX} + G_Y^2.D_{YY} + G_Z^2.D_{ZZ} + 2G_X \cdot G_Y.D_{XY} + 2G_X \cdot G_Z.D_{XZ} + 2G_Y \cdot G_Z.D_{YZ}],$$

which yields $A = A_0 \cdot e^{-t/T_2} \cdot e^{-\sum_{i=1}^{3}\sum_{j=1}^{3} b_{ij} \cdot D_{ij}}$. When the gradients are applied in the primary directions of the diffusion tensor whose non-diagonal terms are null – which is a rare situation – the exponent takes the form $\gamma^2\delta^2(\Delta - \frac{\delta}{3}) \cdot [G_X^2.D_{XX} + G_Y^2.D_{YY} + G_Z^2.D_{ZZ}]$. More generally, we need to consider the determination of the six parameters D_{ij}, rather than simply three parameters D_{ii}. Thus, we have to consider methods for determining the diffusion tensor, voxel by voxel, in 2D images (pixel × slice thickness) or 3D images, and the results are *a priori* different from one voxel to the next. Thus, the orientation of the major axis, when we move from one voxel to its neighbors, may reveal the existence of a prevailing direction of diffusion, propagating throughout the sample, e.g. along tissular fibers. This gives rise to DTI fiber tracking.

With regard to the terms b_{ij} entering into the expression of the ratio A / A_0, where the exponent assumes the form:

$$b_{XX}.D_{XX} + b_{YY}.D_{YY} + G_{ZZ}.D_{ZZ} + 2b_{XY}.D_{XY} + 2b_{XZ}.D_{XZ} + 2b_{YZ}.D_{YZ},$$

we have $b_{ij} = \gamma^2\delta^2(\Delta - \frac{\delta}{3}) \cdot G_i.G_j,\ i, j = X, Y, Z$ These expressions must be modified if the effective form of the gradients is not the ideally rectangular form shown in Figures 2.84, 2.85 and 2.86.

The method for calculating the coefficients b_{ij} proposed by Mattiello, Basser and Le Bihan is able to take account of all the gradients involved over the course of an imaging sequence and their temporal sequencing. To estimate the values of the six coefficients D_{ij}, we have to take at least six images with six different sets of parameters b_{ij}, as well as a reference image obtained without a diffusion gradient.

The gradients for imaging without a specific gradient dedicated to diffusion are organized in such a way that diffusion affects the signals in that image as little as possible. In a 128 × 128 image, determination of the coefficients D_{ij} pixel by pixel requires 16,384 evaluations of six parameters with a diffusion tensor in each of the 16,384 voxels involved. The stacking, voxel after voxel, of the ellipsoids of the diffusion coefficients along their main axis which yielded the image in Figure 2.88 (a) is a morphological representation. By establishing connections between the anatomical regions of the brain, this technique, known as "fiber tracking" shows us the organization of the bundles of cerebral axons (Figure 2.88 (c)).

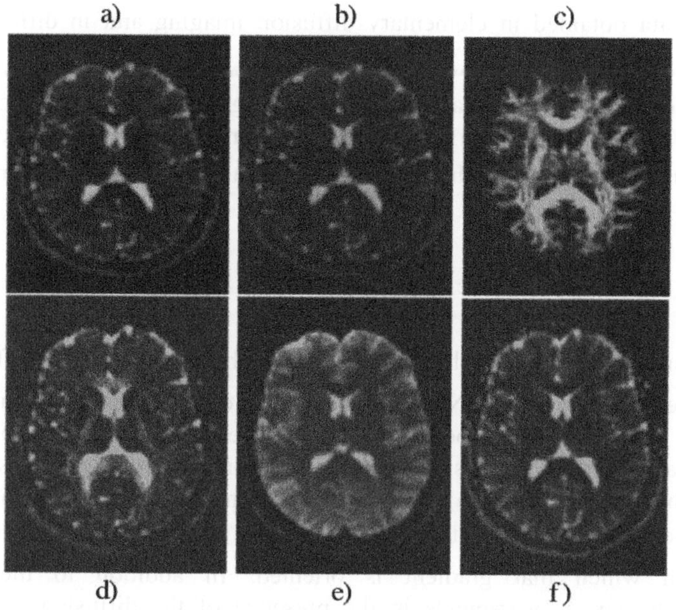

Figure 2.87. *Representations of parametric images obtained by diffusion tensor MRI of the human brain: a) reference image (the term b = 0), b) image of apparent diffusion coefficient; c) image of the fractional anisotropy (FA); (d-e-f) images of the three eigenvalues of diffusivity, respectively:* λ_1, λ_2 and λ_3 *(see section 2.6.3.2.3). Images obtained at 1.5 teslas, with a 128 × 128 matrix size, and a digital resolution of 2.5 × 2.5 × 2.5 mm*^{3}*).Images provided by D. Sappey-Marinier, Cermep, Creatis*

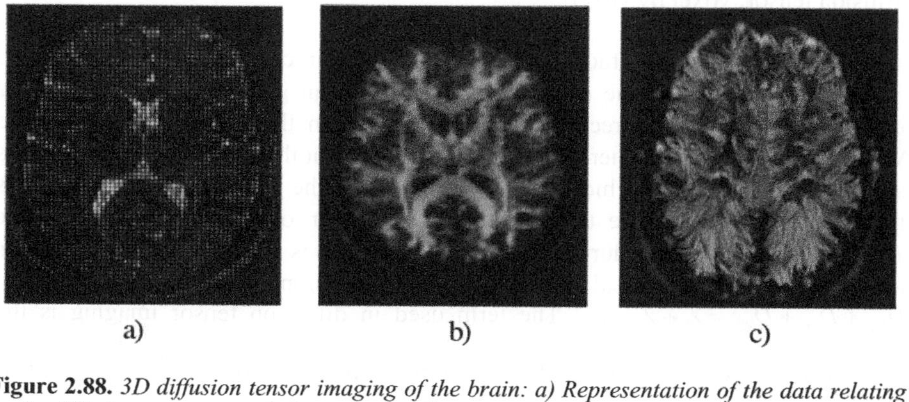

Figure 2.88. *3D diffusion tensor imaging of the brain: a) Representation of the data relating to the diffusion tensor encoded conventionally in a 128 × 128 matrix. b) Image of fractional anisotropy color coded. c) Image of white-matter fibers where the blue fibers are in the coronal plane (up-down); the red ones in the transverse plane (left-right) and the green ones in the sagittal plane (back-front). Images provided by D. Sappey-Marinier, Cermep, Creatis. For a color version of this figure, see www.iste.co.uk/fanet/medimagnet.zip*

2.6.3.2.3. Data obtained in elementary diffusion imaging and in diffusion tensor imaging

The exponential damping caused by diffusion, given by the expression obtained for b above, enables us to write relations between the amplitude A_n of the echo with a diffusion gradient and the amplitude A_0 in the absence of that gradient (the reference image), so that:

$$Log_e(\frac{A_n}{A_0}) = \sum_k b_{kk}^{(n)} D_{kk} + 2\sum_{k \neq l} b_{kl}^{(n)} D_{kl} \quad ; n = 1,..,n_{MAX} \ ; k,l = X,Y,Z$$

where the smallest possible value of n_{MAX} is 6. The values of the coefficients b_{ij} are controlled by the operator. Note that the dimension of these coefficients is that of the inverse of the diffusion coefficients; their order of magnitude is therefore 10^{-9} m^2/s. For a value of the diffusion gradient kept constant from one measurement to the next (changing the index n), with the measurement obtained by modifying the sense of the diffusion gradient, the values of the coefficients b_{ij} depend on the direction in which that gradient is oriented. In addition to the reference measurements, six measurements in the presence of the diffusion gradient may suffice to calculate the eigenvalues of the tensor, as there are six unknowns $(D_{XX}, D_{YY}, D_{ZZ}, D_{XY}, D_{XZ}, D_{YZ})$. As these measurements are performed using noised data, it may be necessary – still with the constant gradient – to use more orientations of the gradient, creating an over-abundant system of equations. Thus, which pieces of information can we deduce from knowing the diffusion coefficients, in cases ranging from that of diffusion-weighting of the image to that of determination of the diffusion tensor, voxel by voxel?

The use of a diffusion gradient applied in a given direction, in MRI, enables us to evaluate an ADC along the direction of that diffusion gradient. By repeating the previous experiment in three orthogonal directions in the space, we obtain three ADC values which are generally distinct from one another, unless the diffusion is isotropic. The image in which we take account of the mean of the three values illustrates the effect of the trace of the diffusion tensor. Indeed, because of the conservation of the trace during a rotation, this trace is the same as that which is determined by the diagonal elements of the diffusion tensor, and is equal to $D_{XX} + D_{YY} + D_{ZZ} = \lambda_1 + \lambda_2 + \lambda_3$. The term used in diffusion tensor imaging is the mean diffusion coefficient λ equal to one third of the trace, so $\lambda = \dfrac{\lambda_1 + \lambda_2 + \lambda_3}{3}$. It is independent of the frame of representation. Consequently, by averaging three measurements with orthogonal diffusion gradients and equal amplitudes, we obtain the image of the mean diffusion coefficient λ, which we can reconstruct if we know the diagonal elements of the diffusion tensor. Knowing the eigenvalues λ_i , i=1,2,3 also enables us to give the mean image coefficient.

To proceed further with the use of these data, we introduce the "fractional anisotropy" (FA) coefficient.

$$\text{FA}=\sqrt{\frac{3}{2}}\sqrt{\frac{(\lambda_1-\lambda)^2+(\lambda_2-\lambda)^2+(\lambda_3-\lambda)^2}{\lambda_1^2+\lambda_2^2+\lambda_3^2}} \qquad [2.51]$$

A value of FA=0 corresponds to isotropic diffusion, and if diffusion takes place exclusively in one direction, we have FA=0 As in the case of the trace, it is also possible to give the image of the fractional anisotropy coefficient (Figure 2.87).

Looking at the three eigenvalues gives us information about the main direction of diffusion, given by the largest value (λ_1). If $\lambda_1 \gg \lambda_2, \lambda_3$, the main direction is clearly predominant and, if this result is confirmed in the neighboring voxels, we can determine the prevailing direction of diffusion in all part of an organ observed by MRI. This is the principle behind fiber tracking (Figure 2.89). Finally if λ_3 , the smallest of the three values is close to noise value, this indicates that diffusion is essentially occurring on one particular plane.

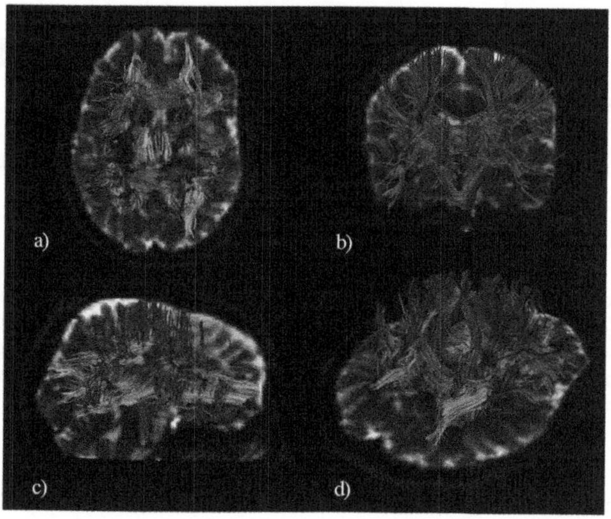

Figure 2.89. *Brain images of fiber tracking using FSL software (FMRIB, Oxford, UK). The axial a), coronal b), sagittal c) and oblique transverse d) views show the orientation of white-matter bundles, the blue one being vertical, the red horizontal and the green running front to back. Images provided by S. Hannoun, Creatis, Cermep. For a color version of this figure, see www.iste.co.uk/fanet/medimagnet.zip*

All of the operations which accompany the determination of a diffusion tensor lead to quite a high number of images of the organ being observed: diffusion-weighted image, mean diffusion coefficient image, fractional anisotropy coefficient image, images (usually 7) for evaluation of the coefficients D_{ij}, images of the eigenvalues of the tensor (usually 3) and images of each component of the eigenvectors (9 images, *a priori*). These images, obtained after the acquisition of a minimum of 7 images, mean the exam will necessarily be lengthy. Thus, the increase of the diffusion gradient number of orientations and number of value intensity will obviously lead to long durations of experiments. These constraints stem from the fact that DTI, on its own, is not capable of indicating whether there are multiple orientations of diffusion within a given voxel.

As illustrated by Figure 2.90, DTI is incapable of resolving the crossing or branching of fibers, and only gives a definite result if the direction of diffusion is unique. This is what happens in fiber tracking where diffusion of water molecules is channeled along a main direction that can be tracked, voxel by voxel. The low angular intra-voxel power of spatial resolution means we need to greatly increase (multiply) the number of values of the vector $\vec{q}$, whilst modifying the orientations of the diffusion gradient associated with the PGSE sequence.

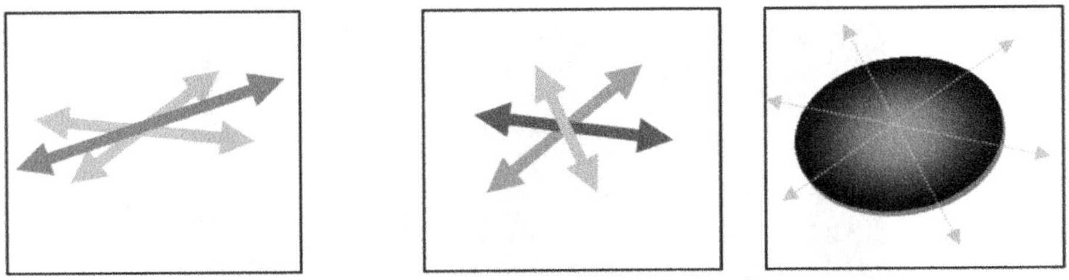

a) b)

Figure 2.90. *Schematic illustration of the difficulty of discriminating crossing directions using DTI. a) When there are two mean directions of diffusion in the voxel (represented by the light gray arrows), diffusion tensor analysis can only give a single, average main direction (the dark gray arrow). b) The superposition of multiple orientations of diffusion (left) in fact leads DTI to propose an average ellipsoid of diffusion (right)*

We now come to one of the modalities of the so-called "HARDI" approach (High Angular Resolution Diffusion Imaging), where we measure an apparent coefficient for a large number of orientations of the vector $\vec{q}$, equally distributed amongst the vertices of a regular icosahedron. We are then led, by the constraint of the Gaussian diffusion model connected to the notion of an ellipsoid, to use statistical analysis methods which are capable of extracting the dominant orientations of diffusion in a voxel.

2.6.3.3. *Improvements to diffusion imaging*

2.6.3.3.1. *q*-space imaging (QSI)

The PGSE sequence enables us to explore the q-space by assigning the diffusion gradient different values for each of its directions occupied during different executions of the sequence. We introduce the vector $\vec{q}$ in the following manner: $\vec{q} = (2\pi)^{-1}\gamma\int_{t'}\vec{G}(t')dt'$, where the variable t' is counted during the application of a lobe of the diffusion gradient. Its duration δ is short so the diffusion during that time interval is hardly perceptible, and so we can take $\vec{q} = (2\pi)^{-1}\gamma\vec{G}\delta$. Therefore, the duration of diffusion is essentially governed by the time interval Δ; hence, in principle with the condition $\delta \ll \Delta$. The ratio of the amplitude of the spin echo with respect to the case where $\vec{q} = \vec{0}$ is given by:

$$E(\vec{q},\Delta) = \int d^3\vec{r_2}\int d^3\vec{r_1}.P(\vec{r_2},\Delta|\vec{r_1},0)\cdot P(\vec{r_1}).e^{j2\pi\vec{q}(\vec{r_2}-\vec{r_1})} \quad [2.52]$$

if the resulting displacement of matter during the time interval Δ is $\vec{r} = (\vec{r_2} - \vec{r_1})$. The diffusion process is represented here by the volumetric conditional probability density $P(\vec{r_2},\Delta|\vec{r_1},0)$, which modifies the initial distribution of the spins $P(\vec{r_1})$. It is supposed to be "homogeneous", meaning that only the difference $(\vec{r_2} - \vec{r_1})$ counts; this hypothesis is normally accepted. We can show that the expression given by equation [2.52] becomes $E(\vec{q},\Delta) = \int d^3\vec{r}.P(\vec{r},\Delta|\vec{0},0).e^{j\vec{q}.\vec{r}}$ when $P(\vec{r_1})$ is constant and taken as equal to the unit. This result is crucially important, because it shows that $P(\vec{r},\Delta|\vec{0},0)$ can be obtained by the Fourier transform $P(\vec{r},\Delta|\vec{0},0) = \int d^3\vec{q}\cdot E(\vec{q},\Delta)\cdot e^{-j2\pi\vec{q}\cdot\vec{r}}$ with respect to $\vec{q}$ without any hypothesis relating to the diffusion model. This formulation gives rise to the technique known as "*q*-space imaging" (QSI), by contrast to the geometric imaging method in the general sense, which is known as "*k*-space imaging". The technique, which is illustrated by the last expression above, in principle enables us to work back to a knowledge of the process of diffusion through $P(\vec{r},\Delta|\vec{0},0)$. This "probability density function" (PDF) determines the spatio-temporal probability density at $\{\vec{r},\Delta\}$. This analysis, performed on the scale of the voxel, indicates that the technique could be able to specify how diffusion takes place in the space situated within that voxel.

Over a relatively large sample, divided into numerous voxels, this approach necessitates lengthy experiments because, in addition to encoding the parameters q, we must encode the parameter k as well in order to correctly represent the result in the image. Note that using morphological imaging with a single direction of "q-imaging" is tantamount to determining ADC. Another special case, also restrictive, which we shall not discuss in detail, arises when the amplitude of variation of the term $\vec{q}$ is small. In this case we can show that QSI yields similar results to DTI. In summary, low values of $\vec{q}$ go hand in hand with high values of the displacements $\vec{r}$. We can indeed see that in order to obtain as sharp a view as possible of the phenomenon of diffusion, we need to exploit a complete spectrum of values of $\vec{q}$. Fulfilling this condition enables us to reveal small values of $\vec{r}$ and consequently to consider examining restricted diffusion and refining the examination of the processes of intravoxel diffusion of the molecules used as probes.

2.6.3.3.2. The concept of an Orientation Distribution Function (ODF)

Measuring $P(\vec{r},\Delta|\vec{0},0)$ based on PGSE experiments leads us to reconstruct the probability that a nucleus will diffuse in a given direction during the time Δ. The distribution of that probability throughout the solid angle of the space – the necessity of which is illustrated in Figure 2.90 – is called the "Orientation Density Function" or "Orientation Distribution Function" (ODF). This function is defined using the typical spherical coordinates (r,θ,φ) in which the vector $\vec{r}$ can be noted as $r\vec{u}$ and we preserve the notation r to express the length of $\vec{r}$ and $\vec{u}$ is the unit vector $(\cos\theta, \sin\theta\cos\varphi, \sin\theta\sin\varphi)$. If we ignore the formulation of the diffusion time Δ, the probability $d^3\vec{r}.P(\vec{r},\Delta|\vec{0},0)$ that a spin will move from position $\vec{r}$ corresponding to the element $d^3\vec{r}$ can also be written as $P(r\vec{u})r^2 dr \sin\theta d\theta d\varphi$. The ODF is then defined by that probability expressed per unit solid angle, which is $\sin\theta d\theta d\varphi$ for all the values of the length r, so it must be written as:

$$\Psi(\vec{u}) = \int_{r=0}^{\infty} P(r\vec{u})r^2 dr \qquad [2.53]$$

This dimensionless value, expressed per unit solid angle, is therefore correctly called "density". Note that as soon as where $P(\vec{r},\Delta|\vec{0},0) = P(-\vec{r},\Delta|\vec{0},0)$, we have $\Psi(\vec{u}) = \Psi(-\vec{u})$, which corresponds to the direction of diffusion defined by $\vec{u}$.

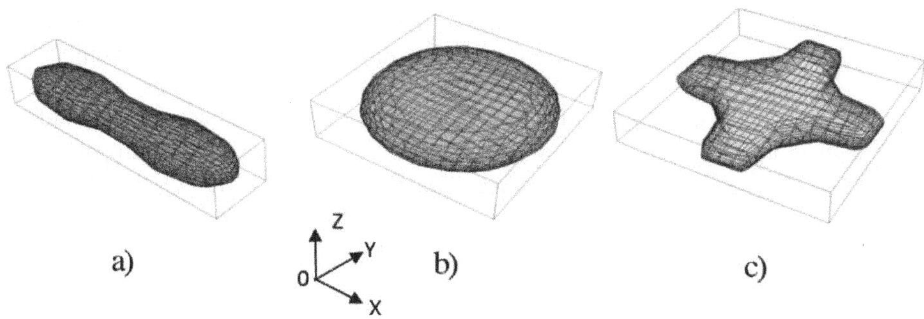

Figure 2.91. *It is possible to analytically determine the ODF using the Gaussian model of the diffusion tensor: a) ODF obtained for a tensor whose main axes lie along* OX *,* OY*,* OZ *with eigenvalues in the respective proportions 8:2:1, defining an element of a fiber in the direction of the* OX *axis. b) Result of the computation of the ODF following the determination of the diffusion tensor when two elements of fibers with properties identical to the previous one cross at a right angle in the plane* XOY *. c) ODF normally expected (Simulations)*

One method for evaluating the ODF is to use the diffusion spectrum. In terms of its principle, it consists of collecting, for a given voxel, a sufficiently abundant set of samples of the value $E(\vec{q},\Delta)$ where, with a fixed value of Δ, the measurements are performed on a 3D Cartesian grid of the values of q . The Fourier transform of this dataset gives us the diffusion spectrum, which is therefore a 3D matrix of the values of $P(\vec{r},\Delta|\vec{0},0)$. These values in turn enable us to determine the ODF by way of the radial summations indicated by relation [2.53]. The drawback to this approach stems from the 6D structure of the overall operation imposed by the sampling of the $\vec{q}$ and $\vec{k}$ spaces, which necessitates the use of a high number of operations for acquisition, not to mention the subsequent digital manipulations, which we shall leave aside in this very brief presentation of the notion of the ODF. However, it must be mentioned that relation [2.53] can be modified by omitting r^2 in the summation. This does not alter the properties of symmetry of the result, which must comprise a re-normalization term to re-establish a correct numerical value. Finally, it should be noted that if we adopt a mathematical expression for the PDF $P(\vec{r},\Delta|\vec{0},0)$, it is possible to determine the corresponding diffusion ODF; this is feasible particularly in the case of the diffusion tensor, where the model of diffusion is Gaussian (whereas with the QSI technique, no model is needed, *a priori*) and for which the ODF appears in the direction given by the long axis of the ellipsoid (Figure 2.91(a)).

The significant duration of the diffusion spectrum method led to the development of another approach, which proves more effective. The "q-ball imaging" method consists of replacing Cartesian sampling by sampling over a sphere (or "ball" in the

q-space. This imposes only a single value of the diffusion gradient reoriented in the directions of the vertices of a regular icosahedron, with a sufficient number of vertices in view of the HARDI principle mentioned above. This method of acquisition has the advantage of being performed in spherical geometry, lending itself well to the reconstruction of the ODF, which presents a structure of radial projection. Consequently, it exhibits a certain flexibility in terms of the choice of radius of the "q-ball" if we need to specifically examine certain spatial frequency bands in the q-space. The measurements of $E(\vec{q},\Delta)$ are taken with a constant $|\vec{q}| = q_0$, on the basis of which the ODF can be reconstructed by referring to the properties of the Funk-Radon transform (a sphere-to-sphere projection technique) or by Fourier analysis, as proposed more recently. In the first case, we determine $\Psi(\vec{u})$ by projection in cylindrical coordinates, with the direction of $\vec{u}$ defining the axis of the cylinders. Then, expression [2.53] must be replaced with $\Psi(\vec{u}) \simeq 2\pi q_0 \int P(r',\theta,z) J_0(2\pi q_0 r') dr' d\theta dz$, which is valid for any orientation $\vec{u}$ of the axis of the cylinder, measured with the variable z, the circulation on the disc of radius r' taking place with the polar angle θ, once the PDF is expressed by changing variables in (r',θ,z). The zero-order Bessel function helps limit the extension of $\Psi(\vec{u})$ in the directions transverse to the direction $\vec{u}$.

2.6.3.4. *Microstructural information obtained by diffusion in the absence of imaging gradients*

The data relating to diffusion, which can be extracted by MRI and considered up until now, are located in a volume of around ten cubic millimeters – the volume of the voxels. However, it is possible to obtain microscopic information on a volumetric scale at best six volumetric orders of magnitude smaller, or two orders of magnitude on a linear scale. The examination of diffusion in such small spaces began with the work in microscopic imaging carried out by P. Callaghan in the 1980s. These works are used for exploration of so-called porous media, exhibiting hollow structures whose dimensions are in the tens of micrometers. These cavities, which are modeled by spheres or cylinders, contain molecules carrying resonating nuclei (e.g. protons) – molecules which are consequently subject to a regime of restricted diffusion. While there is no question of perceiving these pores by the application of imaging gradients, it is possible to observe the overall behavior of the content of the pores under the influence of a PGSE sequence. According to relation [2.52], which is valid when the diffusion gradient pulses are very short in comparison to the duration of diffusion Δ, the relative value of the echo can be written $E(\vec{q},\Delta) = \int d^3\vec{r_2} \int d^3\vec{r_1}.P(\vec{r_2},\Delta|\vec{r_1},0)\cdot\rho(\vec{r_1}).e^{j2\pi\vec{q}(\vec{r_2}-\vec{r_1})}$, where the distribution function for the resonating nuclei in the pores, $\rho(\vec{r_1})$ at the time of the first diffusion gradient pulse, replaces the initial distribution of the spins $P(\vec{r_1})$. The term $\rho(\vec{r_1})$ gives us a sort of 1D projection in the direction of the diffusion gradient, of the internal space of the pores; consequently, it depends on the form of those pores, and its value is zero if we assign $\vec{r_1}$ a value which would lead us to look outside of the pore. Its determination thus provides us with information about the shape and dimension of the pores. The value of Δ, which is high in comparison to δ, is also much higher than the duration of retention of the molecules in the sites that they occupy between two successive diffusion displacements. (This duration was notated τ in the random walk model mentioned at the beginning of this presentation). It is also important that in the PGSE sequence used, the duration Δ is sufficiently high for diffusion to erase any memory effect in terms of the positions of the spins. With Δ considered as infinitely large, the expression of $E(\vec{q},\Delta)$ can be expressed in the form: $E(\vec{q},\Delta \rightarrow \infty) = \int \rho(\vec{r_2}) \cdot e^{j2\pi\vec{q}\cdot\vec{r_2}} d^3\vec{r_2} \int \rho(\vec{r_1}).e^{-j2\pi\vec{q}\cdot\vec{r_1}} d^3\vec{r_1}$, the product of two Fourier transforms, functions of $\vec{q}$ and averaged over all the pores so that $E(\vec{q},\Delta) = \overline{\left|d(\vec{q})\right|^2}$ where $\overline{\mathrm{d}(\vec{q})} = \mathrm{T.F.}\{\rho(\vec{r})\}$. However, the ideal would be to know $d(\vec{q})$, which enables us to use the inverse Fourier transform to work back to $\rho(\vec{r})$, representing a precious piece of information about the microstructure. The approach adopted to solve this problem is to use a new measurement, created by stringing together two PGSE sequences in which the values of the diffusion gradient cannot be identical. We can then show that the term which replaces $E(\vec{q},\Delta)$ with this double sequence becomes $E'(\vec{q},\Delta \rightarrow \infty) = \overline{\left|\mathrm{d}(\vec{q})\right|^2 \mathrm{d}^*(2\vec{q})}$, so that we can deduce $\mathrm{d}(\vec{q})$ with this second experiment to extract $\rho(\vec{r})$ in order to obtain information about the dimensions and the shape of the cavities making up the pores.

Applications to inert media are already very frequently used, because it is not too difficult to work on small samples placed in magnets with a strong field for reasons of sensitivity. With the constant progress being made in MRI hardware in terms of field and performances of the gradients, this type of microstructural determination in living tissues is of interest in the biomedical sphere. In order to move from the concept to these specialized applications, we need to develop the computation methods and test-object models (phantoms) necessary for the simulation of experiments and digital processing of the data. The same is true in the preparation of the protocols of observation compatible with the use of medical installations, and we also need to bring these installations up to date, making them able to handle this new application of diffusion.

2.6.4 *Functional MRI (fMRI) of cerebral activation*

By enabling us to visualize the connections between regions of the brain, diffusion MRI (dMRI) establishes links between the microarchitecture and cerebral connectivity. Cerebral activation fMRI, for its part, enables us to exploit local variations in the image contrast to determine the areas of neuronal activity. Activation fMRI was developed in the 1990s, following the successive discovery of two fundamental phenomena. Firstly, Raichle demonstrates by PET that when presented with a sensory stimulus, there is a "metabolic decoupling" between the two sources of energy in the brain: glucose and oxygen. Indeed, when the brain is stimulated, we observe a local increase of 50% in glucose consumption, whereas the increase in oxygen consumption is limited to 10%. The consequence of this important discovery was the demonstration that the brain could function without using all of the available oxygen, by a mechanism related to anaerobic respiration when dealing with physical stimulation. This notion therefore implied that the brain could produce lactate when faced with a demand for additional energy, as was later demonstrated by MRS. Secondly, Ogawa and his team show, using an animal model, that a decrease in the proportion of oxygen in the air inhaled causes a drop in signal from the blood vessels in brain images obtained with a gradient-echo sequence.

More generally, the variation in signal intensity related to neural activity is a complex problem, which we shall not investigate in detail here. However, the part played by fMRI for the exploration of the brain justifies a number of elementary explanations.

2.6.4.1. *BOLD (Blood Oxygen Level Dependent) effect*

Demonstrated by Ogawa, the BOLD effect is actually caused by the changing concentration of deoxyhemoglobin in the blood. Oxyhemoglobin is diamagnetic and has a negative magnetic susceptibility quite close to that of cerebral tissues. On the other hand, it has been known since 1939 that deoxyhemoglobin is paramagnetic. Thus, the presence of deoxyhemoglobin reduces the apparent transverse relaxation time of deoxygenated blood in comparison to that of oxygenated blood, thus weakening the signal from the blood vessels. Note that for simplicity's sake, the approximate but helpful concept of a time constant T_2^* is used to take account of the local degradation of the uniformity of the static field, which affects the attenuation of the signal during the sequence's echo time. This groundbreaking observation gave rise to the concept of the BOLD effect, which opened the way for human testing by Belliveau and his team. However, the earliest studies to be performed with stimulation of the visual cortex (which did not take place when the BOLD effect was first demonstrated) show not the expected dip in the signal, resulting from an increase in oxygen level in response to the additional neuron activation, but instead a slight increase in the signal. Indeed, according to the results of Raichle's experiment, neuron activation causes a sharp increase in rCBF (regional cerebral blood flow) associated with a slight increase in oxygen consumption. This decoupling between oxygen consumption and CBF causes a relative decrease in the concentration of deoxyhemoglobin, which increases the strength of the signal. As the variation in blood flow is delayed in relation to the neural activity which causes it, it is interesting to be able to observe this delay with an ultra-fast imaging technique exhibiting a good power of temporal resolution (of around a second). The EPI group of sequences (see section 2.5.2.6), which entail the formation of successive gradient echoes sensitive to the variations of the parameter T_2^* and whose execution is indeed very fast, satisfies these conditions. In order to avoid artifacts of susceptibility other than those which are caused by the slight increase (a few percent) in the NMR signal in typical conditions of the SNR, it is useful to adapt a mode of acquisition in several stages rather than a single RF excitation, but the temporal resolution must be less than a second (for example, the EPI sequence "multi-shot" or PRESTO ("PRinciples of Echo-Shifting with a Train of Observations").

2.6.4.2. *Examination with fMRI*

To begin with, the measurement of the local neuronal activation is based on the stimulation of a particular brain function, with the aim of the fMRI exam being to determine the location of that activity in the brain. Thus, in order to attempt to locate the networks activated when performing a given task, we need to define a protocol of activation, called a "paradigm", alternating phases of stimulation and rest. This sequence, which must be repeated sufficient times to give a statistically different signal between the stimulation and rest periods, can be organized in one of two ways. The conventional "block" arrangement is the concatenation of a series of several (plus or minus ten) identical stimulations during a period of twenty to thirty seconds (movement of the fingers, generation of words, mental arithmetic, viewing of an image, etc.) with intervening rest periods of the same duration. If we need to avoid consecutively repeating the same stimulations (which can cause problems of habituation, attention span and so on), we use a different, "event-based" pattern, which is the concatenation of elementary stimulations (or a few stimulations for mini-block) with short rest periods in between (Figure 2.92). Finally, a more recent approach is the observation of brain activity in the absence of stimulation, which is known as *resting state fMRI*. The acquisition is very simple: it involves acquiring EPI images over the course of ten minutes, minimizing external stimulation and asking the subject not to think about anything. Put very simply, the aim of this strategy is to provide information about brain function in the resting state. This technique enables us to extract several networks, including the main "Default Mode Network" connecting the prefrontal and parietal lobes, and to a lesser extent the temporal lobe, but also the networks corresponding to usual brain function such as motor skills, hearing, vision, etc.

To conclude this summary, it should be noted that acquisition by an fMRI exam, although it already appears relatively complex in comparison to a diagnostic exploration, is merely the first step which precedes the data-processing stage.

The analysis of the hundreds of EPI images obtained during an acquisition lasting ten minutes requires numerous software tools, applied in three main stages of pre-processing: retiming, spatial normalization and filtering. (i) The images affected by any major movement of the patient's head are rejected. The images which are kept are "aligned" in order to avoid any mistake in the positioning of voxels whose signals are being monitored. (ii) Then comes the spatial normalization of the images, which involves gathering all the images in a reference space, known as the Talairach atlas. This enables us to compare all the images from one group of "control" subjects against another group of patients, or between two groups of patients, and to define the active regions on the basis of their classification. (iii) Signal- and image-processing: reduction of the noise level in each voxel, leading to a reduction of the initial resolution, correction of the time derivatives of the temporal signal relating to a technical bias or a residual physiological artifact such as breath motion, etc.

After this preparation of the data, statistical analysis is used to compare each voxel in each image, two by two, and detect any significant change in signal strength between conditions of stimulation and rest. In order to obtain a good temporal relation between the paradigm and the data, it is helpful to have an accurate model of the hemodynamic response that is related to the neuron activation. Thus, the hemodynamic response manifested by the mainly positive variation of the MR signal occurs a few seconds (6 seconds, on average) after the beginning of the stimulation, and continues for several seconds before returning to its base state.

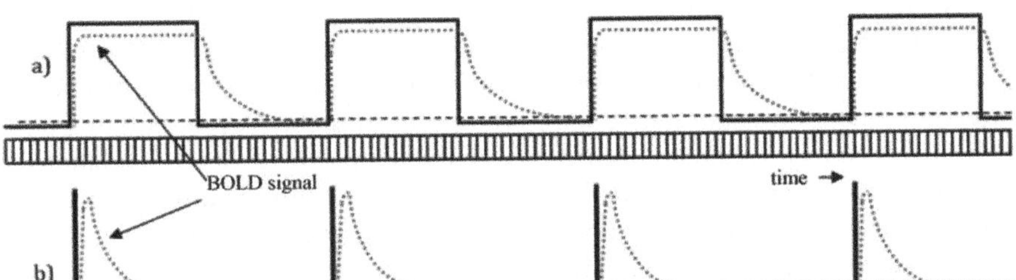

Figure 2.92. *Types of paradigms used in a cerebral activation fMRI exam. a) "Block" mode: the stimulation (represented by rectangular time window) is maintained by repetition for around 30 seconds, followed by an equal rest period, during which the BOLD signal (dotted line) decreases. b) "Event" mode: a single stimulus is used, and repeated at regular intervals for statistical reasons. The pattern shown between diagrams a) and b) reflects the image acquisition using the echo planar imaging sequence*

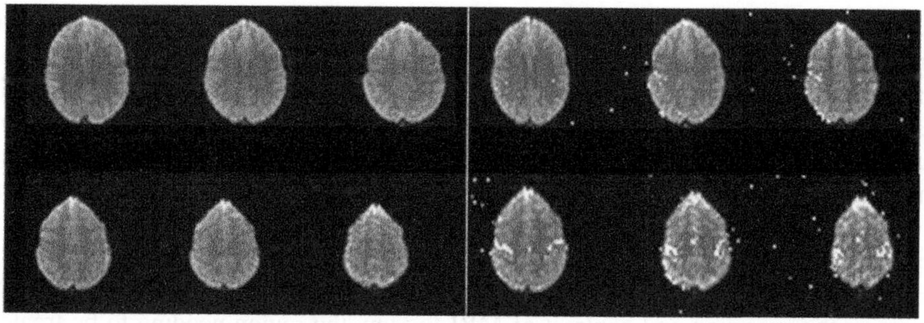

a) b)

Figure 2.93. *Images obtained during an fMRI exam, tapping fingers simultaneously on both hands in "block" mode (30 sec of stimulation followed by 30 sec of rest) (see Figure 2.92). a) The total acquisition leads to a high number (around 10,000) of axial slices. Of the 30 slices obtained every 2 sec for each brain acquisition, only six axial slices are shown, from the middle to the top of the head. b) Statistical analysis, performed during the acquisition, compares images obtained with and without stimulation to identify the regions of activation. Two regions are observed in the primary motor cortical regions for the hands, located on the left- and right-hand sides of the head, as shown here by the bright signal superimposed on the T2*-weighted image, which is obtained by a EPI sequence with a 64 × 64 matrix over a FOV of 240 mm^2 and a slice thickness of 4 mm at 1.5 teslas. Images provided by D. Sappey-Marinier, Cermep, Creatis*

Today, fMRI is becoming more commonly used in clinical practice – particularly in the context of pre-surgical examination of brain tumors or refractory epilepsy to locate the functional motor, language and memory areas. Yet its major use lies in the examination of the most complex cognitive processes. On a technical level, the development of functional and metabolic MRI has been greatly marked by the use of strong fields because of the need for high sensitivity.

2.6.5. *Bi-modal approach to MRI: the example of MR/PET*

Out of all the available imaging techniques, MRI enjoys unparalleled richness of exploratory possibilities, because of its excellent image contrast in conventional imaging, and also because of the extent of the numerous functional and metabolic methods. However, its multiparametric capabilities may prove limited by its low sensitivity, which has led to technological research into how to increase the magnetic field. While this approach seems the most obvious, it is not the easiest, and poses enormous constraints. The development of phased-array antennas, however, has provided a less costly increase in sensitivity, which is more easily transferable to clinical systems, and in particular to 3-tesla systems, for which the technology is now relatively well mastered. However, that of 7-tesla magnets is still very costly and complex to implement. Today, the number of installations worldwide has reached as many as fifty magnets. These centers have been able to exploit strengthened signal and demonstrate their advantage for clinical applications based on the development of new methods for controlling the homogeneity of the static field, new acquisition sequences limiting the energy deposited by the excitation pulses and the development of new multichannel RF receiving and transmitting antennas. The other major axis of development in the area of MRI is that of multimodality, e.g. the implementation of systems combining two modes of acquisition, or combining imaging and therapeutic treatment. We can cite HIFU – "*High-Intensity Focused Ultrasound*". The installation of a HIFU system in an MRI offers us the precision and contrast of MRI to view the region needing to be treated by ultrasounds, and also the ability to control, with very great spatial and thermal precision (within one degree) the warming of the targeted tissues. Today, these systems are applied in clinical practice for treating prostate, liver and breast cancers.

2.6.5.1. *Bimodality of PET and MRI*

The other mode of imaging associated with MRI is PET. The origins of PET, a flagship technique in nuclear medicine, go back thirty years. PET is based on the detection, in coincidence, of γ radiation emitted by the phenomenon of annihilation of the positrons emitted by the transmitting β^+ radionuclide, injected into the patient, and the electrons of surrounding tissues. Thus, PET requires the production by a cyclotron of a positron emitting nucleus, which must then be associated with a tracker molecule by the radiochemist. Depending on its metabolic function, this marker can be used to target certain functions playing a part in the metabolic pathways or in the displacement of neurotransmitters. Since the early 2000s, it has been the favored tool for detection and diagnosis of malignant tumors by a whole-body exploration with the injection of ^{18}FDG (Fluor-deoxy-glucose marked with Fluorine-18). This substance is a radio-tracer which deeply penetrates cancerous cells, whose glycolytic metabolism is faster than that of healthy tissues. Detection of the marker by PET enables us to obtain 3D images of the whole body with very high sensitivity but a poor power of spatial resolution (several mm). This lack of spatial precision led initially to the combination of images obtained with PET with those obtained by CT (computed tomography). More recently, hybrid systems were constructed to combine the submillimetric spatial precision of CT with the detection sensitivity of PET, and produce metabolic images that could be perfectly superposed on the anatomical image. This first approach benefited from the technological facility provided by the combination of two systems, which do not exhibit any mutual contraindications, simply by using them alongside one another. In the case of MRI and PET, the problem becomes more complicated, in that the PET detection systems must be insensitive to the magnetic field of MR. Conversely, the presence of the PET detectors within the magnet leads to a degradation of the quality of the MR images. Various solutions have been put forward by different manufacturers to deal with these technological problems. Philips initially chose to use two separate systems, simply sharing a rotating bed to move the patient smoothly from one system to the other. The approach taken by Siemens was to integrate PET detectors into the magnet; these detectors comprise avalanche photodiodes (APDs) insensitive to the magnetic field and scintillators based on monocrystals of lutetium oxyorthosilicate (LSO). Whilst the first solution is able to preserve all the performances of the independent systems because of the lack of any mutual degradation, the second solution has the immense advantage of being able to perform truly simultaneous acquisition of the two modalities. As the first solution does not present any particular difficulties, we shall concentrate on the Siemens system, whose first machine in France will be installed in 2014 at the CERMEP, the research center for life imaging of Lyon, with others in Britain and Germany. We shall first give an overview of the characteristics and technical performances of each mode, before going on to present the earliest clinical applications.

2.6.5.2. *Technical characteristics, performances and image quality*

The Biograph mMR system from Siemens uses an MRI system (Verio) with a teslas magnet, gradients of 45 mT/m for a rise time of 200T/m/s, and a body antenna offering a maximum field of view of 50 cm.

The PET detection system, installed between the MRI body antenna and the gradient shield, is made up of eight rings of 56 blocks, composed of an 8 × 8 matrix of LSO crystals (4 × 4 × 20 mm), coupled with a 3 × 3 APD network, resulting in 4,032 detection channels. Evaluation of this system's performances shows an energy resolution of 15% for a window of 430-610 KeV, and a temporal resolution of 3 ns for a coincidence window of 6 ns. This system is not capable of the "time-of-flight" (TOF) technique, which calculates the difference in the distance traveled between the two radiations and thus improves the localization of the tracer. This capacity, which is particularly important for full-body applications, is less necessary for cerebral acquisitions. This system, which has a transverse field of view of 60 cm and an axial field of view of 25 cm, can perform full-body acquisitions and record the data in "list" mode – i.e. event by event.

The quality of the anatomical MR images is identical to that obtained with the standard Verio MR system. The MR images show no degradation of the homogeneity of the field, or loss of signal, or increase of noise due to possible electronic coupling between the two systems. The PET images also exhibit characteristics similar to those obtained on the original PET/CT system. The PET acquisition requires correction of the attenuation and the diffusion of γ-rays for the different tissues and compartments, using a density map of the patient. In a conventional PET system or by using the tissue attenuation measured on CT images in a PET/CT system. For MR/PET, this specific mapping is made possible but at the cost of multiple sequences, facilitating good segmentation of the different tissues and compartments such as the air in lungs, muscles and bones.

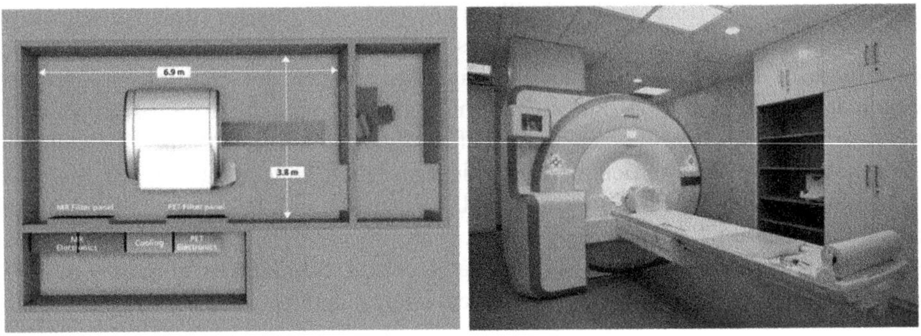

Figure 2.94. *Installation of the MR/PET Biograph mMR device at the University of Erlangen. Information from Siemens Website*

2.6.5.3. *Clinical advantage of MR/PET*

A "clinical" MRI/PET exam can be carried out in thirty minutes by simultaneous recording of the two types of data. For example, a standard protocol for cerebral anatomic MRI today comprises a T_1-weighted millimetric 3D acquisition, a T_2-weighted fast spin-echo sequence and a FLAIR (FLuid Attenuation Inversion Recovery) sequence, which suppress the signal from the cerebrospinal fluid (CSF) of the ventricles, or other media with long relaxation times. These images are taken in an axial orientation. They are supplemented by a DTI acquisition covering the whole brain. All these MRI sequences can be performed whilst recording PET data. During a full-body observation, the MRI and PET data are thus recorded simultaneously and then superposed for the successive positions of the patient (Figure 2.95).

The clinical applications of MR/PET lie mainly in oncology, because of PET's excellent sensitivity for detecting and grading tumors and metastases. However, it is necessary to improve the sensitivity and accuracy of imaging exams in order to personalize the therapeutic monitoring of each patient. While CT provides very precise spatial information, MRI supplements the detection sensitivity of PET by its excellent tissue characterization. In addition, MRI provides a great deal of functional information, about perfusion or diffusion, and about the cerebral metabolism by using MR spectroscopy. Measurement of choline concentration provides a marker for tumor progression, myo-inositol for glial inflammation and lactate for hypoxia, and thus helps tumor grading. Visualization of tumoral heterogeneity, which underlies inflammatory, ischemic and necrotic phenomena, is essential information in order to better target the region to be biopsied or treated by radiotherapy. These metabolic data can be compared against PET images obtained using tumor tracers such as choline marked with ^{11}C or ^{18}F, ^{11}C-methionine or ^{18}F-L-thymidine. They are essential for early evaluation of the response to chemo- and/or radiotherapy, because of their early changes with respect to those in morphology. However, these functional acquisitions, which are probably the future for MR/PET, still need to be evaluated in clinical research.

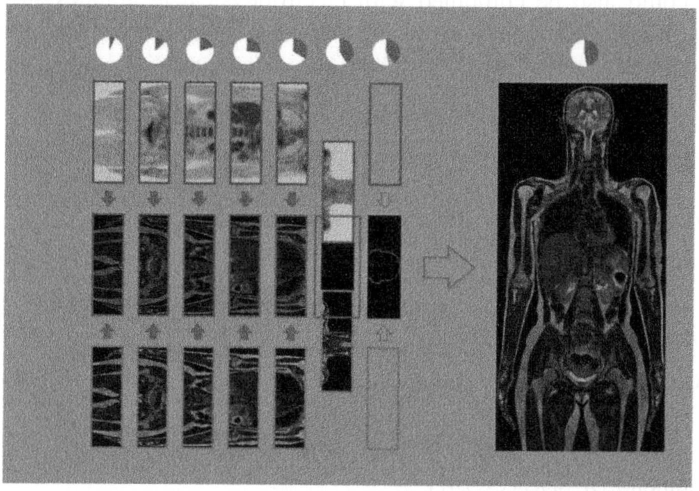

Figure 2.95. *Illustration of simultaneous recording of MRI and PET data for a whole-body acquisition lasting between 20 and 30 minutes, performed in 6 or 7 parts. Information taken from the Siemens Website. For a color version of this figure, see www.iste.co.uk/fanet/medimagnet.zip*

2.6.5.4. *Future applications of MR/PET*

While morphological and functional information provided by MRI offers a physiopathological characterization which is indispensable for diagnosis, the future of MR/PET probably lies in the simultaneous acquisition of additional information such as the kinetics of metabolic mechanisms. For example, the combination of MRI angiography and perfusion techniques with measurements of oxygen consumption and blood flow enables us to pinpoint, with greater temporal and quantitative accuracy, the metabolic processes which are involved in ischemic pathologies, vascular disorders or in tumoral angiogenesis. Similarly, the combination of metabolic information obtained by MRS, about the N-acetyl-aspartate (a neural marker) or myo-inositol (an indicator of glial inflammation) concentration, and with PET thanks to markers of inflammation or amyloid plaques, will help us to gain a better understanding of neurodegenerative diseases. The study of the hemodynamic mechanisms measured by neuronal activation fMRI could be contrasted against the measurements of vascular permeability obtained by the labeling of amino acids in PET. The study of connectivity of neural networks by diffusion tensor MRI could serve as an anatomical support for the monitoring of neurotransmission networks revealed by the labeling of dopamine (^{18}F-DOPA) or serotonin. Also, addiction phenomena, the effects of drugs such as nicotine or the pharmacological mechanisms could simultaneously be pinpointed by marking the synaptic receptors using PET, and quantified by fMRI. Finally, these functional neuroimaging techniques could also be combined with electrophysical techniques such as TMS (transcranial magnetic stimulation) which activates or inhibits certain regions of the brain as and when required.

2.7. Bibliography

General works

[CAL 91] CALLAGHAN P.T., *Principles of NMR Microscopy*, Clarendon Press, Oxford, 1991.

[CAN 02] CANET D., BOUBEL J.C., CANET-SOULAS E., *et al.*, La RMN, *Concepts, Méthodes et Applications*, 2nd ed. Dunod, 2002.

[CHE 89] CHEN C.N., HOULT D.I., *Biomedical Magnetic Resonance Technology*, Adam Hilger, New York, 1989.

[DEC 11] DÉCORPS M., *Imagerie de Résonance Magnétique, Bases Physiques et Méthodes*, CNRS Edition-EDP Sciences, 2011.

[DIL 09] DILLENSEGER J.P., MOERSCHELL E., *Guide des technologies de l'imagerie médicale et de la radiothérapie*, Masson, 2009.

[ERN 87] ERNST R.R., BODENHAUSEN G., WOKAUN A., *Principles of Magnetic Resonance in One and Two Dimensions*, Clarendon Press, Oxford, 1987.

[FUK 81] FUKUSHIMA E., ROEDER S.B.W., *Experimental Pulse NMR: A Nuts and Bolts Approach*, Addison-Welsey, New York, 1981.

[GRA 02] GRANGEAT P., *Tomography*, ISTE, London, John Wiley & Sons, New York, 2002.

[HAA 99] HAACKE E.M., BROWN R.W., THOMPSON M.R., *et al.*, *Magnetic Resonance Imaging, Physical principles and sequences design*, John Wiley & Sons, New York, 1999.

[JAV] Java Magnetic Resonance User Interface (jMRUI)) and Matlab Magnetic Resonance User Interface (mMRUI), www.MRUI.uab.es.

[JOH 09] JOHANSEN-BERG H., BEHRENS T.E.J., *Diffusion MRI: From Quantitative Measurement to In Vivo Neuroanatomy*, Elsevier, 2009.

[LEB 95] LE BIHAN D., *Magnetic Resonance Imaging of Diffusion and Perfusion: Application to Functional Imaging*, Lippincott-Raven Press, New York, 1995.

[LIA 00] LIANG Z.P., LAUTERBUR P.C., *Principles of Magnetic Resonance Imaging. A Signal Processing Perspective*, IEEE Press, Series in Biomedical Imaging, 2000.

[LIN 05] LINET G., "MRI from A to Z, a definitive guide for Medical Professionals", www.cambridge.org/9780521606387, 2005.

[MAT] Mathematica™, Wolfram Research Europe Ltd, Long Haran Borough, Oxfordshire, OX29 8FD (UK).

[MCR 07] MCROBBIE D.W., MOORE E.A., GRAVES M.J., *et al.*, *MRI From Picture to Proton*, Cambridge University Press, 2nd ed., 2007.

[MOR 86] MORRIS P.G., *Nuclear Magnetic Resonance Imaging in Medicine and Biology*, Oxford University Press, 1986.

[OPP 05] OPPELT A., *Imaging Systems for Medical Diagnostics*, Publishing Corporate Editing, 2005.

[SCH 98] SCHMITT F., STEHLING M.K., TURNER R., *Echo-planar Imaging; Theory, Technique and Application*, Springer, 1998.

[SLI 90] SLICHTER C.P., *Principles of Magnetic Resonance*, 3rd ed., Springer Verlag, New York, 1990.

[VLA 96] VLAARDINGERBROEK M.T., DEN BOER J.A., *Magnetic Resonance Imaging*, Springer, 1996.

Pioneering work and foundations

[BLO 46a] BLOCH F., "Nuclear induction", *Phys. Rev.*, vol. 70, pp. 460–473, 1946.

[BLO 46b] BLOCH F., HANSEN W., PACKARD M.E, "The nuclear induction experiment", *Phys. Rev.*, vol. 70, pp. 474–485, 1946.

[BLO 46c] BLOCH F., HANSEN W., PACKARD M.E., "Nuclear induction", *Phys. Rev.*, vol. 69, p. 127, 1946.

[CAR 54] CARR H.Y., PURCELL E.M., "Effects of diffusion on free precession in nuclear magnetic resonance experiments", *Phys Rev.*, vol. 94, pp. 630–638, 1954.

[DIL 09] DILLENSEGER J.P., "Bases de l'IRM", in J.P. DILLENSEGER, E. MOERSCHELL, *Guide des technologies de l'imagerie médicale et de la radiothérapie*, Masson, Chapter 4, pp. 151–214, 2009.

[HAH 50] HAHN E.L., "Spin echoes", *Phys. Rev.*, vol. 80, pp. 580–601, 1950.

[HOU 00] HOULT D.I., "The principle of reciprocity in signal strength calculations – a mathematical guide", *Concepts in Magn. Reson.*, vol. 12, pp. 173–187, 2000

[KUM 75] KUMAR A., WELTI D., ERNST R.R., "NMR fourier zeugmatography", *J. Magn. Reson.*, vol. 18, pp. 69–83, 1975.

[LAU 73] LAUTERBUR P.C., "Image formation by induced local interactions, example using magnetic resonance", *Nature,* vol. 242, pp. 190–191, 1973.

[MAN 73] MANSFIELD P., GRANNEL P.K., "NMR "diffraction" in solids?", *J. Phys. C. Solid State Phys.*, vol. 6, pp. L422–L426, 1973.

[MAN 76] MANSFIELD P., MAUDSLEY A.A., BAINES T., "Fast scan proton density imaging by NMR", *J. Phys E Scientific Instrum.*, vol. 9, pp. 271–278, 1976.

[MAN 77] MANSFIELD P., "Multi-planar image formation using NMR spin echoes", *J. Phys. C Solid,* vol. 10, pp. L55–L58, 1977.

[MEA 99] MEADOWS M., "Precession and Sir Joseph Larmor", *Concepts in Magn. Reson.*, vol. 11, pp. 239–241, 1999.

[MEI 58] MEIBOOM S., GILL D., "Modified spin echo method for measuring nuclear relaxation times", *Rev. Sci. Instrum.*, vol. 29, pp. 688–691, 1958.

[PUR 46] PURCELL E.M., TORREY H.C., POUND R.V., "Resonance absorption by nuclear magnetic resonance in a solid", *Phys. Rev.*, vol. 69, pp. 37–38, 1946.

[RAB 39] RABI I.I., MILLMAN S., KUSCH P., *et al.*, "The molecular beam resonance method for measuring nuclear magnetic moments", *Phys. Rev. 55:526–535*, 1939.

[STE 07] STEINMETZ W.E., MAHER C.R., "Magnetic resonance imaging on an NMR Spectrometer. An experiment for the Physical Chemistry Laboratory", *Concepts in Magn. Reson.*, vol. 30A, pp. 133–139, 2007.

Pulses

[DEC 11] DÉCORPS M.,. "Les impulsions en spectroscopie et imagerie", *Imagerie de Résonance Magnétique, Bases Physiques et Méthodes*, CNRS Edition-EDP Sciences, pp. 95–186, 2011.

[EMS 90] EMSLEY L., BODENHAUSEN G., "Gaussian pulses cascades for rectangular selective excitations: new analytical functions for rectangular selective inversion and in-phase excitation in NMR", *Chem. Phys. Letters*, vol. 165, pp. 469–476, 1990.

[HOU 79] HOULT D.I., "The solution of the Bloch equations in the presence of a varying B1 field. An approach to selective pulse analysis", *J. Magn. Reson.*, vol. 35, pp. 69–86, 1979.

[PAU 89] PAULY J.M., NISHIMURA D.G. , MACOVSKI A., "A k-space analysis of small tip angle excitation", *J. Magn. Reson.*, vol. 81, pp. 43–56, 1989.

Reciprocal space

[LJU 83] LJUNGREN S., "A simple graphical representation of Fourier-based imaging methods", *J. Magn. Reson.*, vol. 54, pp. 338–343, 1983.

[TWI 83] TWIEG D.B., "The k-trajectory formulation of the NMR imaging process with applications in analysis and synthesis of imaging methods", *Med. Phys.*, vol. 10, pp. 610–621, 1983.

Multi-parameter techniques

Perfusion

[DET 92] DETRE J.A., LEIGH J.S., WILLIAMS D.S., *et al.*, "Perfusion imaging", *Magn. Reson. Med.*, vol. 23, pp. 37–45, 1992.

[GRA 07] GRAND S., LEFOURNIER V., KRAINIK A., *et al.*, "Imagerie de perfusion: principes et applications cliniques", *J. Radiol.*, vol. 88, pp. 444–471, 2007.

[WIL 92] WILLIAMS D.S., DETRE J.A., LEIGH J.S., *et al.*, "Magnetic resonance imaging of perfusion using spin inversion of arterial water", *Proc. Natl. Acad. Sci.*, vol. 89, pp. 212–216, 1992.

Angiography, fluid motions

[DUM 86] DUMOULIN C.L., HART H.R., "Magnetic resonance angiography", *Radiology*, vol. 161, pp. 717–721, 1986.

[MOR 82] MORAN P.R., "A flow velocity zeugmatographic interlace for NMR imaging in humans", *Magn. Reson. Imaging,* vol. 1, pp. 197–203, 1982.

[SAL 87] SALONER D., MORAN P.R., "Velocity imaging by rapid cycle tagging", *Med. Phys.*, vol. 14, pp. 167–171, 1987.

Chemical shift imaging

[BRO 82] BROWN T.R., KINCAID B.M., UGURBIL K., "NMR Chemical shift imaging in three dimensions", *Proc. Natl. Acad. Sci.*, vol. 79, pp. 3523–3526, 1982.

[DIX 84] DIXON W.T., "Simple proton spectroscopic imaging", *Radiology*, vol. 153, pp. 189–194, 1984.

[HAA 86] HAACKE E.M., PATRICK J.L., LENZ G.W., *et al.*, "The separation of water and lipid components in the presence of field inhomogeneities", *Rev. Magn. Reson. Med.*, vol. 1, pp. 123–154, 1986.

[PAN 98] PAN J.W., TWIEG D.B., HETHERINGTON H.P., "Quantitative spectroscopic imaging of the human brain", *Magn. Reson. Med.*, vol. 40, pp. 363–369, 1998.

Particular contrasts

[LOU 00] LOUIE A.Y., HÜBER M.M., AHRENS E.T., *et al.*, "In vivo visualization of gene expression using magnetic resonance imaging", *Nature Biotechnology.*, vol. 18, pp. 321–325, 2000.

Instrumentation

[VLA 96] VLAARDINGERBROEK M.T., DEN BOER J.A., "MRI and its hardware", *Magnetic Resonance Imaging*, Springer, pp. 1–43, 1996.

Detection

[BOD 04] BODURKA J., LEDDEN P.J., VAN GELDEREN P., *et al.*, "Scalable Multichannel MRI data acquisition system", *Magn. Reson. Med.*, vol. 51, pp. 165–171, 2004.

[DUY] DUYN J., "Receiver technology", http://www.mr.ethz.ch/parallelmri04/abstracts/pub/Duyn.pdf.

[GIO 06] GIOVANNETI G., HARTWIG V., VITTI V., *et al.*, "Application of undersampling technique for the design of an NMR signals digital receiver", *Concepts Magn. Reson.*, vol. 29B, pp. 107–114, 2006.

[REY 06] REYKOWSKI A., "Overview of signal detection and the RF Chain", *ISMRM Ann. Meeting [R.F. system Engineering Session]*, Stockholm, 2006.

Resonators

[ALD 79] ALDERMAN D.W., GRANT D.M., "An efficient decoupler coil design which reduces heating in conducting samples in superconducting spectrometers", *J. Magn. Reson.*, vol. 36, pp. 447–451, 1979.

[CAS 05] CASSIDY P.J., CLARKE K., EDWARDS D.J., "Determining the tuning and matching requirements of RF coils using electromagnetic simulation and electric circuit analysis", *Concepts in Magn. Reson.*, vol. 25B, pp.27–41, 2005.

[DAR 03] DARRASSE L., GINEFRI J.C., "Perspectives with cryogenic RF probes in biomedical MRI", *Biochimie.*, vol. 85, pp. 915–937, 2003.

[DEC 85] DÉCORPS M., BLONDET P., REUTENAUER H., *et al.*, "An inductively coupled, series tuned NMR probe", *J. Magn. Reson.*, vol. 65, pp. 100–109, 1985.

[DOT 98] DOTY F.D., ENTZMINGER G., YANG Y.A., "High-resolution NMR probe design. I: general methods", *Concepts in Magn. Reson.*, vol. 10, pp. 133–156, 1998.

[GIN 01] GINEFRI J.C., DARRASSE L., CROZAT P., "High-temperature superconducting surface coil for in vivo microimaging of the human skin", *Magn. Reson. Med.*, vol. 45, pp. 376–382, 2001.

[GLO 85] GLOVER G.H., HAYES C.E., PELC N.J., *et al.*, "Comparison of linear and circular polarization for magnetic resonance imaging", *J. Magn. Reson.*, vol. 64, pp. 255–270, 1985.

[HAA 00] HAASE A., ODOJ F., VON KIENLIN M., *et al.*, "NMR Probeheads for in vivo applications", *Concepts in Magn. Reson.*, vol. 12, pp. 361–388, 2000.

[HAL 91] HALL A.S., ALFORD N.M., BUTTON T.W., *et al.*, "Use of high temperature superconductor in a receiver coil for magnetic resonance Imaging", *Magn. Reson. Med.*, vol. 20, pp. 340–343, 1991.

[HAY 09] HAYES C.E., "The development of the birdcage resonator: a historical perspective", *NMR Biomed.*, vol. 22, pp. 908–918, 2009.

[HAY 85] HAYES C.E., EDELSTEIN W.A., SCHENCK J.F., *et al.*, "An efficient highly homogeneous radiofrequency coil for whole-body NMR imaging at 1.5 T", *J. Magn. Reson.*, vol. 63, pp. 622–628, 1985.

[KOR 83] KORETSKY A.P., WANG S., MURPHY-BOESCH J., *et al.*, "31P NMR spectroscopy of rat organs, in situ, using chronically implanted radiofrequency coils", *Proc. Natl. Acad. Sci.*, vol. 80, pp. 7491–7497, 1983.

[MIS 06] MISPELTER J., LUPU M., BRIGUET A., *NMR Probe-heads for biophysical and biomedical experiments*, Imperial Publishing Company, London, 2006.

[PEC 95] PECK T.L., MAGIN R.L., LAUTERBUR P.C., "Design and analysis of micro coils for NMR microscopy", *J. Magn. Reson.*, vol. B108, pp. 114–124, 1995.

[TRA 09] TRAKIC A., LI B.K., WEBER E., *et al.*, "A Rapidly rotating RF coil for MRI", *Concepts in Magn. Reson.*, vol. 35B, pp. 59–66, 2009.

[TRO 89] TROPP J., "The theory of the birdcage resonator", *J. Magn. Reson.*, vol. 82, pp. 51–62, 1989.

[VAU 94] VAUGHAN J.T., HETHERINGTON H.P., PAN J.W., *et al.*, "High frequency volume coils for clinical NMR imaging and spectroscopy", *Magn. Reson. Med.*, vol. 32, pp. 206–218, 1994.

Phased-array coils

[HAR 08] HARDY C.J., GIAQUINTO R.O., PIEL J.E., *et al.*, "128-channel body MRI with a flexible high-density receiver-coil array", *J. Magn. Reson. Imaging*, vol. 28, pp. 1219–1225, 2008.

[HOU 02] HOULT D.I., TOMANEK B., "Use of Mutually inductive coupling in probe design", *Concepts in Magn. Reson.*, vol. 15, pp. 262–285, 2002.

[KUM 06] KUMAR A., BOTTOMLEY P.A., "Optimizing the intrinsic signal-to-noise ratio of MRI strip detectors", *Magn. Reson. Med.*, vol. 56, pp. 157–166, 2006.

[LEE 01] LEE R.F., WESTGATE C.R., WEISS R.G., *et al.*, "Planar strip array (PSA) for MRI", *Magn. Reson. Med.*, vol. 45, pp. 673–693, 2001.

[LEE 04] LEE R.F., HARDY Ch.J., SODICKSON D.K., *et al.*, "Lumped-element planar strip array (LPSA) for parallel MRI", *Magn. Reson. Med.*, vol. 51, pp. 172–183, 2004.

[OHL 06] OHLIGER M.A., SODICKSON D.K., "An introduction to coil array design for parallel MRI", *NMR in Biomedicine,* vol. 19, pp. 300–315, 2006.

[ROE 90] ROEMER P.B., HAYES C.E., SOUZA S.P., *et al.*, "The NMR phased array", *Magn. Reson. Med.*, vol. 16, pp. 192–225, 1990.

[YIN 10] YING L., LIANG Z.P., "Parallel MRI using phased array coils", *IEEE Signal Processing Mag.*, vol. 27, pp. 90–98, 2010.

[YIN] YING L., LIANG Z.P., "Parallel MRI using phased array coils: Multichannel sampling theory meeting spin physics", https://pantherfile.uwm.edu/leiying/public/SPM_binder.pdf.

[ZHU 04] ZHU Y., HARDY C.J., SODINCKSON D.K., *et al.*, "Highly parallel volumetric imaging with a 32-element RF coil array", *Magn. Reson. Med.*, vol. 52, pp. 869–877, 2004.

Integration into the measuring session

[HAZ 01] HAZIZA N., Conception d'une électronique de réception numérique modulable pour l'acquisition multi-antennes en IRM, Thesis, Université Paris 11, 2001.

[POS 12] POSSANZINI C., HARVEY P.R., HAM K., *et al.*, "Architecture dStream", Documentation Philips™, Koninklijke Philips Electronics N.V., 2012.

[REY 95] REYKOWSKI A., WRIGHT S.M., PORTER J.R., "Design of matching networks for low noise preamplifiers", *Magn. Reson. Med.*, vol. 33, pp. 848–852, 1995.

Signal-to-noise ratio (SNR)

[EDE 86] EDELSTEIN W.A., GLOVER G.H., HARDY C.J., *et al*, "The intrinsic signal-to-noise ratio in NMR imaging", *Magn. Reson. Med.*, vol. 3, pp. 604–618, 1986.

[HOU 76] HOULT D.I., RICHARDS R.E., "The signal-to-noise ratio of the nuclear magnetic resonance experiment", *J. Magn. Reson.*, vol. 24, pp. 71–85, 1976.

[HOU 79] HOULT D.I., LAUTERBUR P.C., "The sensitivity of the zeugmatographic experiment involving human samples", *J. Magn. Reson.*, vol. 34, pp. 414–433, 1979.

[OCA 98] OCALI O., ATALAR E., "Ultimate intrinsic signal-to-noise ratio in MRI", *Magn. Reson. Imaging.*, vol. 39, pp. 462–473, 1998.

Signal dynamics

[BEH 05] BEHIN R., BISHOP J., HENKELMANN R.M., "Dynamic Range requirements for MRI", *Concepts in Magn. Reson.*, vol. 26B, 28–35, 2005.

[GAB 09] GABR R.E., SCHÄR M., EDELSTEIN A.D, *et al.*, "MRI dynamic range and its compatibility with signal transmission media", *J. Magn. Reson.*, vol. 198, pp. 137–145, 2009.

Magnetic fields and gradients

[FRA 62] FRANZEN W., "Generation of uniform magnetic fields by means of air core coils", *Rev. Sci. Instrum.*, vol. 33, pp. 933–938, 1962.

[GAR 67] GARRET M.W., "Thick cylindrical coil systems for strong magnetic fields with field or gradient homogeneities of the 6th to 20th order", *J. Appl. Phys.*, vol. 38, pp. 2563–2586, 1967.

[MCK 36] MCKEEHAN L.W., "Combinations of circular currents for producing uniform magnetic fields", *Rev. Sci. Instrum.*, vol. 7, pp. 150–153, 1936.

[ROM 82] ROMÉO F., HOULT D.I., "Magnet field profiling: analysis and correcting coil design", *Magn. Reson. Med.*, vol. 1, pp. 44–65, 1982.

[SCH 87] SCHENCK J.F., HUSSAIN M.A., EDELSTEIN W.A., Transverse gradient field coils for nuclear magnetic resonance imaging, US Patent #4646024, http://radaris.com/p/John/Schenck/, 1987.

[TUR 93] TURNER R., "Gradient coils design: a review of methods", *Magn. Reson. Imaging.*, vol. 11, pp. 903–920, 1993.

Safety

[ARM 04] ARMENEAN C., PERRIN E., ARMENEAN M., *et al.*, "RF-induced temperature elevation along metallic wires in clinical magnetic resonance imaging: influence of diameter and length", *Magn. Reson. Med.*, vol. 52, pp. 1200–1206, 2004.

[BOT 85] BOTTOMLEY P.A., REDINGTON R.W., EDELSTEIN W.A., *et al.*, "Estimating radiofrequency power deposition in body in NMR imaging", *Magn. Reson. Med.*, vol. 2, pp. 336–349, 1985.

[COL 09] COLLINS C.M., "Numerical field calculations considering the human subject for engineering and safety assurance in MRI", *NMR in Biomedicine.*, vol. 22, pp. 919–926, 2009.

[COL 98] COLLINS C.M., LI S., SMITH M.B., "SAR and B1 field distribution in a heterogeneous human head model within a birdcage coil", *Magn. Reson. Med.*, vol. 40, pp. 847–856, 1998.

[HOU 00] HOULT D.I., "The sensitivity and power deposition of the high field imaging experiment", *J. Magn. Reson. Imaging.*, vol. 12, pp. 46–67, 2000.

[SCH 98] SCHMITT F., IRNICH W., FISHER H., *et al.*, "Physiological side effects of fast gradient switching", *Echoplanar Imaging: Theory, Technique and Application*, Springer, pp. 201–252, 1998.

[WAN 07] WANG Z., LIN J.C., MAO W., *et al.*, "SAR and temperature: simulations and comparison to regulatory limits for MRI", *J. Magn. Reson. Imaging,* vol. 26, pp. 437–441, 2007.

[YEU 02] YEUNG C.J., SUSIL R.C., ATALAR E., "RF heating due to conductive wires during MRI depends on the phase distribution of the transmit field", *Magn. Reson. Med.*, vol. 48, pp. 1096–1098, 2002.

[YEU 07] YEUNG C.J., KARMARKAR P., MCVEIGH E.R., "Minimizing RF heating of conducting wires in MRI", *Magn. Reson. Med.*, vol. 58, pp. 1028–1034, 2007.

Image properties

Spatial resolution

[PLE 12] PLENGE E., POOT D.H., BERNSEN M., *et al.*, "Super-resolution methods in MRI: Can they improve the trade-off between resolution, signal to noise ratio, and acquisition time?", *Magn. Reson. Med,* vol. 68, pp. 1983–1993, 2012.

[WAT 00] WATTS R., WANG Y., WINCHESTER P.A., *et al.*, "Rose model in MRI: noise limitation on spatial resolution and implications for contrast enhanced MR angiography", *Proc. Intl. Soc. Mag. Reson. Med.*, vol. 8, p. 463, 2000.

[WAT 02] WATTS R., WANG Y., "k-Space interpretation of the Rose model: noise limitation on the detectable resolution in MRI", *Magn. Reson. Med.*, vol. 48, pp.550–554, 2002.

Contrast and noise

[CON 91] CONSTABLE R.T., HENKELMAN R.M., "Contrast, resolution and detectability in MR imaging", *J. Comput. Ass. Tom.*, vol. 15, pp. 297–303, 1991.

[EDE 83] EDELSTEIN W.A., BOTTOMLEY P.A., HART H.R., *et al.*, "Signal, noise and contrast", *J. Comp. Assisted Tom.*, vol. 7, pp. 391–401, 1983.

[GUD 95] GUDBJARTSSON H., PATZ S., "The Rician distribution of noisy MRI data", *Magn. Reson. Med.*, vol. 34, pp. 910–914 ,1995.

[KOP 11] KOPANOGLU E., ERTURK V.B., ATALAR E., "Analytical expressions for the ultimate intrinsic signal to noise ratio and ultimate intrinsic specific absorption rate in MRI", *Magn. Reson. Med.*, vol. 66, pp. 846–858, 2011.

[WIE 04] WIESENGER F., BOESIGER P., PRUESSMANN K.P., "Electrodynamics and ultimate SNR in parallel imaging", *Magn. Reson. Med.*, vol. 52, pp. 376–390, 2004.

Contrast agents

[BUR 08] BURTEA C., LAURENT S., VAN DER ELST L., *et al.*, "Contrast agents: magnetic resonance", *Molecular Imaging: Handbook of Experimental Pharmacology*, vol. 185–1, pp. 135-165, 2008.

[COR 06] COROT C., ROBERT P., IDÉE J.M., *et al.*, "Recent advances in iron oxide nanocrystal technology for medical imaging", *Adv. Drug Deliv. Rev.*, vol. 58, pp. 1471–504, 2006.

[DE 06] DE MARCO G., ROCH A., PERETTI I., *et al.*, "Agents de contraste en IRM: approche théorique des mécanismes de relaxation", *J. Med. Nucl.*, vol. 30, pp. 645–658, 2006.

[MAR 13] MARINESCU M., CHAUVEAU F., DURAND A., *et al.*, "Monitoring therapeutic effects in experimental stroke by serial USPIO-enhanced MRI", *Eur. Radiol.*, vol. 23, pp. 37–47, 2013.

[MEN 06] MENUEL C., GUILLEVIN R., VALLÉE J.N., *et al.*, "L'IRM de perfusion", *Neurologies*, vol. 86, pp. 375–381, 2006.

[MUL 86] MULLER R.N., "Contrast agents in whole body MR: operating mechanisms", *J. Wiley, Encyclopedia of NMR*, 1986.

[VAN 03] IVANCEVIC M.K., "Tissue perfusion", section 1.4 in MR tissue perfusion quantification: MR sequence optimization and in flow effect correction, Thesis, 2003.

[WAN 01] WANG Y.X., HUSSAIN S.M., KRESTIN G.P., "Superparamagnetic iron oxide contrast agents: physicochemical characteristics and applications in MR imaging", *Eur Radiol.*, vol. 11, pp. 2319–31, 2001.

[WEH 84] WEHRLI F.W., MARCHAL J.R., SHUTTS D., *et al.*, "Mechanisms of contrast in NMR imaging", *J. Comp. Assisted Tom.*, vol 8, pp. 369–380, 1984.

[WEI 90] WEISSLEDER R., ELIZONDO G., WITTENBERG J., *et al.*, "Ultrasmall superparamagnetic iron oxide: characterization of a new class of contrast agents for MR imaging", *Radiology.*, vol. 175, pp. 489-93, 1990.

[WU 10] WU J.C., ABRAHAM M.R., KRAITCHMAN D.L., "Current perspectives on imaging cardiac stem cell therapy", *J. Nucl. Med.* 51, pp. 128S–135S, 2010.

Sequences and modes of reconstruction

[HEN 99] HENNIG J., "k-Space sampling strategies", *European Radiology*, vol. 9, pp. 1020–1031, 1999.

Spin echo sequences

[EDE 80] EDELSTEIN W.A., HUTCHINSON J.M.S., JOHNSON G., *et al.*, "Spin warp NMR imaging and applications to human whole body imaging", *Phys. Med. Biol.*, vol. 25, pp. 751–756, 1980.

[HUT 78] HUTCHINSON J.M.S., SUTHERLAND R.J., MALLARD J.R., "NMR imaging: image recovery under magnetic fields with large non uniformities", *J. Phys. E, Sci. Instrum.* vol. 11, pp. 217–221, 1978.

[KIE 94] KIEFER B., GRÄSSNER J., HAUSMANN R., "Image acquisition in a second with half-Fourier acquired single shot turbo spin echo", *J. Magn. Reson. Imaging.*, vol. 4, pp. 86–87, 1994.

Gradient-echo sequences; fast and very fast sequences

[AHN 86] AHN C.B., KIM J.H., CHO Z.H., "High speed spiral-scan echo planar imaging", *IEEE Trans. Med. Imaging,* vol. 52, pp. 2–5, 1986.

[BRA 92] BRANT-ZAWADZKI M., GILLAN G.D., NITZ W.R., "MP RAGE: a three-dimensional, T1-weighted, gradient-echo sequence – initial experience in the brain", *Radiology*, vol. 182, pp. 769–775, 1992.

[COH 91] COHEN M.S., WEISSKOFF R.M., "Ultra-fast imaging", *Magn. Reson. Imaging.*, vol. 9, pp.1–37, 1991.

[HAA 85] HAASE A., FRAHM J., MATTHAEI D., *et al.*, "Rapid NMR imaging using stimulated echoes", *J. Magn. Reson.*, vol. 65, pp.130–135, 1985.

[HAA 89] HAASE A., FRAHM J., MATTHEI D., "FLASH imaging, rapid NMR imaging using low flip angle pulses", *J. Magn. Reson.*, vol. 67, pp. 388–397, 1989.

[HAA 91] HAACKE E.M., WEILOPOLSKI P.A., TKACH J.A., "A comprehensive technical review of short TR fast magnetic resonance imaging techniques", *Rev. Magn. Reson. Med.*, vol. 3, pp. 53–170, 1991.

[HEN 86] HENNIG J., NAUERT A., FRIEDBURG H., "RARE imaging. A fast imaging method for clinical MR", *Magn. Reson. Med.*, vol. 3, pp. 823–833, 1986.

[HEN 91] HENNIG J., "Echoes – How to generate, recognize, use or avoid them in MR imaging sequences. Part I", *Concepts in Magn. Reson.*, vol. 3, pp. 125–143, 1991.

[HEN 91] HENNIG J., "Echoes in imaging sequences. Part II", *Concepts in Magn. Reson.*, vol. 3, pp. 179–192, 1991.

[LIU 93] LIU G., SOBERING G., DUYN J., *et al.*, "A functional MRI technique combining principles of echo-shifting with a train of observations (PRESTO)", *Magn. Reson. Med.*, vol. 30, pp. 764–768, 1993.

[OPP 86] OPPELT A., GRAUMANN R., BARFUSS H., *et al.*, "FISP – a new fast MRI sequence", *Electromedica*, vol. 3, pp. 15–18, 1986.

[PAU 06] PAUL D., MARKL M., FAUTZ H.P., *et al.*, "T2-weighted balanced SSFP Imaging (T2-TIDE) using variable flip angles", *Magn. Reson. Med.*, vol. 56, pp. 82–93, 2006.

[VAN 85] VAN DER MEULEN P., CROEN J.P., CUPPEN J.J.M., "Very fast MR imaging by field echoes and small angle excitation", *Magn. Reson. Imaging.*, vol. 3, pp. 297–299, 1985.

Echo Planar Imaging

[BAN 98] BANDETTINI P.A., WONG EC., SCHMITT F., *et al.*, "Echo-planar magnetic resonance imaging of human brain activation", *Echo planar Imaging; theory, technique and application,* Springer, pp. 493–530, 1998.

[KIE 98] KIEFER B. SCHMITT F., STEHLING M.K., *et al.*, "Turbo spin-echo imaging", *Echo-planar Imaging; theory, technique and application,* Springer, pp. 583–604, 1998.

[MAN 77] MANSFIELD P., "Echo planar imaging", *J. Phys. C 10:L55-L58*, 1977.

[MEY 98] MEYER C.H. SCHMITT F., STEHLING M.K., *et al.*, "Spiral echo-planar imaging", *Echo planar Imaging; theory, technique and application,* Springer, pp. 633–658, 1998.

[STE 91] STEHLING M.K., TURNER R., MANSFIELD P., "Echo-planar imaging: magnetic resonance imaging in a fraction of a second", *Science,* vol. 254, pp. 43–50, 1991.

Reconstructions

[BOU 01] BOURGEOIS M., WAJER F.T.A.W., VAN ORMONDT D., *et al.*, "Reconstruction of MRI images from non-uniform sampling: application to intrascan motion correction in functional MRI", *Modern Sampling theory: Mathematics and Applications, Birkhauser*, pp. 343–362, 2001.

[BRI 74] BRINGHAM E.O., *The Fast Fourier Transform*, Prentice Hall, 1974.

[BRO 76] BROOKS R.A., DI CHIRO G., "Principles of computer assisted tomography (CAT) in radiographic and radio-isotopic imaging", *Phys. Med. Biol.*, vol. 21, pp. 689–732, 1976.

[COR 73] CORMACK A.M., "Reconstructions of densities from their projections with applications in radiological physics", *Phys. Med. Biol.*, vol. 18, pp. 195–207, 1973.

[ERN 66] ERNST R.R., ANDERSON W.A., "Application of Fourier transform spectroscopy to magnetic resonance", *Rev. Sci. Instrum.*, vol. 37, pp. 93–102, 1966.

[GLO 92] GLOVER G.H., PAULY J.M., "Projection reconstruction techniques for reduction of motion effects in MRI", *Magn. Reson. Med.*, vol. 28, pp. 275–289, 1992.

[JAC 01] JACCO A., DE ZWART F.C., VIMEUX F., *et al.*, "On-line correction and visualization of motion during MRI-controlled hyperthermia", *Magn. Reson. Med.,* vol. 45, pp. 128–137, 2001.

[LAI 81] LAI C.M., LAUTERBUR P.C., "Automatic correction of nuclear magnetic resonance zeugmatographic projections", *J. Phys E Sci. Instrum.*, vol. 14, pp. 874–879, 1981.

[LAI 81] LAI C.M., LAUTERBUR P.C., "True dimensional image reconstruction by nuclear magnetic resonance", *Phys. Med. Biol.*, vol. 26, pp. 851–856, 1981.

[LEH 03] LEHTMATE R., RATINEY H., WAJER F.T.A.W., *et al.*, "Dynamic magnetic resonance imaging with radial scanning: a post acquisition keyhole approach", *MAGMA*, vol. 16, pp. 21–28, 2003.

[PIP 00] PIPE J.G., "Reconstructing MR images from undersampled data: Data weighting considerations", *Magn. Reson. Med.*, vol. 43, pp. 867–875, 2000.

[SON 04] SONG H.K., DOUGHERTY L., "Dynamic MRI with projection reconstruction and KWIC processing for simultaneous high spatial and temporal resolution", *Magn. Reson. Med.*, vol. 52, pp. 815–824, 2004.

Reconstruction and Point Spread Function (PSF)

[BRO 87] BROOKER H.R., MARECI T.H., MAO J.T., "Selective Fourier transform localization", *Magn. Reson. Med.*, vol. 5, pp. 417–33, 1987.

[LAU 96] LAUZON M.L., RUTT B.K., "Effects of polar sampling in k-space", *Magn. Reson. Med.*, vol. 36, pp. 940–949, 1996.

[POH 97] POHMANN R., VON KIENLIN M., HAASE A., "Theoretical evaluation and comparison of fast chemical shift imaging methods", *J. Magn. Reson.*, vol. 129, pp 145–160, 1997.

[ROB 97] ROBSON M.O., GORE J.C., CONSTABLE R.T., "Measurement of the point spread function in MRI using constant time imaging", *Magn. Reson. Med.*, vol. 38, pp. 733–740, 1997.

[SCH 98] SCHEFFER K., HENNIG J., "Reduced circular field of view imaging", *Magn. Reson. Med.*, vol. 40, pp. 474–480, 1998.

Parallel imaging

[BLA 04] BLAIMER M., BREUER F., MUELLER M., *et al.*, "SMASH, SENSE, PILS, GRAPPA, How to choose the optimal method", *Topics in Magn. Reson. Imaging.*, vol. 15, pp. 223–236, 2004.

[GRI 00] GRISWOLD M.A., JAKOB P.M., NITTKA M., *et al.*, "Partially parallel imaging with localized sensitivities (PILS)", *Magn. Reson. Med.*, vol. 44, pp. 602–609, 2000.

[GRI 02] GRISWOLD M.A., JAKOB P.M., HEIDEMANN R.M., *et al.*, "Generalized autocalibrating partially parallel acquisitions (GRAPPA)", *Magn. Reson. Med.*, vol. 47, pp. 1202–1210, 2002.

[KAT 06] KATSCHER U., BÖRNERT P., "Parallel RF transmission in MRI", *NMR Biomed.*, vol. 9, pp. 393–400, 2006.

[LAR 07] LARKMANN D.J., NUNES R.G., "Parallel magnetic resonance imaging", *Phys. Med. Biol.*, vol. 52, pp. R15–R55, 2007.

[PRE] Premières images IRM in vivo en transmission parallèle à 7 Teslas: http://irfu.cea.fr/Phocea/Vie_des_labos/Ast/ast.php?t=fait_marquant&id_ast=2864.

[PRU 06] PRUESSMANN K.P., "Encoding and Reconstruction in Parallel MRI", *NMR Biomed.*, vol. 19, pp. 288–299, 2006.

[PRU 99] PRUESSMANN K.P., WEIGER M., SCHEIDEGGER M.B., *et al.*, "SENSE: sensitivity encoding for fast MR", *Magn. Reson. Med.*, vol. 42, pp. 953–962, 1999.

[SOD 00] SODICKSON D.K., "Tailored SMASH image reconstructions for robust in vivo parallel MR Imaging", *Magn. Reson. Med.*, vol. 44, pp. 243–251, 2000.

Application of MRI: uses and evolution in the biomedical field

Complementarity of MRI and MRS techniques

Technical aspects

[BOT 83] BOTTOMLEY P.A., HART H.R., EDELSTEIN W.A., *et al.*, "NMR imaging/spectroscopy system to study both anatomy and metabolism", *Lancet*, vol. 2, no. 8344, pp. 273–274, 1983.

[CHA 84] CHANCE B., "Applications of P-31 NMR to clinical biochemistry", *Ann. N.Y. Acad. Sci.*, vol. 428, pp. 318–332, 1984.

[DUY 93] DUYN J.H., MOONEN C.T., "Fast proton spectroscopic imaging of human brain using multiple spin-echoes", *Magn. Reson. Med.*, vol. 30, pp. 409–414, 1993.

[DYD 01] DYDAK U., WEIGER M., PRUESSMANN K.P., *et al.*, "Sensitivity-encoded spectroscopic imaging", *Magn. Reson. Med.*, vol. 46, pp. 713–722, 2001.

[HAA 85] HAASE A.. FRAHM J., HANICKE W., *et al.*, "1H NMR chemical shift selective (CHESS) imaging", *Phys. Med. Biol.*, vol. 30, pp. 341–344, 1985.

[HAS 83] HASELGROVE J.C., HARIHARA SUBRAMANIAN V., LEIGH JR J.S., *et al.*, "In vivo One-dimensional imaging of phosphorus metabolites by Phosphorus-31 Nuclear Magnetic Resonance", *Science*, vol. 220, 1170–1173, 1983.

[HEN 88] HENNIG J. FRIEDBURG H., "Clinical applications and methodological developments of the RARE technique", *Magn. Reson. Imaging.*, vol. 6, pp. 391–395, 1988.

[KRE 93] KREIS R., ERNST T., ROSS B.D., "Development of the human brain: in vivo quantification of metabolite and water content with proton magnetic resonance spectroscopy", *Magn. Reson. Med.*, vol. 30, pp. 424–437, 1993.

[MAU 94] MAUDSLEY A.A., MATSON G.B., HUGG J.W., *et al.*, "Reduced phase encoding in spectroscopic imaging", *Magn. Reson. Med.*, vol. 31, pp. 645–651, 1994.

[PFE 99] PFEUFFER J., TKAC I., PROVENCHER S., *et al.*, "Toward an in vivo neurochemical profile quantification of 18 metabolites in a short echo time 1H spectra of the rat brain", *J. Magn. Reson.*, vol. 141, pp. 104–120, 1999.

[PON 94] PONDER S.L., TWIEG D.B., "A novel sampling method for 31P spectroscopic imaging with improved sensitivity, resolution and sidelobe suppression", *J. Magn. Reson. B.*, vol. 104, pp. 85–88, 1994.

[POS 95] POSSE S., TEDESCHI G., RISINGER R., *et al.*, "High speed 1H spectroscopic imaging in human brain by echo planar spatial-spectral encoding", *Magn. Reson. Med.*, vol. 33, pp. 34–40, 1995.

MRI and MRS of different nuclei

[ACK 80] ACKERMAN J.J., GROVE T.H., WONG G.G., *et al.*, "Mapping of metabolites in whole animals by ^{31}P NMR using surface coils", *Nature.*, vol. 283, pp. 167–170, 1980.

[BEN 06] BENDAHAN D., MATTEI J.P., GUIS S., *et al.*, "Explorations de la fonction musculaire par spectroscopie et imagerie de résonance magnétique", *Rev. Neurol. (Paris).*, vol. 162, pp. 467–484, 2006.

[CAL 01] CALLOT V., CANET E., VIALLON M., *et al.*, "MR perfusion imaging using encapsulated polarized ^{3}He", *Magn. Reson. Med.*, vol. 46, pp. 535–540, 2001.

[CAL 08] CALLOT V., GALANAUD D., LE FUR Y., *et al.*, "1H MR spectroscopy of human brain tumors: a practical approach", *Eur. J. Radiol.*, vol. 67, pp. 268–274, 2008.

[CLA 84] CLARK JR L.C., ACKERMAN J.L., THOMAS S.R., *et al.*, "Perfluorinated organic liquids and emulsions as biocompatible NMR imaging agents for 19F and dissolved oxygen", *Adv. Exp. Med. Biol.*, vol. 180, pp. 835–845, 1984.

[GAL 03] GALANAUD D., NICOLI F., LE FUR Y., *et al.*, "Multimodal Magnetic Resonance imaging of the central nervous system", *Biochimie*, vol. 85, pp. 905–914, 2003.

[GAL 07] GALANAUD D., NICOLI F., CONFORT-GOUNY S., *et al.*, "Spectroscopie par résonance magnétique du proton. Quelles indications neurologiques en 2007?", *Rev. Neurol. (Paris).*, vol. 163, pp. 287–303, 2007.

[GOV 00] GOVINDARAJU V., YOUNG K., MAUDSLEY A., "Proton NMR chemical shifts and coupling constants for brain metabolites", *NMR in Biomedicine.*, vol. 13, pp. 129–153, 2000.

[LHO 03] LHOSTE J.M., MISPELTER J., COZZONE P.J., "Metabolic NMR", *Biochimie.*, vol. 85, pp. 821–822, 2003.

[POU 98] POUWELS P.J., FRAHM J., "Regional metabolite concentrations in human brain as determined by quantitative localized proton MRS", *Magn. Reson. Med.*, vol. 39, pp. 53–60, 1998.

[THO 09] THOMAS C., MAILLARD P., LOOCK B., *et al.*, "23Na MRI longitudinal follow-up of PDT in a xenograft model of human retinoblastoma", *Photodiagnosis and Photodynamic Therapy.*, vol. 6, pp. 214–220, 2009.

[TKA 03] TKAC I., RAO R., GEORGIEFF M.K., *et al.*, "Development and regional changes in the neurochemical profile of the rat brain determined by in vivo 1H spectroscopy", *Magn. Reson. Med.*, vol. 50, pp. 24–32, 2003.

[WOL 87] WOLF W., ALBRIGHT M.J., SILVER M.S., *et al.*, "Fluorine-19 NMR spectroscopic studies of the metabolism of 5-fluorouracil in the liver of patients undergoing chemotherapy", *Magn. Reson. Imaging.*, vol. 5, pp. 265–169, 1987.

Fundamentals in diffusion

[COO 74] COOPER R.L., CHANG D.B., YOUNG A.C., *et al.*, "Restricted diffusion in biophysical systems", *Biophys. J.*, vol. 14, pp. 161–177, 1974.

[FEL 58] FELLER W., *Probabilities and Statistics.*, vol. 1 (Chap. XIV), 1958.

[LE 06] LE BIHAN D., POUPON C., AMADON A., *et al.*, "Artifacts and pitfalls in diffusion MRI", *J. Magn.Reson. Imaging.*, vol. 24, pp. 478–488, 2006.

[MAT 94] MATTIELLO J., BASSER P.J., LE BIHAN D., "Analytical expressions for the *b* matrix in NMR diffusion imaging and spectroscopy", *J.Magn. Reson. A.*, vol. 108, 131–141, 1994.

[MIT 95] MITRA P.P., HALPERIN B.I, "Effects of finite gradient-pulse widths in pulsed-field gradient diffusion measurements" *J. Magn. Reson. A.*, vol. 113, pp. 94–101, 1995.

[PRI 97] PRICE W.S., "Pulsed-field gradient nuclear magnetic resonance as a tool for studying translational diffusion: Part 1 Basic theory", *Concepts Magn. Reson.*, vol. 9, pp. 299–336, 1997.

[STE 65] STEJSKAL E.O., "Use of spin echoes in a pulsed magnetic-field gradient to study anisotropic restricted spin diffusion and flow", *J. Chem. Phys.*, vol. 43, pp. 3597–3603, 1965.

[STE 65] STEJSKAL E.O., TANNER J.E., "Spin diffusion measurements: Spin echoes in the presence of a time-dependant field Gradient", *J. Chem. Phys.*, vol. 42, pp. 288–292, 1965.

[TAN 68] TANNER J.E., STEJSKAL E.O., "Restricted self-diffusion of protons in colloidal systems by the pulsed-gradient, spin-echo method", *J. Chem. Phys.*, vol. 49, pp. 1768–1777, 1968.

[TOR 56] TORREY H.C., "Bloch equations with diffusion terms", *Phys. Rev.*, vol. 104, pp. 563–565, 1956.

Diffusion tensor imaging

[BAS 01] BASSER P.J., JONES D.K., "Diffusion-tensor MRI: theory, experimental design and data analysis – a technical review", *NMR Biomed.*, vol. 15, pp. 456–467, 2001.

[BAS 94] BASSER P.J., MATTIELLO J., LE BIHAN D., "Estimation of the effective self-diffusion tensor from the NMR spin echo", *J. Magn. Reson. B.*, vol. 103, pp. 247–254, 1994.

[BAS 94] BASSER P.J., MATTIELLO J., LE BIHAN D., "MR diffusion tensor spectroscopy and imaging", *Biophys. J.*, vol. 66, pp. 259–267, 1994.

[LEB 01] LE BIHAN D., MANGIN J.F., POUPON C., *et al.*, "Diffusion tensor imaging: concepts and applications", *J. Magn. Reson. Imaging.*, vol. 13, pp. 534–546, 2001.

The q-space imaging technique

[AGA 10] AGANI I., LENGLET Ch., SAPIRO G., *et al.*, "Reconstruction of the orientation distribution function in single- and multiple-shell q-ball imaging within constant solid angle", *Magn. Reson. Med.*, vol. 64, pp. 554–566, 2010.

[BAS 02] BASSER P.J., "Relationships between diffusion tensor and q-space MRI", *Magn. Reson. Med.*, vol. 47, pp. 392–397, 2002.

[PER 05] PERRIN M., POUPON C., RIEUL B., *et al.*, "Validation of q-ball imaging with a diffusion fibre crossing phantom on a clinical scanner", *Phil. Trans. R. Soc. B.*, vol. 360, pp. 881–891, 2005.

[TUC 02] TUCH D.S., REESE T.G., WIEGELL M.R., *et al.*, "High angular resolution diffusion imaging reveals intravoxel white matter fiber heterogeneity", *Magn. Reson. Med.*, vol. 48, pp. 577–582, 2002.

[TUC 04] TUCH D.S., "Q-ball imaging", *Magn. Reson. Med.*, vol. 52, pp. 1358–1372, 2004.

[WEE 05] WEEDEN V.J., HAGMANN P., TSENG W.Y.I., *et al.*, "Mapping complex tissue architecture with diffusionspetrum magnetic resonance imaging", *Magn. Reson. Med.*, vol. 54, pp. 1377–1386, 2005.

Applications of diffusion to porous media

[ALE 08] ALEXANDER D.C., "A general framework for experiment design in diffusion MRI and its application in measuring direct tissue microstructure features", *Magn. Reson. Med.*, vol. 60, pp. 439–448, 2008.

[CAL 91] CALLAGHAN P.T., COY A., MCGOWAN D., *et al.*, "Diffraction-like effects in NMR diffusion studies of fluids in porous solids", *Nature.*, vol. 351, pp. 467–469, 1991.

[LAU 12] LAUN F.B., KUDER T.A., WETSCHEREK A., *et al.*, "NMR-based diffusion pore imaging", *J. Phys. E.*, vol. 86, no. 021906, pp. 1–23, 2012.

[SHE 12] SHEMESH N., WESTIN C.F., *et al.*, "Magnetic resonance imaging by synergistic diffusion-diffraction patterns", *Phys. Rev. Letters*, vol. 108, no. 058103, pp. 1–5, 2012.

Applications of diffusion to living tissues

[ALE 02] ALEXANDER D.C., BARKER G.J., ARRIDGE S.R., "Detection and modeling of non-Gaussian apparent diffusion coefficient profiles in human brain data", *Magn. Reson. Med.*, vol. 48, pp. 331–340, 2002.

[BEA 02] BEAULIEU C., "The basis of anisotropic water diffusion in the nervous system – a technical review", *NMR Biomed.*, vol. 15, pp. 435–455, 2002.

[BER 03] BERG J., SAILER A., RAND T., *et al.*, "Diffusivity and T2 imaging at 3 teslas for the detection of degenerative change in excised tissue with high resolution: atherosclerotic arteries", *Invest. Radiol.*, vol. 38, pp. 452–459, 2003.

[CAL 10] CALLOT V., DUHAMEL D., LE FUR Y., *et al.*, "Echo planar diffusion tensor imaging of the mouse spinal cord at thoracic and lumbar levels", *Magn. Reson. Med.*, vol. 63, pp. 1125–1134, 2010.

[HAN 12] HANNOUN S., BAGORY M., DURAND-DUBIEF F., *et al.*, "Correlation of diffusion and metabolic alterations in different clinical forms of multiple sclerosis", *PLoS One.*, vol. 7, no. 032525, 2012.

[HAN 12] HANNOUN S., DURAND-DUBIEF F., CONFAVREUX C., *et al.*, "Diffusion tensor-MRI evidence for extra-axonal neuronal degeneration in caudate and thalamic nuclei of patients with multiple sclerosis", *Am. J. Neuroradiol.*, vol. 33, pp. 1363–8, 2012.

[HAR 06] HARSAN L.A., POULET P., GUIGNARD B., *et al.*, "Brain dysmyelination by noninvasive in vivo diffusion tensor magnetic resonance imaging", *J. Neuroscience Research.*, vol. 83, pp. 392–402, 2006.

[HAS 05] HASAN K.M., NARAYANA P.A., "DTI parameter optimization at 3.0 T: potential application in entire normal human brain mapping and multiple sclerosis research", *Medica Mundi*, vol. 49, pp. 30–45, 2005.

[SHM 07] SHMAHMANN J.D., PANDYA D.N., WANG R., *et al.*, "Association fibre pathways of the brain: parallel observations from diffusion spectrum imaging and autoradiography", *Brain.*, vol. 130, pp. 639–653, 2007.

[TUC 03] TUCH D.S., REESE T.G., WIEGELL M.R., *et al.*, "Diffusion MRI of complex neural architecture", *Neuron*, vol. 40, pp. 885–895, 2003.

[WIE 00] WIEGELL M.R., LARSSON H.B.W., WEDEEN V.J., "Fiber crossing in human brain depicted with diffusion tensor MR imaging", *Radiology.*, vol. 217, pp. 897–903, 2000.

BOLD effect

[BEL 91] BELLIVEAU J.W., KENNEDY D.N., MCKINSTRY R.C., *et al.*, "Functional mapping of the human visual cortex by magnetic resonance imaging", *Science.*, vol. 254, pp. 716–719, 1991.

[GOR 03] GORE J.C., PROST R.W., "Functional MRI is fundamentally limited by an inadequate understanding of the origin of fMRI signals in tissue", *Med. Phys., and Controversies in Med. Phys.*, vol. 30, pp. 2859–2861, 2003.

[LOG 02] LOGOTHETIS N.K., "The neural basis of the blood-oxygen-level-dependent functional magnetic resonance imaging signal", *Phil. Trans. R. Soc.*, London B., vol. 357, pp. 1003–1037, 2002.

[OGA 90a] OGAWA S., LEE T.M., KAY A.R., *et al.*, "Brain magnetic resonance imaging with contrast dependent on blood oxygenation", *Proc. Natl. Acad. Sci.*, USA., vol. 87, pp. 9868–9872, 1990.

[OGA 90b] OGAWA S., LEE T.M., NAYAK A.S., *et al.*, "Oxygenation-sensitive contrast in magnetic resonance image of rodent at high magnetic fields", *Magn. Reson. Med.*, vol. 14, pp. 68–78, 1990.

[PRI 91] PRICHARD J., ROTHMAN D., NOVOTNY E., *et al.*, "Lactate rise detected by 1H NMR in human visual cortex during physiologic stimulation", *Proc. Natl. Acad. Sci.*, U.S.A., vol. 88, pp. 5829–5831, 1991.

[SAP 92] SAPPEY-MARINIER D., CALABRESE G., FEIN G., *et al.*, "Effect of photic stimulation on human visual cortex lactate and phosphates using 1H and 31P magnetic resonance spectroscopy", *J. Cereb. Blood Flow Metab.*, vol. 12, pp. 584–592, 1992.

Combining MRI and PET

[DEL 11] DELSO G., FÜRST S., JAKOBY B., *et al.*, "Performance measurements of the Siemens mMR integrated whole-body PET/MR scanner", *J. Nucl. Med.*, vol. 52, pp. 1914–1922, 2011.

[DRZ 12] DRZEZGA A., SOUVATZOGLOU M., EIBER M., *et al.*, "First clinical experience with integrated whole-body PET/MR: comparison to PET/CT in patients with oncologic diagnoses", *J. Nucl. Med.*, vol. 53, pp. 845–855, 2012.

[HER 10] HERZOG H., PIETRZYK U., SHAH N.J., *et al.*, "The current state, challenges and perspectives of MR-PET", *Neuroimage.*, vol. 49, pp. 2072–2082, 2010.

[HOE 08] HOEHN M., HIMMELREICH U., KRUTTWIG K., *et al.*, "Molecular and cellular MR imaging: potentials and challenges for neurological applications", *J. Magn. Reson. Imaging.*, vol. 27, pp. 941–954, 2008.

[MAR 09] MARTINEZ-MÖLLER A., SOUVATZOGLOU M., DELSO G., *et al.*, "Tissue classification as a potential approach for attenuation correction in whole-body PET/MRI: evaluation with PET/CT data", *J. Nucl. Med.*, vol. 50, pp. 520–526, 2009.

[SCH 09] SCHLEMMER H.P., PICHLER B.J., KRIEG R., "An integrated MR/PET system: prospective applications", *Abdominal Imaging.*, vol. 34, pp. 668–674, 2009.

[SCH 12a] SCHWENZER N.F., SCHMIDT H., CLAUSSEN C.D. "Whole-body MR/PET: applications in abdominal imaging", *Abdominal Imaging.*, vol. 37, pp. 20–28, 2012.

[SCH 12b] SCHWENZER N.F., STEGGER L., BISDAS S., *et al.*, "Simultaneous PET/MR imaging in a human brain PET/MR system in 50 patients – current state of image quality", *Eur. J. Radiol.*, vol. 81, pp. 3472–3478, 2012.

[TOW 08] TOWNSEND D.W., "Dual-modality imaging: combining anatomy and function", *J. Nucl. Med.*, vol. 49, pp. 938–955, 2008.

[ZAI 11] ZAIDI H., DEL GUERRA A., "An outlook on future design of hybrid PET/MRI systems", *Med. Phys.*, vol. 38, pp. 5667–5689, 2011.

[ZAI 11] ZAIDI H., OJHA N., MORICH M., J *et al.*, "Design and performance evaluation of a whole-body Ingenuity TF PET-MRI system", *Phys. Med. Biol.*, vol. 56, pp. 3091–3106, 2011.

Temperature monitoring by MRI

[BOS 07] BOSS A., GRAF H., BERGER A., *et al.*, "Tissue warming and regulatory responses induced by radio frequency energy deposition on a whole-body 3-Tesla magnetic resonance imager", *J. Magn. Reson. Imaging.*, vol. 26, pp. 1334–1339, 2007.

[EHS 08] EHSES P., FIDLER F., NORDBECK P., *et al.*, "MRI thermometry: fast mapping of RF-induced heating along conducting wires", *Magn. Reson. Med.*, vol. 60, pp. 457–461, 2008.

[PAR 84] PARKER D.L., "Applications of NMR Imaging in Hyperthermia: An evaluation of the potential for localized tissue heating and noninvasive temperature monitoring", *IEEE Trans. Biomed. Eng. BME.*, vol. 31, pp. 161–167, 1984.

[QUE 00] QUESSON B., DE ZWART J.A., MOONEN C.T.W., "Magnetic Resonance temperature Imaging for guidance of thermotherapy", *J. Magn. Reson. Imaging.*, vol. 12, pp. 525–533, 2000.

[QUE 02] QUESSON B., VIMEUX F., SALOMIR R., *et al.*, "Automatic control of hyperthermia therapy based on real-time Fourier analysis of MR temperature maps", *Magn. Reson. Med.*, vol. 47, 1065–1072, 2002.

[SAL 00a] SALOMIR R., PALUSSIÈRE J., VIMEUX F. C., *et al.*, "Hyperthermia by MR-guided focused ultrasound: spiral trajectory of the focal point optimized for temperature uniformity in the target region", *J. Magn. Reson. Med.*, vol. 12, pp. 571–583, 2000.

[SAL 00b] SALOMIR R., VIMEUX F.C., JACCO A., *et al.*, "Hyperthermia by MR-guided focused ultrasound: accurate temperature control based on fast MRI and a physical model of local energy deposition and heat conduction", *J. Magn. Reson. Med.*, vol. 43, pp. 342–346, 2000.

List of Authors

Adrian BASARAB
IRIT
Toulouse
France

Olivier BASSET
CREATIS
University of Lyon
France

Olivier BEUF
CREATIS
University of Lyon
France

André BRIGUET
University Claude Bernard Lyon 1
France

Elisabeth BRUSSEAU
CREATIS
University of Lyon
France

Christian CACHARD
CREATIS
University of Lyon
France

Philippe DELACHARTRE
CREATIS
University of Lyon
France

Valérie DETTI
CREATIS
University of Lyon
France

Hervé FANET
CEA LETI
France

Guy FRIJA
Hôpital Européen Georges Pompidou
Paris
France

Hervé LIEBGOTT
CREATIS
University of Lyon
France

François VARRAY
CREATIS
University of Lyon
France

Dominique SAPPEY-MARINIER
University Claude Bernard Lyon 1
France

Didier VRAY
CREATIS
University of Lyon
France

Index

Printed and bound by CPI Group (UK) Ltd, Croydon, CR0 4YY
13/09/2026
14961081-0004